Arbitrage and Rational Decisions

This unique book offers a new approach to the modeling of rational decision-making under conditions of uncertainty and strategic and competition interactions among agents. It presents a unified theory in which the most basic axiom of rationality is the principle of no-arbitrage, namely that neither an individual decision maker nor a small group of strategic competitors nor a large group of market participants should behave in such a way as to provide a riskless profit opportunity to an outside observer.

Both those who work in the finance area and those who work in decision theory more broadly will be interested to find that basic tools from finance (arbitrage pricing and risk-neutral probabilities) have broader applications, including the modeling of subjective probability and expected utility, incomplete preferences, inseparable probabilities and utilities, nonexpected utility, ambiguity, noncooperative games, and social choice. Key results in all these areas can be derived from a single principle and essentially the same mathematics.

A number of insights emerge from this approach. One is that the presence of money (or not) is hugely important for modeling decision behavior in quantitative terms and for dealing with issues of common knowledge of numerical parameters of a situation. Another is that beliefs (probabilities) do not need to be uniquely separated from tastes (utilities) for the modeling of phenomena such as aversion to uncertainty and ambiguity. Another over-arching issue is that probabilities and utilities are always to some extent indeterminate, but this does not create problems for the arbitrage-based theories.

One of the book's key contributions is to show how noncooperative game theory can be directly unified with Bayesian decision theory and financial market theory without introducing separate assumptions about strategic rationality. This leads to the conclusion that correlated equilibrium rather than Nash equilibrium is the fundamental solution concept.

The book is written to be accessible to advanced undergraduates and graduate students, researchers in the field, and professionals.

Robert Nau is a Professor Emeritus of Business Administration in the Fuqua School of Business, Duke University. He received his Ph.D. in Operations Research from the University of California at Berkeley. Professor Nau is an internationally known authority on mathematical models of decision-making under uncertainty. His research has been supported by the National Science Foundation, and his papers have been published in journals such as *Operations Research, Management Science, Annals of Statistics, Journal of Economic Theory*, and the *International Journal of Game Theory*. He was a co-recipient of the Decision Analysis Society Best Publication Award. One of the themes in Professor Nau's research is that models of rational decision-making in various fields are linked by a single unifying principle, namely the principle of no-arbitrage, i.e., avoiding sure loss at the hands of a competitor. This principle is central to modern finance theory, but it can also be shown to be the fundamental rationality concept that underlies Bayesian statistics, decision analysis, and game theory. Professor Nau has taught the core MBA courses on Decision Models and Statistics in several programs, and he developed an MBA elective course on Forecasting, which he has taught throughout his career. He also teaches a course on Rational Choice Theory in the Ph.D. program that draws students from other departments and schools at Duke University.

Chapman & Hall/CRC Financial Mathematics Series

Series Editors

M.A.H. Dempster
Centre for Financial Research
Department of Pure Mathematics and
Statistics
University of Cambridge, UK

Rama Cont
Mathematical Institute
University of Oxford, UK

Dilip B. Madan
Robert H. Smith School of Business
University of Maryland, USA

Robert A. Jarrow
Samuel Curtis Johnson School of Management
Cornell University, USA

For more information about this series please visit: Chapman and Hall/CRC Financial Mathematics
Series - Book Series - Routledge & CRC Press

Arbitrage and Rational Decisions

Robert Nau

CRC Press
Taylor & Francis Group
Boca Raton London New York

CRC Press is an imprint of the
Taylor & Francis Group, an **informa** business

A CHAPMAN & HALL BOOK

First edition published 2025
by CRC Press
2385 Executive Center Drive, Suite 320, Boca Raton, FL 33431

and by CRC Press
4 Park Square, Milton Park, Abingdon, Oxon, OX14 4RN

CRC Press is an imprint of Taylor & Francis Group, LLC

Library of Congress Cataloging-in-Publication Data
Names: Nau, Robert, author.
Title: Arbitrage and rational decisions / Robert Nau.
Description: Boca Raton : CRC Press, 2024. | Includes bibliographical references and index.
Identifiers: LCCN 2024033616 | ISBN 9781032863511 (hardback) | ISBN 9781032859095 (paperback) | ISBN 9781003527145 (ebook)
Subjects: LCSH: Arbitrage. | Decision making. | Rational choice theory.
Classification: LCC HG4521 .N38 2024 | DDC 332.6--dc23/eng/20240829
LC record available at https://lccn.loc.gov/2024033616

ISBN: 978-1-032-86351-1 (hbk)
ISBN: 978-1-032-85909-5 (pbk)
ISBN: 978-1-003-52714-5 (ebk)

DOI: 10.1201/9781003527145

Typeset in Nimbus Roman font
by KnowledgeWorks Global Ltd.

Contents

Preface

This book presents a unified theory of rational choice in which the most basic axiom of rationality is the principle of no-arbitrage, namely that neither an individual decision-maker nor a small group of strategic competitors nor a large group of market participants should behave in such a way as to provide a riskless profit opportunity to an outside observer. All of the most fundamental theorems of rational choice theory can be framed as optimization models (linear programs) in which the problem of finding an arbitrage opportunity is the mirror image (the dual) of the problem of finding a probability distribution or a utility function or a state-dependent utility function or a game-theoretic equilibrium or a market equilibrium with respect to which the behavior of all the participants is optimal. As such, the different branches of rational choice theory are more tightly integrated than they are usually portrayed to be.

The geometrical representation of such an optimization problem involves a plane (more generally a hyperplane) that separates the arbitrage opportunities (sure losses) from the set of bets or trades that decision-makers are willing to accept. The parameters that determine the hyperplane's orientation in space are the numbers (probabilities, utilities, equilibria) that quantify the rational decision-maker's states of mind.

Duality theorems and separating hyperplanes are the hallmarks of optimization models in general, and they are used throughout microeconomics, although not always in plain view. The use of a no-arbitrage requirement on one side of them is very familiar in several branches of rational choice theory—particularly in subjective probability (where it is called coherence and also the Dutch Book Argument), in welfare economics (where it is called Pareto efficiency), and in finance (where it is the basis of option pricing methods)—but its role in other branches is not so well known, especially in the large middle ground now occupied by game theory, where Nash equilibrium dominates the landscape. The broader concept of correlated equilibrium, illustrated on the cover of this book, fits more seamlessly into the spectrum of rationality models.

Even where its relevance is implicitly understood, the no-arbitrage principle is often obscured in formal models of rational behavior that are vertically constructed in the tradition of methodological individualism. In this tradition, a numerical

representation of the rational individual's beliefs and tastes is first derived from half-
a-dozen separate axioms imposed on preferences, and that representation then serves
as a platform on top of which higher-order knowledge assumptions and equilibrium
concepts are erected to describe, or predict, or recommend how two or more rational
individuals should interact in games and markets. Here, a single rationality principle
rather than a heirarchy of principles will be invoked.

In the first half of the book, it will be shown that there are two parallel sets of axioms for models of individual rationality, one that refers to binary preferences which
satisfy independence and transitivity conditions and another that refers to acceptable
bets which satisfy convexity and linear extrapolation conditions. The former is more
conventional, while the latter is more easily visualized and better suited for extensions to multi-agent settings and for dealing with common knowledge issues and
the determination of strategic and competitive equilibria, which are discussed in the
second half of the book.

In most of the models considered here, it will be assumed that agents communicate their preferences to each other through small bets they are willing to take, while
an observer seeks to exploit any arbitrage opportunities they may create. What the
observer knows is by definition what the agents commonly know. This scenario is
only a thought experiment in settings that do not involve real markets or games
played for money, but to the extent that rational choice models assume that agents
possess mutually consistent probabilities and commonly known utilities and that
they exhibit equilibrium behavior toward each other, it is as if such communication
occurs and as if suitable markets exist and as if arbitrage opportunities are exploited.
Wherever the arbitrage story strains the imagination, it also raises questions about
the quantitative precision with which rational agents can be expected to articulate
their preferences (even to themselves), about the degree to which utilities can be
commonly known and probabilities can be commonly held, and about the uniqueness and refinement of equilibria they can be expected to seek, even in an as-if sense.

When agents are uncertainty-averse or otherwise have state-dependent marginal
utility for money, the units of revealed beliefs and equilibria become so-called "risk-
neutral probabilities" as in the theory of asset pricing by arbitrage. These have the
interpretation of products of subjective probabilities and relative marginal utilities
for money, but probabilities and utilities are not uniquely separated, nor do they
need to be separated in order for Bayes' rule to apply. Neither do they need to be
separated for modeling aversion to uncertainty and ambiguity. The derivatives of the
risk-neutral probabilities provide a tool for measuring those aspects of local preferences. The use of risk-neutral probabilities in the modeling of beliefs and knowledge
and interactions among agents will be a recurring theme.

The rationality standard of no-arbitrage remains applicable in situations where many of the usual assumptions do not apply, particularly the completeness axiom that guarantees the uniqueness of subjective probabilities and utility functions, the independence axioms that yield additive representations of preferences in which beliefs and tastes are cleanly separated, and the restrictions on reciprocal beliefs that are typically invoked in game theory. In these respects, the book extends a literature of generalized choice theory dealing with issues such as incomplete preferences, state-dependent utility, non-expected-utility, ambiguity, and coarsenings of Nash equilibrium. No-arbitrage is a principle of social rationality as well as individual rationality, so it can be used to directly characterize rational group behavior even when individual behavior is only quasi-rational. It implies that strategic and competitive equilibria will exist (at least approximately) even in the absence of well-defined personal probabilities and utilities.

The book's emphasis on the many uses of money may appear to be very limiting, even old-fashioned (I hesitate to say too-business-oriented). Most applications of rational choice theory deal with situations in which a decision-maker's experiences may have non-monetary physical or psychological attributes. Indeed, the path-breaking work of von Neumann and Morgenstern and Savage was aimed squarely at the problem of modeling preferences among acts whose outcomes could be consequences of any kind, such as getting rained upon or vaporized in a nuclear explosion. I will argue that the role of money—real money—in *quantitative* reasoning and quantitative interpersonal knowledge has become greatly underappreciated as the field has expanded in those directions.

In the chapters on expected utility and subjective expected utility, the units of money will be taken to be objective probabilities attached to consequences, as is conventional, but the operational and epistemic differences between probability currency and real currency will be emphasized. The alternative modeling framework of state-preference theory is used in later chapters. It generalizes the subjective probability model to allow nonlinear and state-dependent utility for money, and it does not attach intrinsic importance to the independence axiom of expected utility theory. As such it is a good environment in which to model non-expected-utility phenomena such as aversion to ambiguity, where it leads to variants of the so-called "smooth model."

The chapter on noncooperative game theory contains the results that are most novel (for most readers), and it is the key link in the chain of models that are unified by the no-arbitrage principle. It will show that in games with finite numbers of players and outcomes (the most elementary examples of noncooperative games), special kinds of side bets for money can be used to elicit information about the players' payoff functions, which need not themselves be monetary, and this method provides an operational standard of common knowledge of that part of the rules of the game that determines personal incentives. The parameters of such bets just happen to be the minimal data needed to calculate Nash equilibria and correlated equilibria. What it means for the numerical parameters of a game to be common knowledge is usually not directly addressed in game theory, at least not at a basic level. Rather, the values of the players' payoffs, in units of utility, are merely assumed to be part of the description of the game that is shared by everyone. Here the modeling of common knowledge of payoffs will be the very starting point of analysis.

Outcomes of a game that are rational, as seen by observers, are those which do not lead to arbitrage in the side bets, and by the separating hyperplane theorem they are precisely the outcomes that have positive probability in correlated equilibria, the set of which is a convex polytope. That is to say, an outcome of the game is commonly known to be rational in the no-arbitrage sense if and only if it is one that could have arisen from implementation of a correlated equilibrium (not necessarily a Nash equilibrium), which is a strategic version of the Dutch Book Argument. In light of this result, correlated equilibrium may be viewed as the more fundamental solution concept: it is the natural game-theoretic extension of the same principles of Bayesian rationality that apply to individuals (also to markets), as previously claimed by Aumann.

The correlation in a correlated equilibrium does not necessarily mean that the players have used correlated randomization (a common and unfortunate misportrayal), although that is a sometimes-desirable option to have. It also does not require Aumann's assumption of a common prior distribution. Rather, it reflects the natural limits of an observer's knowledge—and the players' common knowledge—based only on information that is incentive-compatible for them to reveal unilaterally through betting. In general, such knowledge is represented by a convex set of probability distributions that rationalize the acceptable bets, as in the theories of imprecise subjective probabilities and incomplete financial markets, and in the game setting those probability distributions are correlated equilibria. There is no simple additional rationality principle which requires those probability distributions to further satisfy the nonlinear independence constraints of Nash equilibria, which may lead to constellations of isolated points or to curves in space or to unique solutions that involve irrational numbers. Meanwhile, solution concepts that are (sometimes) weaker than correlated equilibrium, such as rationalizability and correlated rationalizability, do not protect the players from being collectively exploited by an arbitrageur.

In the case where players have diminishing marginal utility for money, the parameters of correlated equilibria are risk-neutral probabilities. A stylized fact is that the set of such equilibria is typically larger than what it would be if the players were risk-neutral, i.e., risk aversion tends to make the players less candid in revealing their preferences and this reduces the precision of common knowledge and the determinacy of solutions.

The models discussed here are a set of variations on a theme rather than a comprehensive and even-handed treatment of the theory. It is intended to serve as a complement rather than a substitute for other texts. To some readers the book may seem at first like an overly simplistic and polemical treatment of a field in which giants have walked (and still do), but I hope that you will suspend judgment for long enough to see where the story leads. It is written to be accessible to advanced undergraduates, graduate students, and researchers in fields where rational choice theory is applied, as well as interested readers from other disciplines. The only required tools are multivariable calculus (derivatives of functions), linear algebra (vectors and matrices), and elementary probability theory.

Robert Nau
Professor Emeritus, Decision Sciences
Fuqua School of Business, Duke University
November 12, 2024, © 2024

Acknowledgments: I am indebted to a great many colleagues for helpful discussions over many years, especially Jean Baccelli, Edi Karni, Peter Klibanoff, Mark Machina, Mark Schneider, Teddy Seidenfeld, and Peter Wakker, as well as my co-authors Kevin McCardle, Jim Smith, and Bob Winkler.

The cover picture illustrates the set of rational solutions of some classic two-player noncooperative games: chicken, stag hunt, and battle-of-the-sexes. The saddle is the set of randomized strategies that are independent between players. The hexahedron is the set of no-arbitrage solutions: correlated equilibria. Their three points of intersection are Nash equilibria. See Chapter 8 for details.

Dedicated to Becca

Chapter 1

Introduction

1.1 Social physics

Rational choice theory is social physics, a search for universal mathematical principles to explain social phenomena in terms of lawful interactions among their atoms, who are human decision-makers. In rational choice models, *individuals* strive to satisfy their *preferences* for the expected *consequences* of their *actions* given their *beliefs* about events and their *tastes* for risk and consumption. Their beliefs are quantified by *probability distributions*, their tastes are quantified by *utility functions*, and their equations of motion and resting points are determined by *equilibrium conditions* expressed in terms of their probabilities and utilities. This paradigm is the foundation of microeconomic theory, game theory, finance theory, statistical decision theory, and applied decision analysis, as well as mathematical subdisciplines in fields such as philosophy, political science, law, and sociology.

The label "social physics" has been used by detractors[1] of rational choice theory as well as its adherents, but it is historically accurate. It was coined by Quetelet (*Sur l'homme et le développement de ses facultés, ou Essai de physique sociale* (1835)),

---

[1]"More heat than light" in the words of Mirowski (1989). See also Green and Shapiro (1994) and Friedman (1996).

DOI: 10.1201/9781003527145-1

the astronomer who introduced statistical methods to the social sciences.[2] The flip term "mathematical psychics," which didn't catch on, was later proposed by Edgeworth (1881). Many who have studied the laws of mathematics and physics have looked for analogous laws of cognition and behavior. Descartes developed principles of analytic geometry and is considered to be the founder of the rationalist school of thought. Pascal made important contributions to the theory of fluid dynamics as well as probability and philosophy. Laplace used Bayesian methods to solve inferential problems in astronomy and later showed their applications to jurisprudence. Daniel Bernoulli (1738) explored the principle of conservation of energy as well as the principle of maximizing expected utility. Pareto's (1906) theory of economic equilibrium was shaped by his earlier studies of equilibrium in mechanical systems. Von Neumann axiomatized both quantum mechanics and expected utility. Samuelson and many others have explored the obvious parallels between economics and thermodynamics, and training in physics is literally the best preparation for a career as a rocket scientist on Wall Street. A time line of these developments and other milestones in physics and rational choice theory is shown in Figure 1.1.

Modern rational choice theory dates back nearly a century to a string of seminal publications that began with Ramsey's and de Finetti's papers on subjective probability in the 1920s and 1930s, von Neumann's minimax theorem for zero-sum games (1928) and his collaboration with Morgenstern on *Theory of Games and Economic Behavior* (1944), Nash's papers on bargaining and noncooperative and cooperative games (1950ab, 1951, 1953), Arrow's book *Social Choice and Individual Values* (1951b) and his papers on risk bearing (1951c, 1953),[3] Arrow and Debreu's papers on welfare economics and existence of competitive equilibria (Arrow (1951a), Debreu (1951), Arrow and Debreu (1954)), Blackwell and Girshick's *Theory of Games and Statistical Decisions* (1954), and Savage's formalization of Bayesian decision theory in his *Foundations of Statistics* (1954). These works were part of a general ferment of interdisciplinary research in economics, statistics, and applied mathematics that began in the early post-WWII era and laid the foundation for a steady growth in the use of formal models in the social sciences—especially axiomatic methods, concepts of measurable utility and personal probability, and tools of optimization

---

[2]Quetelet studied the statistical profile of the "average man" in diverse settings, and he looked for principles of equilibrium and conservation in social systems that were analogous to those of cosmological systems (Jahoda (2007)). The term "social physics" had also been proposed by Comte, who later invented "sociology" to distinguish his own search for abstract social laws from the more empirical approach of Quetelet. Ironically, the field that became known as sociology evolved more along Quetelet's lines than Comte's (Turner et al. (2011)). More recently the term has been warmly embraced by physicists working on "big data" applications in the social sciences, e.g., spatial analysis of activity in cities, internet social media traffic, etc., in the spirit of Quetelet's original work.

[3]A number of his papers from this period are collected in Arrow (1971).

Descartes: analytical geometry	Descartes: rationalism (learning from intuition and logic, vs. empiricism, learning from experience)
Pascal: projective geometry, hydrodynamics	
	Pascal, Fermat: probability in games of chance, "Art of Persuasion" (via axioms)
Newton: calculus; laws of motion, optics, gravitation; "rational mechanics" **1700**	
	D. Bernoulli: calculus applied to expected utility maximization in games of chance
D. Bernoulli: hydrodynamics, Bernoulli principle	Condorcet: theory of voting
Bayes, Price: Bayes' theorem	Laplace: "moral expectation", Bayesian inference applied to jurisprudence
1800	
Laplace: general theory of probability, celestial mechanics, Bayesian inference in astronomy	Bentham: "felicific calculus," utilitarianism
	Quetelet: statistical theory of the average man, "social physics"
Lagrange, Hamilton: equations of mechanics	
Maxwell: equations of electrodynamics	Cournot: theory of monopoly and duopoly
	Jevons, Menger, Walras: marginal utility analysis, general equilibrium
Gibbs: equilibria of thermodynamic systems	Edgeworth, Pareto: indifference curve analysis
Einstein: special relativity theory, Brownian motion model of particle movements **1900**	Bachelier: Brownian motion model of stock price movements
	Slutsky: fundamental equation of value theory
Farkas' lemma (separating hyperplane theorem)	
	Ramsey: axiomatization of subjective expected utility calibrated with ethically neutral propositions
Brouwer: fixed point theorem	
Hilbert, Russell, Whitehead: axiomatization of mathematics	von Neumann: minimax theorem for zero-sum games
Heisenberg, Schrödinger: quantum theory, uncertainty principle	de Finetti: subjective theory of probability based on coherent bets for money
von Neumann: axiomatization of set theory, quantum mechanics	von Neumann/Morgenstern: axiomatization of expected utility, game-theoretic foundations of economics
Bonnesen/Fenchel: theory of convex bodies	Samuelson: thermodynamic methods in economics
von Neumann: duality theorem of linear programming based on Farkas' lemma	Nash: equilibria of noncooperative games
1950	Arrow: axiomatic social choice
Dantzig, Gale, Kuhn, Tucker: algorithms for linear and nonlinear programming, applications to zero-sum game theory	Arrow/Debreu: existence of general equilibrium
	Savage: SEU foundations for Bayesian statistics

FIGURE 1.1: Milestones in physics/mathematics and rational choice theory (to early 1950s)

theory, general equilibrium theory, and game theory.[4,5] In subsequent decades new theoretical frontiers were opened by refining and generalizing the classic models to expand the range of applications they cover. New developments in game theory, particularly in the modeling of incomplete-information games and designed mechanisms, have made it a universal wrench in the toolkit of microeconomics. Finance theory was revolutionized by the development of models for arbitrage-free pricing of derivative securities. Models of "non-expected utility" preferences were developed to explain robust patterns of behavior that had formerly been considered paradoxical. By now the field is broad, deep, and mature, but it is far from unified.

---

[4]Much of this research was sponsored by the RAND Corporation. Poundstone (1992) describes the environment there: "It was at RAND rather than the groves of academia that game theory was nurtured in the years after von Neumann and Morgenstern's book. In the late 1940s and early 1950s, few of the biggest names in game theory and allied fields *didn't* work for RAND, either full-time or as consultants. Besides von Neumann, RAND employed Kenneth Arrow, George Dantzig, Melvin Dresher, Merrill Flood, R. Duncan Luce, John Nash, Anatol Rapoport, Lloyd Shapley, and Martin Shubik—nearly all of whom were there at the same time. It is difficult to think of any other scientific field in which talent was concentrated so exclusively at one institution." In the late 1950s, Daniel Ellsberg became a nuclear strategy analyst at RAND, and his observations on military decision-making provided the inspiration for his Ph.D. research on the topic of ambiguity (1961) as well as his later release of classified documents known as the Pentagon Papers that helped to turn public sentiment against the Vietnam War and brought about the downfall of President Richard Nixon.

[5]Several watershed conferences between 1949 and 1952 showcased the pioneering work on expected utility, subjective probability, mathematical programming, and their applications in modeling games and competitive economies. A conference on "Activity Analysis of Production and Allocation" was held at the University of Chicago in 1949, sponsored by the Cowles Commission on Economic Research and organized by its director, T.C. Koopmans. The participants included mathematicians G. Brown, G. Dantzig, M. Flood, M. Gerstenhaber, D. Gale, H. Kuhn, A. Tucker, and C. Tompkins, as well as economists K. Arrow, R. Dorfman, N. Georgescu-Roegen, L. Hurwicz, A. Lerner, J. Marschak, O. Morgenstern, S. Rieter, P. Samuelson, and H. Simon. In the parts of the meeting devoted to mathematics and computation, the papers included "Convex Polyhedral Cones and Linear Inequalities" by Gale, "Theory of Convex Polyhedral Cones" by Gerstenhaber, "Linear Programming and the Theory of Games" by Gale, Kuhn and Tucker, "A Proof of the Equivalence of the Linear Programming Problem and the Game Problem" by Dantzig, "Application of the Simplex Method to a Game Theory Problem" by Dorfman, and "Iterative Solution of Games of Fictitious Play" by Brown. The Second Berkeley Symposium on Mathematical Statistics and Probability was held in 1950, and its participants included Arrow ("Extensions of the Basic Theorems of Welfare Economics"), Marschak ("Why 'should' Statisticians and Businessmen Maximize 'Moral Expectation'?"), Kuhn and Tucker ("Nonlinear Programming"), de Finetti ("Recent Suggestions for the Reconciliation of Theories of Probability"), Girshik and Savage ("Bayes and Minimax Estimates for Quadratic Loss Functions"), as well as R. Feynman ("The Concept of Probability in Quantum Mechanics") and many other scientists from other fields. In 1952, an international conference on "The Foundations and the Applications of the Theory of Risk" was held in Paris, organized by M. Allais and including K. Arrow, B. de Finetti, M. Frechet, M. Friedman, R. Frisch, J. Marschak, E. Malinvaud, P. Masse, P. Samuelson, L. Savage, and H. Wold. By this time the theory of expected utility was well enough entrenched that Allais stood out for his criticism of it. More history of this period is given by Dantzig (1963) and Debreu (1984).

Von Neumann and de Finetti drew explicit analogies between their theoretical ambitions and those of modern physics. Von Neumann compared the project of developing a quantitative theory of utility with the project of developing a quantitative theory of heat, which also started from subjective experience. De Finetti appealed to the spirit of operationalism in physics—a focus on what is measurable by real observers with real instruments—following the examples of Einstein, Mach, and Bridgman. The contributions of von Neumann and de Finetti to several main branches of rational choice theory—expected utility, subjective probability, and game theory—are intertwined in ways that will be emphasized throughout this book. A brief outline is the following. In 1928, von Neumann proved the minimax theorem for 2-player zero-sum games, which shows that such a game has a unique rational solution, and it instructs each player to optimize against the most damaging strategy of her opponent. His elegant and complex proof relied on showing the existence of a saddle point in independently randomized strategies, using topological methods. He later discovered that the same result could be proved very easily using algebraic methods based on a separating hyperplane theorem for convex polyhedral sets. In the 1944 book, von Neumann and Morgenstern presented a special case of the separating hyperplane theorem, which they called the "theorem of the alternative," and they used it to give a simple proof of the minimax theorem.[6] This same theorem can be used to prove the duality theorem of linear programming, as von Neumann explained to George Dantzig in 1947, who was developing the first computational algorithm for solving such problems on a large scale.[7] Meanwhile, de Finetti pursued the idea that subjective probabilities should be operationally defined in terms of rates at which agents are willing to make conditional and unconditional *monetary bets* on events, and he used linear equations to establish that *coherent* betting rates—those which avoid arbitrage—must satisfy the additive and multiplicative laws of probability. Later authors (e.g., Scott (1964), Heath and Sudderth (1972)) showed that this result can be framed more neatly as an application of the separating hyperplane theorem. The subjective probability distribution is obtained in one step this way, and thus, von Neumann and Morgenstern implicitly provided the right tool for a simple proof of de Finetti's theorem as well as the minimax theorem. Somewhat ironically they did not apply the same tool to the proof of existence of a

---

[6]Soltan (2021) discusses the history of the separating hyperplane theorem in general. Minkowski (1911) formulated it for the 3-dimensional case. Bonneson and Fenchel (1934) presented the *n*-dimensional version with the terminology "separating hyperplane" without reference to Minkowski. Farkas' lemma (1902) is a special case that is the basis of the duality theorem of linear programming. Ville (1938) gave the first proof of the minimax theorem using algebraic methods, which provided inspiration to von Neumann and Morgernstern for their own proof, as discussed by Kjeldsen (2001) and Dimand and Ben-El-Mechaiekh (2010). A simpler proof is given by Ben-El-Mechaiekh and Dimand (2011).

[7]See footnote 22 in Chapter 2 for the details of von Neumann's meeting with Dantzig.

von Neumann-Morgenstern utility function in games of chance. Later authors did so (e.g., Fishburn (1975b), Gilboa (2009)), and it will be used here for that purpose in Chapter 4.

Another irony in the importation of tools and concepts from physics into social physics is the general absence of any analog of the uncertainty principle in the latter. The uncertainty principle in physics, which is associated with the name of Heisenberg (of whom von Neumann was a protégé), was a landmark development in 20th-century physics and it can be interpreted to represent the fact that an act of physical measurement perturbs the object whose properties are being measured. For example, a measurement of the position of a particle perturbs its momentum, and vice versa. One might think that this general principle would also resonate deeply in social physics through the fact that psychological measurements inevitably perturb the mind of the subject (e.g., via conversations with others or being asked to contemplate fictional opportunities in a laboratory setting). The closest approach of rational choice theory to the incorporation of a *quantitative* uncertainty principle is to be found in the theory of imprecise probabilities (discussed in the later parts of Chapter 3), where lower and upper bounds on revealed subjective probabilities might be attributed to information that is expected to be gained from interactions with a betting opponent who plays the role of a measuring instrument. This representation is obtained when preferences among bets are allowed to be "incomplete," i.e., when the decision-maker is allowed to be undecided. More broadly a social-physics uncertainty principle ought to be acknowledged by allowing preferences to be incomplete not only for bets but for things in general. The importance of addressing the problem of incomplete preferences is a recurring theme in the book: it will come up in nearly every chapter.

Nash (1950a, 1951) used a fixed-point theorem (another topological tool) to establish the existence of an equilibrium point of an *n*-player non-zero-sum game in which expected-utility-maximizing players may use independently randomized strategies, a result which generalizes the minimax theorem. (In a 2-player zero-sum game, the minimax solution is the unique Nash equilibrium.) Several decades later, Aumann (1974, 1987) proposed that players in a noncooperative game should be allowed to seek an equilibrium in *correlated* randomized strategies, and he showed that this solution concept is the natural one from the viewpoint of subjective probability theory, assuming that the players begin with a common prior distribution on private signals received from an external device which may be used to correlate their strategies. In 1990 Kevin McCardle and I showed that when acceptable bets a-la-de-Finetti are used to elicit the rules of the game, an ex post application of the requirement of no-arbitrage ("joint" coherence) leads to the result that the rational outcomes of the game are those that have positive probability in some correlated

equilibrium,[8] without separately assuming a common prior and without any requirement for a correlated randomization device. The as-if existence of a common prior,
as seen from the perspective of an observer, is itself a requirement of no-arbitrage.
The proof uses the separating hyperplane theorem to show that exactly one of the
following two alternatives must hold: (i) the outcome of the game leads to an arbitrage loss for the players as a group, or (ii) the outcome is one that has positive
probability in a correlated equilibrium. This result directly generalizes de Finetti's
theorem to a setting in which events are under the control of game players with reciprocal subjective beliefs, and it leapfrogs over Nash equilibrium in extending the
minimax theorem to more general noncooperative games. As such it provides a direct integration of game theory with single-agent decision theory and many-agent
market theory. It also calls attention to some fundamental issues of interpersonal
measurement: who knows what about whom, how do they know it, and if they were
behaving irrationally, *so what*?

1.2 The importance of having money

It's good to have money. It's what makes economic behavior possible, and it
has been central to the functioning of great societies for thousands of years.[9] It's
a tool of thinking as well as transacting. It's portable and divisible, it is subject to
exact measurement in public, and it is a conserved quantity at the microeconomic
level. It directly extends principles of physics to principles of social physics: it is
(almost) on a par with length, mass, and time as a physical variable of the natural

---

[8]The set of *all* correlated equilibria of a game is a convex polytope, analogous to convex polytopes
of subjective probabilities that represent incomplete beliefs of individuals. Vertices of the polytope can
be found via linear programming. The cover of this book shows a picture of the correlated equilibrium
polytope for the battle-of-sexes game (and also the games of chicken and stag hunt). Given that there
was already intense interest in the properties of convex polytopes and the use of linear programming in
the context of zero-sum noncooperative games by the late 1940s, it is a remarkable oddity in the history
of economic theory that the concept of correlated equilibrium was not even formalized until 1974 (by
Aumann). Roger Myerson has remarked: "If there is intelligent life on other planets, in a majority of
them, they would have discovered correlated equilibrium before Nash equilibrium." (Quoted in Solan
and Vohra (2002).)

[9]Money in the form of weighed amounts of precious metals was used in the Babylonian empire c. 2000
BCE, and the Code of Hammurabai established laws to regularize the banking industry. Coined money
was introduced in the 7th or 8th century BCE, and Galbraith (1975) quotes a passage from Herodotus
about its use as a savings medium in the business of prostitution by marriageable young women of Lydia.
A treatise on "how finance made civilization possible" has been written by Goetzmann (2016).

world, and its units combine with theirs.[10] A layperson might guess that a book on mathematical models of rational choice would typically emphasize the importance of money in *quantitative* reasoning and communication and action with respect to beliefs and values, because without it the numbers may be fuzzy and hard to agree upon and act upon, but she would be wrong.

The idea that a decision-maker ought to perform expected-value calculations in units of utility rather than money, and that money should have diminishing marginal utility, was proposed by Daniel Bernoulli (1738) as an explanation for risk-averse behavior in the context of gambling and insurance. The proposition that a decision-maker's values for things in general might be numerically measured in units of utility was pursued by 19th-century economists who sought to bring quantitative methods to bear on the analysis of consumer behavior, where it resonated with contemporary concepts of unobservable fields and forces in physics. The (literal) cardinal role that objectively measurable monetary risk had played in Bernoulli's analysis was not appreciated. By the early 20th century the use of utility functions in the context of demand theory was discredited due to a lack of sound foundations for their measurement (Cassel (1924), Stigler (1950)), but the utilitarian program had set a precedent for using a numeric variable to represent the preferences of a rational individual.

Utility functions were resurrected and re-animated by von Neumann and Morgenstern (1944), who introduced a system of axioms for preferences that implies the existence of a cardinal index of choice under risk, providing a foundation for modeling the use of randomized strategies in games where payoffs are not necessarily monetary. It is too bad that the baggage-laden term "utility" was recycled as the name for this new construct,[11] but it served the same purpose as the earlier one insofar as it shifted the focus away from money and toward a subjective measure of value as the unit of account. Strictly speaking, the von Neumann-Morgenstern utility of an outcome is not a measure of its ex post value to the decision-maker. Rather, it is only an index to be used ex ante in the context of an expectation, as Bernoulli had done. But it behaves just like a universal cardinal measure of value, one that also has the very convenient property of being subject to multiplication by probabilities.[12]

---

[10]Price per unit of length, mass, time, etc.

[11]Ellsberg (1954) described it this way: "...von Neumann and Morgenstern, in their famous aside to the economics profession, announced they had succeeded in synthesizing 'measurable utility.' That feat split their audience along old party lines. It appeared that a mathematician had performed some elegant sleight-of-hand and produced, instead of a live rabbit, a dead horse. ... It is unfortunate that old terms have been retained, for their associations arouse both hopes and antagonisms that have no real roots in the new context."

[12]It is possible to axiomatize a cardinal measure of value that does not necessarily bear any relation to the von Neumann-Morgenstern utility function by introducing additional psychological measurements of relative strength of preference, e.g., "the degree of preference for w over x is greater than or equal to the

Savage (1954) introduced an axiomatic model of preferences under uncertainty that grafted together de Finetti's subjective probability model and von Neumann and Morgenstern's expected-utility model,[13] providing a foundation for a theory of statistical inference in which the role of data is to update subjective probabilities in accordance with Bayes' rule in completely general decision settings where consequences may be arbitrary rewards or personal experiences. The Bayesian model was slow to gain acceptance because of its computational demands[14] and its personalistic interpretation of probability, but it was revolutionized by the introduction of Monte Carlo Markov Chain methods (Gelfand and Smith (1990)) that made it possible to calculate posterior distributions for virtually unlimited numbers of parameters, stimulating the development of "objective" Bayesian methods that use impersonal, non-informative priors. In its broader role as a model of the rational economic person, Savage's axiom system (and the simpler variant due to Anscombe and Aumann (1963)) aims for a clean separation of beliefs from tastes: *subjective probabilities are measures of beliefs about events, which do not depend on the consequences to which they may lead, and utilities are measures of tastes for consequences, which do not depend on the events that may have led to them.*[15]

It has become conventional for textbooks of decision theory and microeconomic theory and their fields of application to present the subjective-expected-utility model as the bedrock of rational behavior under uncertainty (with a few inconvenient paradoxes for which more parameters may be needed) and to model the cognition of the individual in terms of her utility function and subjective probabilities. It may even be asserted that the model provides a "guide to introspection" (Mas-Colell et al. (1995) p. 178). *Virtually no one introspects in this way.* The exception that proves this rule is provided by the field of Bayesian decision analysis, which uses the subjective

---

degree of preference for y over z." (Krantz et al. (1971), Dyer and Sarin (1979)) However, these sorts of measurements do not correspond to choices, so the value/utility distinction is not often raised except in models of multiattribute utility.

[13] Savage spent two years as a statistical assistant to von Neumann from 1941 to 1942 at the Institute for Advanced Study (Lindley (1980)) . He struck up a relationship with de Finetti during the latter's U.S. visit in 1950 (Fulvia de Finetti (2011)) and became strongly influenced by his subjectivistic views.

[14] My own interest in Bayesian decision theory was sparked by a survey course on Bayesian statistics that Dennis Lindley taught in the IEOR department at Berkeley in 1979. He ended the course by giving us the disappointing news that Bayesian methods would never be practical because it would never be feasible to perform numerical integration in as many as ten dimensions.

[15] The separation of beliefs and tastes is also emphasized in contemporary models of non-expected utility preferences that address the issue of aversion to ambiguity as well as aversion to risk, such as the max-min expected utility model of Gilboa and Schmeidler (1989), the smooth model of Klibanoff et al. (2005), and the variational preference model of Maccheroni et al. (2006). These models retain the feature of a scalar function that maps consequences to units of measurable utility, and they combine it with a richer description of beliefs such as a set of probability distributions, or a second-order probability distribution, or an additive penalty function applied to probability distributions.

expected utility model for prescriptive purposes (Howard (1968), Raiffa (1968)) but which emphasizes that probabilities and utilities must be constructed, a process that often requires well-paid expert guidance. This paradigm has proved to be successful but not world-conquering, and significantly the most important benefits in decision analysis applications generally do not flow from thinking in terms of subjective probability and expected utility per se. Rather, they flow from a disciplined approach to decision modeling in which objectives are clarified, alternatives are generated, risks are prioritized, data is collected, financial and statistical models are built, and, ideally, new options are discovered along the way that are obviously better than the original ones. Introspection occurs mainly in the early problem-structuring phase of this process. Where probabilities are needed, they are often obtained from data analysis or outside experts. When risk attitudes are formally modeled at all, an exponential[16] utility function is typically used for analytical convenience, but introspection still is not carried out in the language of utility. Rather, the decision-maker's risk attitude is assessed in terms of a single parameter–her *risk tolerance*, which is a quantity of money–and alternatives are compared in terms of their *certainty equivalents*, which are also quantities of money. Log and power utility functions are widely used in financial economics, but there too the decision-maker's risk attitude is modeled with a single parameter, which measures her risk tolerance relative to her level of wealth.

In game theory, some texts begin with the assumption that outcomes are measured in units of von Neumann-Morgenstern utility and the game is played as if the players are in separate laboratory cubicles where they see each others' utilities displayed on computer screens (Fudenberg and Tirole (1991), p. 4) . Utility functions have also re-emerged in the standard treatment of consumer theory, subject to the caveat that they are unique only up to increasing (not necessarily linear) transformations. In models of situations where money plays a role, it is often not referred to as money: it is merely a consequence whose range is an interval of real numbers, or the linear variable in a quasilinear utility function, or the numeraire for a budget constraint.

---

[16] An exponential utility function can be written in the form $u(x) = 1 - \exp(x/\tau)$, where the parameter τ is the decision-maker's risk tolerance. Its value can be assessed (to a very close approximation) by asking the following question: what is the value of τ for which you are just barely willing to accept a gamble offering a 50% chance of winning $\$\tau$ and a 50% chance of losing $\$\frac{1}{2}\tau$? The certainty equivalent of a risky alternative is the amount of money which, if received with certainty, would be equally preferred to it. Under an exponential utility function, the 50-50 gamble between winning $\$\tau$ and losing $\$\frac{1}{2}\tau$ has a certainty equivalent of almost exactly zero when τ is the decision-maker's risk tolerance. In applications of this model to high-level corporate decision-making, the appropriate risk tolerance is typically some characteristic fraction of equity. Howard (1988) suggests a value in the vicinity of one-sixth based on interviews with corporate executives. In applications of this model to personal decision-making under uncertainty, such as investing for retirement or insuring against catastrophe, risk tolerances are also generally elicited or prescribed relative to the decision-maker's level of wealth.

The measurement of outcomes in units of utility and beliefs in units of subjective probability confers a sweeping generality on the theory of rational choice. Any personal decision or social interaction, no matter what the setting, plausibly lies within the scope of the theory, and the same mathematical models of rational behavior—Bayesian inference and expected utility maximization by individuals, equilibrium-seeking by players in games and markets and political systems—can be applied everywhere. Exact predictions are achieved by using highly refined solution concepts requiring not only that utilities and probabilities should be uniquely determined but also that the utilities should be commonly known and the probabilities should be mutually consistent in situations where there are multiple actors, and the solution concepts themselves must be commonly embraced by the actors. Such assumptions are qualified by "as-if" provisos, but compelling reasons why it should be as if these things are true are often lacking. Rather, it is assumed that by definition rational actors will behave as if these things are true.

What's wrong with any of this? There's nothing wrong in principle with using subjective probabilities and utilities (or generalizations thereof) to parameterize a tractable psychological model of an individual whose real and hypothetical choices are determined as if by preferences that satisfy normatively appealing or descriptively suggestive rationality axioms. After all, the individual has the right to think in whatever terms she wants and satisfy whatever axioms she wants in her own small world.[17] The psychological and behavioral models are merely primal and dual representations of the same person: an inner self and an outer self. In fact, they are literally primal and dual in the sense of linear programming, which is "the" fundamental theorem of modern rational choice theory and one of the central themes of this book. The inner-self representation is more convenient to use for purposes of modeling because it requires only the specification of probabilities and utilities. The parsimony of this parameterization follows from the fact that it exploits the constraints imposed by the preference axioms, which are assumed to be satisfied exactly.[18]

---

[17] On the other hand, she might choose to satisfy her preferences as they are and "let the axioms satisfy themselves" as Samuelson (1950) famously remarked.

[18] In this respect social physics is perhaps better characterized as social mathematics, along the lines of the distinction drawn by Brillouin (1964): "Open a book of pure mathematics and consider a theorem. It is always built on a typical scheme: given certain conditions A, B, C, which are assumed to be exactly fulfilled, it can be proven rigorously that conclusion Q must be true. Here the physicist starts wondering: how can we know that A, B, and C are *exactly* fulfilled? No observation can tell us that much. The only thing we know is that A, B, and C are approximately satisfied with certain limits of error. Then, what does the theorem prove? Very small errors of A, B, or C may result in a very small error of the final statement Q, or they may destroy it completely. The discussion is not complete until the stability of the theorem has been investigated, and that is another story!" (p. 33)

The difficulty arises when the parameters of the inner self rather than those of the outer self are treated as if they have an objective reality and are asked to bear the additional weight of common knowledge assumptions, inter-agent consistency requirements, and equilibrium conditions expressed in numerically exact terms. Different agents may not discriminate the events and alternatives and outcomes in the same way, they cannot really know each others' preferences among hypothetical alternatives, and they satisfy the preference axioms only approximately at best. The situation is even worse if the outcomes are intrinsically unquantifiable. As we get farther away from settings whose parameters are concrete, the probabilities and utilities become fuzzier and the observability and consistency requirements they should be expected to satisfy become correspondingly weaker. Universally respected scales of distance and weight and time—and money—exist because the things to which they refer really do behave lawfully enough that stable and precise quantitative measurements are possible. Moreover, *public* measurements are possible, and a quantity that is publicly measured with an objective yardstick is de facto common knowledge. Subjective probabilities and utilities do not have this property.

1.3   The impossibility of measuring beliefs

In the course of spending my entire career working in the field of Bayesian decision theory, I have gradually come to the dismal conclusion (shared by a few others) that it is not possible in general to observe subjective probabilities that measure only the decision-maker's beliefs about events, irrespective of the consequences that may be attached to them, particularly in situations where the events are of personal importance to the decision-maker (Nau (2001)). The reason for this is that there is no way to be sure a priori that a risk-averse decision-maker has no significant prior stakes in events or that her marginal utilities for money or her utilities of consequences in general are not intrinsically state-dependent. The unknown state-dependent utilities would interact with subjective probabilities in the determination of observable preferences, because expected utility depends (only) on the product of the two. So, when the decision-maker's subjective probabilities are elicited by methods such as asking her to bet money on events or to compare hypothetical lotteries in which prizes are staked on real events against hypothetical lotteries in which the same prizes are staked on spins of a roulette wheel, it may be impossible for observers to disentangle her subjective probabilities for events from her possibly-state-dependent marginal utilities for money or possibly-state-dependent utilities for prizes. And if it is impossible for others to disentangle them, then it may also be impossible (or at least, unhelpful) for the decision-maker to try to disentangle them for herself by reflecting on her own preferences. A half-century of work in behavioral decision theory has shown that people tend to think in many other ways under conditions of uncertainty. Theoretical approaches to the separation of probability and utility have been widely explored (especially by Karni (1996, 2003, 2011ab), Karni et al. (1983), Karni and Mongin (2000), Karni and Schmeidler (2016); also Seidenfeld et al. (1990a), Baccelli (2017, 2019) and many others), and the intricacy of the modeling that is required, which generally involves higher levels of counterfactualism in the definitions of acts over which preferences are assumed to be observable, is a testament to the profound nature of the problem.

Savage (1954) defined a consequence in his model to be an outcome that (at least in the imagination) can occur and be experienced in the same way in every state of the world, in terms of which an act is defined as an arbitrary mapping from states to consequences. Thus, consequences exist independently of states and acts, and this is what allows subjective probabilities of states to be uniquely separated from utilities of consequences through observations of preferences among all such acts. The profession has largely followed him over the last 70-odd years in this line of thinking, notwithstanding the literature cited above. There is a strong economic and philosophical tradition of viewing beliefs about events as things that exist unto themselves, independently of consequences to which those events may lead, and the pursuit of this idea leads to parsimonious and flexible models for analysis. But strictly speaking this is a contradiction in terms: *a consequence is literally that which "follows from" some particular course of actions and events,*[19] *so it can't be logically divorced from them, and even an outcome with an objective definition may have a situationally dependent value.* A famous example is given by Aumann (1971) in a letter to Savage, in which a man whose beloved wife is about to undergo a dangerous medical operation is asked whether he would prefer to bet $100 on her survival or on the flip of a coin. If his enjoyment of everything in life would be diminished by her loss, then for him there is no such thing as a consequence that is the same under both outcomes, and the contemplation of bets on the operation's success or hypothetical acts in which they live happily together after her death might not help him to articulate his subjective probabilities. He might also feel better or worse about the outcome depending on the decision that preceded it. Savage himself had no trouble with this, and he famously wrote back to Aumann:

---

[19]Definition dating to the 15th Century according to https://www.etymonline.com/word/consequence. Also at that link, see the description of the entertaining 1796 parlor game called "consequences."

"In particular, I can contemplate the possibility that the lady dies med-
ically and yet is restored in good health to her husband. Put a little
differently, I can ask Mr. Smith how he would bet on the operation if
the continuance of his family life were not dependent on its outcome.
Make believe is certainly involved, and indeed it is extremely difficult
to make believe to the required extent. Yet, it does seem to be a helpful goal. Incidentally, it would not be nonsensical, though unmannerly,
for the experimenter to guarantee to execute Mrs. Smith if she recovers
from the operation. And I see no real objection to Mr. Smith imagining
this cruel situation if it helps him assess his own probabilities."

Judge for yourself. Other examples of a similar nature readily come to mind: judg-
ing the probability that your house will burn down in the next year, or that you will
suffer a debilitating illness or gruesome injury, or that the airplane you are flying
on will crash, or that your expensive new car will break down just after the manu-
facturer's warranty expires, or that another stock market plunge will wipe out your
retirement portfolio. Is it a worthwhile exercise to assess your own probabilities for
such events by contemplating scenarios that are not only unlikely but logically or
socially impossible? The situations in which individuals might be expected to make
informative and unbiased subjective probability judgments about important events
tend to be ones in which they have expert knowledge about the events but do not
have personal stakes in them, and where they routinely make many such judgments
that can be calibrated over the long run, as in the case of weather forecasting. But
those are the exceptions that prove this rule.

That's the dismal news. But there's a bright side for the decision-maker: in most
of the important situations where personal probability assessment is most difficult
(insuring against catastrophe, investing in financial markets, choosing among unreliable products, etc.) she has no need to do it by herself in any detail. The market
has already done it for her. She can buy the customary amount of insurance for her
car or her house or her life at the market rates, she can buy index funds rather than
solve a portfolio selection problem, she can use prices as a guide to relative quality, she can search for testimonials on web sites, and she can observe what others
around her are doing. She will make some adjustments for her own resources and
idiosyncrasies and whatever private knowledge she may possess, but she doesn't
need to go to the trouble of articulating her beliefs in units of subjective probability.[20] A boundedly rational consumer makes rational choices under uncertainty by

---

[20]The individual also faces a stream of less-important risks and uncertainties on a day-to-day or even
minute-to-minute basis, and these generally do not require precise anticipation or quantification either:
they are typically dealt with as they arise by relying partly on heuristics and partly on as-needed support
from other individuals and from institutions whose business is to deal with the unexpected.

letting the market generate an efficient set of risky alternatives and price them out relative to each other and then try to manipulate her awareness and desires for them. She only has to make fairly high-level decisions about the risks she wishes to take relative to what is conventional for her peer group, within the limits of her own budget, and that is her own incremental contribution to the rational market. And, as with most other consumer choices, her preferences for the alternatives that she is offered will be most readily quantifiable when she is able to think of them directly in terms of money.

1.4 Risk-neutral probabilities

There is also a bright side for the theorist in giving special attention to the uses of money in decision-making, namely that in many situations it suffices for modeling phenomena such as aversion to uncertainty and ambiguity without separating probabilities from utilities. In the most fundamental models of choice under uncertainty to be discussed in the latter half of the book, the parameters of belief are "risk-neutral probabilities," the local marginal rates at which an individual is willing to bet money on events. Risk-neutral probabilities are central to the theory of asset pricing by arbitrage in finance. Here they will also be applied across other domains of choice under uncertainty. They can equally well be called "risk-adjusted probabilities" or "relative state prices," and they were dubbed "util-probs" by Samuelson and Merton (1969). (See Rubinstein (2006), pp. 115-116 for some history.) The term is perhaps misleading insofar as it may seem to imply that the decision-maker or the representative agent is risk-neutral, which she isn't. Rather, her risk-neutral distribution is the probability distribution with respect to which she evaluates the acceptability of small bets *given* her current portfolio of investments and her degree of aversion to uncertainty, and as such it is the subjective probability distribution that an observer would attribute to her on the naive assumption that she is risk-neutral.

If the decision-maker has subjective-expected-utility preferences in her own mind (which in general she need not have), then her risk-neutral probabilities are the product of her subjective probabilities and her relative marginal utilities for money in the status quo, but these parameters of her cognition may not be separately distinguishable by observers of her behavior. Drèze ((1970), p. 152) states: "It may come as a surprise to the reader that, *while the prices for contingent claims to the numeraire ... have all the properties of a probability distribution, they are to be interpreted as the product of a probability by a marginal utility*. There is, however, nothing paradoxical about such a situation. In the more familiar context of expected utility analysis for the individual... the marginal expected utilities of contingent claims to the numeraire also behave like probabilities; still, they are not to be confused with probabilities."[21]

Risk-neutral probabilities are (I will argue) one of the most foundationally important tools in the modeling of choice under uncertainty in general. They are especially useful in settings where agents interact with each other in public and money exchanges hands, which are best-case scenarios for building quantitative models of rational behavior whose parameters can be measured with precision and which do not require the decision-maker to assess or reveal preferences among acts that are counterfactual. Strangely, this tool is not often mentioned in the decision theory literature outside of the field of finance, and its properties and potential uses are underexplored. Risk-neutral probabilities will be discussed in great detail in the second half of the book, but here is a brief introduction in the meantime.

Suppose that there are two states of the world for which the decision-maker's subjective probabilities are p_1 and p_2, suppose that her attitude toward risk is represented by a state-independent Bernoulli[22] utility function u, and let u' denote the first derivative of u, i.e., $u'(x)$ is the decision-maker's marginal utility of money when her wealth is x. Then a total utility function U that represents her preferences can be written as $U(x_1, x_2) = p_1 u(x_1) + p_2 u(x_2)$, where x_1 and x_2 denote hypothetical wealth in states 1 and 2. Now let w_1 and w_2 denote her status quo wealth in the two states. The decision-maker's risk-neutral probability for state m in the status quo is $\pi_m = (p_m u'(w_m))/(p_1 u'(w_1) + (p_2 u'(w_2))$, because that is the marginal price of a unit lottery ticket on state m that yields zero marginal change in expected utility if she buys or sells it. (A unit lottery ticket pays \$1 if the event occurs and \$0 otherwise.) If her utility for money is also intrinsically state-dependent, represented by utility functions u_1 and u_2 in states 1 and 2, then her risk-neutral probability for state m is just $\pi_m = (p_m u'_m(w_m))/(p_1 u'_1(w_1) + (p_2 u'_2(w_2))$. If she has non-expected utility preferences represented by a utility function $U(x_1, x_2)$ that is not additive across states, then her risk-neutral probability for state m is $(\partial U/\partial x_m)/((\partial U/\partial x_1) + (\partial U/\partial x_2))$ evaluated at $(x_1, x_2) = (w_1, w_2)$. In higher dimensions, the risk-neutral probability distribution at the status quo is just the

---

[21] The version of this paper published in *European Economic Review* in 1970 was a slightly updated version of the paper presented at the First World Congress of the Econometric Society held in Rome in 1965. There it was presented in a session entitled "Economics of Uncertainty" organized by Arrow and chaired by de Finetti. It builds upon results in Arrow's 1953 paper on "The role of securities in the optimal allocation of risk-bearing." It was later reprinted in Drèze (1987).

[22] A Bernoulli utility function measures utility for money, i.e., it is a von Neumann-Morgenstern utility function in the special case where the outcomes are levels of wealth or monetary gains with respect to a known origin of coordinates.

normalized gradient of U. So, risk-neutral probabilities exist and (as will be shown) they are useful parameters for the modeling of decisions even in settings where subjective or objective probabilities may *not* exist.[23] In general, risk-neutral probabilities are unique only up to the choice of a currency (dollars, euros, ounces of gold...) because of the possibility that rates of exchange between currencies could be state-dependent (e.g., if the states under consideration are *defined* by future values of exchange rates),[24] but this is not a problem if the rates are the same for everyone in all states. The decision-maker and her observers can make the same conversions of risk-neutral probabilities from one denomination to another.

To the extent that relative marginal utilities for money are warped by unknown prior stakes in events or state-dependent utility or state-dependent exchange rates or non-expected utility preferences, risk-neutral probabilities will not be interpretable as subjective probabilities that measure beliefs alone. And, for the same reason, preferences among risky gambles will not reveal a von Neumann-Morgenstern utility function that measures attitudes toward risk alone: those preferences will also be contaminated by unmeasurable background uncertainty and state-dependence. It may appear as though this inseparability of subjective probability and utility must have dire consequences for both Bayesian decision theory and game theory, inasmuch as the former is concerned with issues such as how subjective probabilities ought to be updated upon receipt of information and how risk attitudes are determined by the curvature of the utility function, while the latter is concerned with issues such as how game players should endogenously form beliefs about each others' strategy choices based on the assumption that their von Neumann-Morgenstern utility functions are common knowledge. This will turn out to be less of a problem than is commonly imagined. *The unique separation of probability from utility is not an essential requirement of Bayesianism.*

Risk-neutral probabilities are updated by Bayes' rule in exactly the same way as true probabilities when information is obtained from an impersonal source with agreed-upon reliability, so attempting to disentangle an individual's subjective probabilities from her state-dependent marginal utilities for money is no more productive than asking her to explain why her prior probabilities are what they are. For modeling decisions in which money plays a role, it is more useful to know an individual's risk-neutral probabilities than to know her true subjective probabilities, and

---

[23] In problems of choice under certainty, the corresponding parameters of the individual are her relative marginal rates of substitution between commodities. Indeed, risk neutral probabilities are a special case of relative marginal rates of substitution insofar as lottery tickets are a special form of commodity.

[24] The state-dependent exchange rate issue is discussed by Seidenfeld et al. (1990b). It arises with reward– or loss-based methods of probability elicitation in general, as discussed by Schervish et al. (2013).

the former exist under conditions where the latter may not. The fact that risk-neutral probabilities generally depend on the status quo is not very problematic either. The manner in which risk-neutral probabilities vary in response to changes in wealth reveals (among other things) the individual's local attitude toward risk and uncertainty, and the Arrow-Pratt measure of risk aversion can be generalized in these terms. The epistemic assumptions of game theory also make more sense when applied to risk-neutral probabilities, and the familiar solution concepts of game theory can be similarly generalized, although they turn out to be variants of correlated equilibrium rather than variants of Nash equilibrium.

The modeling of not-quite-pure beliefs in terms of risk-neutral probabilities is consistent with the state-preference approach to modeling choice under uncertainty that was pioneered by Arrow and Debreu and expanded upon by Drèze (1987). It is basically an application of neoclassical consumer theory to goods and monetary payments that are state-dependent and perhaps also time-dependent. It has long been used in models of markets under uncertainty and is especially popular among theorists who are agnostic about the existence of subjective probabilities. Here it will serve as a platform on which to model a wide range of other issues.

1.5   No-arbitrage as common knowledge of rationality

Money is not only a portable and durable and divisible and transferable unit of account, but it also directly addresses the problem of common knowledge of a numerically precise measure of belief or value. A posted price is common knowledge between buyers and sellers, the quantity of money that one person hands to another is common knowledge between them, and the terms of a financial contract are common knowledge among those who sign it. What makes posted prices common knowledge is that they are not merely verbal assertions of preference and they do not refer to merely hypothetical (possibly even counterfactual) alternatives leading to consequences whose units of evaluation—utilities—may be uniquely personal. They create real opportunities for reciprocal action with material consequences whose units are understood in the same way by everyone. If Alice announces in front of Bob that she is willing to sell her apples to him for $0.83 apiece, at his discretion, this is equivalent to saying "I would rather have $0.83 than an apple (because I am committing myself to making this trade if you want to take me up on it, which would materially affect both of us in a way that we both understand), and you know it (because you know that you have the opportunity to take me up on it, which would materially affect both of us in a way that you know that I know that we both understand), and I know you know it (because I know that you know that I have offered you this opportunity, and I am still offering it), and you know that I know you know it (because you know I am continuing to offer it)..." and so on. Alice and Bob may have different personal needs for dollars or apples, but at least they can agree on what the price is and what it means. Its two-digit value in this scenario is common knowledge in the "specular" sense of two or more people who observe themselves in the same mirror at the same time, which needs no elaboration. Definitions of common knowledge of the occurrence of discrete events have been given by Lewis (1969) and Aumann (1976), but it is hard to apply them to the problem of common knowledge of utility functions.

And not only is money the unit in terms of which reciprocal knowledge is quantified, it provides the most basic standard of economic rationality itself: *you should not throw it away*. That is, you (and your fellow players, if any) should not behave in such a way as to expose yourself to arbitrage at the hands of another actor in the same scene. This is not only a standard of rationality, it is a standard of common knowledge of rationality. An arbitrage opportunity is created when one or more persons publicly offer to buy or sell goods or contingent claims in such a way that it is possible for an observer to earn a riskless profit. If Alice is offering to sell apples for $0.83 apiece and Bob (or Alice herself) is simultaneously offering to buy them for $0.95 apiece, in front of everyone, they are jointly behaving irrationally: one of them should shut up and take the deal that the other is offering, or they should strike a deal that benefits both of them rather than allowing someone else to walk off with the surplus. It is important that the offers be public and simultaneously available: if Alice would prefer to have $0.83 rather than an apple and Bob would prefer to have an apple rather than $0.95, this is not an arbitrage opportunity unless they make their preferences common knowledge by advertising their willingness to trade at those prices.

Of course this is an idealization: opportunities for pure arbitrage are rare, while opportunities for near arbitrage (easy profit) arise continually but are not noticed by everyone, and their exploitation by a few smart agents is one of the forces that keeps the market nearly efficient. The occurrence of unexpected events in time and space creates a steady stream of price imbalances that enable those with the sharpest eyes and fastest computers and lowest transaction costs to earn modest profits that are almost riskless, thus capturing a small portion of the surplus, and this leads to price adjustments that diffuse information throughout the system so that everyone can behave somewhat rationally, but no one knows everything at any given moment. These observations on markets are not new ("Adam Smith said that"). The important point is that this fundamental principle of rationality in markets is also the fundamental principle of rationality in games and individual decisions, and it applies from the top down as well as from the bottom up. It is inherently a principle of social rationality that forces individuals to be made rational, up to a point, whether or not they are born that way. It resonates with the view expressed by Arrow (1986, 1987):

> "[R]ationality is not a property of the individual alone, although it is usually presented that way. Rather, it gathers not only its force but also its very meaning from the social context in which it is embedded. It is most plausible under very ideal conditions. When these conditions cease to hold, the rationality assumptions become strained and possibly even self-contradictory. They certainly imply an ability at information processing and calculation that is far beyond the feasible and that cannot well be justified as the result of learning and adaptation."

A corollary is that not only is it impossible to explain every social phenomenon merely by axiomatizing and equilibrating the behavior of individuals in the right way, it is also unrealistic to expect the predictions of rational choice models to be validated in laboratory experiments which lack a social context. Long lists of novel choice problems are presented to isolated subjects who are not allowed the luxury of reflecting upon them, discussing them with others, and drawing upon relevant social knowledge. Such studies have revealed many fascinating regularities and irregularities in cognition and behavior, which shed their own light on what happens outside the laboratory, but it is not surprising that subjects do not always behave rationally in them in the narrow sense of rational choice theory. Thought experiments such as those of Allais and Ellsberg, in which many theorists (including this one) reject the prescriptions of the standard models even after reflecting deeply on them, are a more direct challenge insofar as they suggest that the models are unsatisfactory on their own terms, even as idealizations of rational behavior.

The no-arbitrage characterization of rational choice more strongly unifies the single- and multiple-actor versions of the theory than the usual expected-utility-plus-equilibrium-concepts approach, and it provides a different grip on issues such as the precision of probabilities and utilities, the separation of beliefs from tastes, the distinction between ambiguity and risk, the characterization and construction of common knowledge, and the existence and uniqueness of equilibria. It might seem that this standard is rather too idealized and limited in scope insofar as it requires the decision problem to be embedded, at least hypothetically, in a financial market in which monetary side bets provide a medium of credible communication in the language of numbers, even if money itself is not the primary objective. I will argue that this idealization is not as radical as it may look, especially in comparison to the make-believe counterfactualism of the acts that must be contemplated in subjective expected utility theory or the heroism of the common knowledge and common prior and equilibrium selection assumptions that must be made in game theory. The harder it is to imagine that the numerical parameters of an individual or collective decision problem could be validated by side bets that would expose the actors to arbitrage if they were not consistent in their thoughts and actions, the less likely it is that they will display optimizing behavior or equilibrium behavior that can be quantified with precision and the more likely it is that alternative methods of explanation drawing upon other assumptions and sources of information will useful.

1.6 A road map of the book

Figure 1.2 shows a road map for the theories to be discussed in the remainder of the book, arranged according to the strength of their assumptions and the number of actors that are involved (one, a few, many). At the top is the principle of *no-arbitrage,* which is the foundation of all the theories. A slightly stronger requirement, which will turn out to be more useful in the context of game theory, is that of *no-ex-post-arbitrage,* under which a individual is judged irrational after-the-fact if she offered to accept bets that resulted in a loss, given the way events turned out, without having admitted the possibility of a gain had events turned out otherwise.

There are three main vertical branches, dealing with the cases of *individual rationality* (decision theory), *strategic rationality* (game theory), and *competitive rationality* (market theory). Each branch extends through several domains in which models are grouped according to the representations of beliefs that they use. From top to bottom, in decreasing order of generality, are the domains of *risk-neutral probabilities, subjective probabilities,* and *objective probabilities.* The more special cases of *non-additive probabilities* and *certainty* are treated along a side branch.

Classic models of individual rationality include *ordinal utility* (for choice under conditions of certainty), *expected utility* (for choice under conditions of risk, where probabilities for events are objectively determined), and *subjective expected utility* and *state-preference theory* (for choice under conditions of uncertainty, where probabilities for events are subjectively determined or perhaps undetermined). Models of strategic rationality include various equilibrium concepts of game theory, of which the most fundamental is *correlated equilibrium,* not Nash equilibrium. Models of competitive rationality include the general equilibrium model and particular forms of it that are used in finance theory. The focus here will be on asset markets, where the payoffs are all monetary, but the same general principles apply in more general

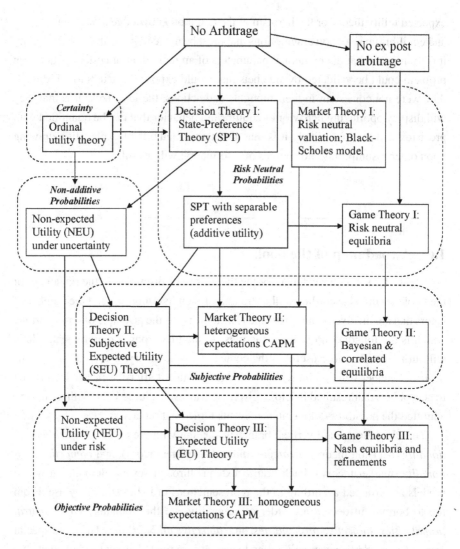

FIGURE 1.2: Rationality theories and associated uncertainty concepts (arrows indicate direction of stronger assumptions and/or narrower applications)

kinds of markets in which money is present. If it isn't, then all bets are off, so to speak, and assumptions about measurability and consistency of preferences must be weaker.

Much of the research in rational choice theory in the last few decades has been devoted to explorations of weaker alternatives to the standard theories of expected utility and subjective expected utility, creating a host of forms of *non-expected utility* in which one or more axioms of individual rationality are relaxed. Most of these generalized utility models have focused on relaxations of the *independence* axiom, which is used for obtaining representations of preferences that are linear in the probabilities or additive across states but which is directly violated in the Allais and Ellsberg paradoxes. Non-expected-utility preferences will be discussed here mainly in the context of the state-preference model, where the independence axiom plays no important role in modeling risk aversion (hence the passage from risk aversion to ambiguity aversion does not require any extra machinery there) and where the separability of probability from utility is not particularly important.

The most problematic of the standard axioms is that of *completeness*, which does not permit the decision-maker to ever be undecided between two alternatives or to ever be unwilling to take one side or the other of a bet. Despite its lack of realism, most theorists are reluctant to give it up because that path leads to models that don't make sharp predictions. Although it is nice to say that a rational decision-maker always ought to be able to make up her mind, this is a ponderous assumption when it is applied to large sets of abstract acts that are far from the decision-maker's experience and mostly counterfactual. In practice, preferences are constructed as needed and with finite effort, a process that does not always converge, particularly under novel conditions. Completeness also strains all the other axioms because it requires them all to be satisfied in hairsplitting terms, which makes it fairly easy to construct situations in which they are violated. Dropping the completeness axiom leads to representations of preferences in terms of *convex sets* of probability distributions or utility functions or state-dependent utility functions rather than unique values. Correspondingly, probabilities of events and utilities of consequences have imprecise values described in terms of lower and upper bounds. Geometrically and mathematically this does not change the model by very much. In fact, it doesn't affect the method of proof of the fundamental theorems: equality constraints are merely relaxed to inequality constraints. The "imprecise" versions of the fundamental theorems are included in the chapters dealing with subjective probability, expected utility, and subjective expected utility, and the fundamental solution concept for games is also expressible in terms of convex sets of probabilities.

Chapter 2

Preference axioms, fixed points, and separating hyperplanes

2.1   The axiomatization of probability and utility

By the end of the 19th century, probability theory was an established branch of applied mathematics (but still largely classical in its interpretation), the principle of maximizing expected utility in the context of gambling and insurance had been floated (but had not gained wider traction), the idea that utility-maximizing agents should seek an equilibrium in a competitive market had been explored (although utility had fallen into disrepute in consumer theory), and the framing of economic competition as a game had been proposed (in the theory of duopoly). The extension and unification of these concepts that gave rise to the modern theory of rational choice was carried out with the help of additional tools—and toolmakers—from mathematics and physics.

In the late 19th and early 20th centuries, axiomatic methods spread through the fields of mathematics and physics. Peano (1889) axiomatized the natural numbers, Hilbert axiomatized geometry (1902) and the real numbers. Zermelo, under Hilbert's influence, axiomatized set theory (1908). Quantum mechanics was axiomatized by Dirac (1930) and von Neumann (1932). Russell and Whitehead (1910, 1912, 1913) attempted to axiomatize all of mathematics in terms of principles of logic, which Goedel (1931) later showed to be impossible. Some axiom systems

DOI: 10.1201/9781003527145-2

(those of mathematics) are purely formal while others (those of physics) are based on observations or conjectures in regard to natural phenomena. Following in the latter tradition, axioms for rational choice under risk and uncertainty and in competitive games and markets began to develop in the social-physics literatures of statistical decision theory and economics in works such as those by Frisch (1926), Ramsey (1926), de Finetti(1931, 1937), Koopman (1940), von Neumann and Morgenstern (1944), Marschak (1950), Nash (1950b, 1953), and Savage (1954). Axiomatization of decision-theoretic models provides a clear statement of their assumptions for purposes of debate and empirical testing, it permits the use of very powerful mathematics for theorem-proving, and it allows new theories to be spun off by tweaks of the axioms. However, there are important distinctions between natural physics and social physics. In the former, particles and collections of them are really identical, they plausibly obey the same laws inside and outside the laboratory, and experiments are reproducible within limits of error that are themselves theoretically grounded. At least, these things are true at the centers of fields, and they justify the elegance of the mathematics. In social physics, by comparison, the particles are heterogeneous boundedly rational individuals whose decision processes are shaped by idiosyncratic personal experience, limits of cognition, literacy, degrees of attention and effort, advice from others, and whatever they may see on television and read online. They do not naturally think as-if in terms of numbers in most situations they face.

The axiomatic movement in rational choice theory drew together several new strands of thought. One was the recognition that the concept of *uncertainty* needed to be given a prominent place in the foundations of economic theory, as it had achieved in contemporary theoretical physics, and that this would require modeling the beliefs of economic agents in terms of probabilities. The concept of uncertainty played essentially no role in classical or neoclassical economics outside the context of insurance (a topic that Bernoulli had addressed), but in the early 20th century it began to be appreciated that risk and uncertainty are central to the understanding of phenomena such as financial investments and entrepreneurship. The earliest theory of continuous-time random processes was developed by Louis Bachelier (1900) for modeling the behavior of stock prices, several years before Einstein (1905) independently developed a similar model for explaining Brownian motion,[1] and John Maynard Keynes (1921) and Frank Knight (1921) wrote treatises on probability that were directed at economists and philosophers.

---

[1] The existence of a continuous-time stochastic process having the properties of the models proposed by Bachelier and Einstein was later proved by the mathemetician Norbert Wiener (1923), hence it is now known as a Wiener process. Wiener also ventured into the social physics arena with his writings on "cybernetics" (1948), and at MIT he recruited a research team that did pioneering work in cognitive science and artificial intelligence.

There are many conceptually and historically distinct interpretations of probability, including the classical interpretation, the quantum-mechanical interpretation, the frequentist interpretation, the measure-theoretic interpretation, the logical interpretation, and the subjective interpretation.[2] Under the classical interpretation, developed over several centuries by mathematicians such as Pascal, Fermat, Bernoulli(s), and Laplace, an uncertain situation is decomposed into equally likely cases and the probability of an event is defined as the ratio of the number of its favorable (supporting) cases to the total number of cases. (This definition is obviously limited in its generality insofar as most uncertainties aren't naturally decomposable in this way.) Under the quantum-mechanical interpretation, subatomic events are intrinsically indeterminate with probabilities determined by physical laws and universal constants. (This interpretation may describe the uncertainty that exists in the mind of an observer of a physical system, although it is not subjective.) According to the frequentist interpretation, the basis of statistical methods of hypothesis testing, the probability of an event is its frequency of occurrence in a large—ideally infinite—number of repeated trials. (All observable sequences are finite, though, and most events that are of personal importance to a decision-maker are not repeatable at all, which limits the relevance of this view for modeling decisions.) In the measure-theoretic interpretation, axiomatized by Kolmogorov, probabilities are measures of the relative masses of sets of events. (This is the standard model of modern probability theory insofar as mathematics is concerned, but it is agnostic about the meaning of probabilities and how their values should be determined a priori.) According to the logical (or "necessary") interpretation, which was embraced by Keynes, probabilities are attached to sentences or propositions rather than events, and they are deduced by logical reasoning from the truth values assigned to their premises. (But it is not clear how the truth values of the premises are to be determined, as Ramsey pointed out in his criticism of Keynes. Keynes admitted that some probabilities were indeterminate under this approach, and he especially felt that it would be impossible to describe the behavior of the stock market in terms of logical probabilities.)

According to the subjective interpretation—which was advocated by Knight, Ramsey, de Finetti, and Savage, and which underlies the Bayesian methods of statistics[3] that are dominant today—probability is a degree of personal belief that can be assigned to any event (repeatable or not), which can be measured by psychometric

---

[2]Some good discussions of the history of the different schools of thought are given by Kyburg and Smokler (1964), Fine (1973), Hacking (1975), and Weatherford (1982).

[3]Bayesian statistical models nowadays often have such large numbers of parameters that it would be impossible to subjectively elicit probability distributions for all of them, and the results of the analysis usually do not depend sensitively on the prior anyway, so various standard forms of "reference prior" are used.

methods (such as eliciting preferences among bets), which can be systematically updated upon the receipt of further information by application of Bayes' rule, and which can be neatly integrated into a theory of decision-making. The subjective view has been criticized by those who feel that personal probabilities are vague and unreliable as well as by those who feel that they have determinate values that should be agreed upon by everyone who has access to the same information. Much of the literature in economics that deals with choice under uncertainty assumes that agents have mutually consistent and correct beliefs about the relevant events, an issue to which we shall return periodically.

A second new strand of thought was the idea that *preferences* are the true fundamental particles of rational thought and behavior and that they should be expressed in *binary* terms. Toward the end of the marginalist era the idea of utility as a primitive psychological concept was abandoned because of doubts about the necessity or even the possibility of measuring it.[4] Choice under certainty could be modeled equally well with demand functions and indifference curves. But, demand functions and indifference curves are high-dimensional psychological objects. Their assessment requires the individual to contemplate tradeoffs among many variables at once or to identify whole equivalence classes of commodity bundles among which she is indifferent. Asserting a direction of preference between two discrete alternatives is a much more elementary mental task. Preference is a qualitative binary relation and as such it lends itself to the same sort of ordering axioms that are employed elsewhere in mathematics to prove the existence of numbers. With the right axioms, the numbers turn out to be interpretable as probabilities and utilities.[5]

A third innovation in the axiomatization of rational choice was the incorporation of probabilities directly into the definitions of alternatives among which decision-makers are assumed to have preferences, which makes possible the construction of a cardinal scale of utility under conditions of risk and uncertainty. In designing a new game-theoretic model of strategic thinking that could be applied to economic behavior in general, von Neumann and Morgenstern needed a cardinal measure of utility

---

[4]"This purely formal [utility] theory, which in no way extends our knowledge of actual processes, is in any case superfluous for the theory of price. It should further be noted that this deduction of the nature of demand from a single principle, in which so much childish pleasure has been taken, was only made possible by artificial constructions and a considerable distortion of reality." (Cassel, 1924, p. 81)

[5]In the 1930's, Paul Samuelson developed an axiomatic theory of "revealed preference" in which the primitive concept is that of a choice function rather than a binary preference relation. A choice function determines the object that will be chosen out of any set of objects that might be available. Depending on the axioms that choice functions and preferences are assumed to satisfy, they may or may not have exactly the same implications for behavior. A brief overview is given by Kreps (1988, pp. 11-17). One of the issues in comparing these approaches is that choices that are *observable* are usually made from sets consisting of more than just two objects. The question of how to observe preferences will be dealt with in later sections of the book in the context of modeling common knowledge in noncooperative games..

to serve as a "single monetary commodity" whose expected value the players would seek to maximize, regardless of the setting. The ability to take expectations is essential for the evaluation of randomized strategies in games, not only for the evaluation of natural risks. If utility is only ordinally measurable, as in consumer theory, then it is meaningless to say something like "the utility of y is exactly halfway between those of x and z" and there is no justification for multiplying utilities by probabilities to determine their expected values. Von Neumann and Morgenstern showed that you *can* do these things meaningfully if you assume that the primitive objects of preference are not the final consequences of decisions but rather "acts" that are lotteries in which those consequences are to be received with objective probabilities. Then you can say that by definition a 50-50 lottery between x and z has a utility that is exactly halfway between the utility of x and the utility of z, and if y is equally preferred to such a lottery, then it too has a utility that is exactly halfway between the utility of x and the utility of z.

2.2   The independence axiom

A fourth key step in developing what has become the standard model of the rational individual was the invention of the independence axiom, which is the consistency condition that beliefs about events or preferences among acts must satisfy in order to be chained together to construct measures of personal probability and utility that are *additive* functions of the separate elements of the alternatives to which they apply. (Wakker (1989) gives a definitive treatment of this topic and Fishburn and Wakker (1995), and Bleichrodt et. al. (2016) discuss its history.) The independence condition for qualitative personal probability, which was introduced as an axiom by de Finetti (1937) and derived as a theorem by Savage (1954), is that if A and B are events, A is more likely than B if and only if, for any other event C that is disjoint from both A and B, A-or-C is more likely than B-or-C. That is, the direction of relative likelihood between two events is preserved when they are conjoined with the same alternative event. The independence condition for expected utility is that x is preferred to y if and only if, for any other act z and any α strictly between 0 and 1, an α chance of x and a $1-\alpha$ chance of z is preferred to an α chance of y and a $1-\alpha$ chance of z. That is, the direction of preference between two acts is preserved when they are probabilistically mixed in the same way with the same alternative act. The latter condition is implicit in von Neumann and Morgenstern's theory, but this was not fully appreciated for a few years. It explicitly appears in later axiomatizations of expected utility (beginning with Marschak (1950)) as well as in Anscombe and Aumann's (1963) axiomatization of subjective expected utility. Other approaches to axiomatizing subjective expected utility, such as those of Savage (1954) and Wakker (1989) also rely on independence axioms to obtain additive representations of preferences.

The effect of the independence axiom in all of these models of decision-making under risk and uncertainty is that when it is combined with the completeness and transitivity axioms it yields a linear/additive measure of total probability or expected utility, so *common terms can be subtracted out when comparing probabilities or expected utilities between two alternative events or acts, and the direction of preference is preserved when two different comparisons of events or acts are linearly combined in the same way.* In the expected utility model this means that if x_1 is preferred to y_1 and x_2 is preferred to y_2, then an α chance of x_1 and a $1-\alpha$ chance of x_2 is preferred to an α chance of y_1 and a $1-\alpha$ chance of y_2. These are the properties that enable the separating hyperplane theorem to be invoked in the representation theorems, as will be shown in the next few chapters.[6]

In parallel with von Neumann and Morgenstern's work, Sono (1943), Samuelson (1947) and Leontief (1947) discovered that the imposition of an independence condition on preferences under *certainty* implies that a consumer must have an additively-separable utility function, i.e., a utility function of the form $U(x_1, x_2, \ldots, x_N) = u_1(x_1) + u_2(x_2) + \ldots + u_N(x_N)$, where x_n is the amount of the nth commodity in a consumption bundle and $u_n(x_n)$ is the contribution to total utility that the consumer attaches to it.[7] A utility function of this form is unique up to an increasing linear transformation and is therefore cardinally measurable. The necessary condition is variously called "coordinate independence" or "factor independence" or "conjoint independence." It requires that, when there are three or more commodities, preferences between commodity bundles do not depend on elements that they have in common. For example, if a bundle consisting of 1 apple and 2 bananas and 3 cantaloupes is preferred to a bundle consisting of 4 apples and 1 banana and 3 cantaloupes, then the direction of preference between the two bundles remains the same if the common number of cantaloupes is changed from 3 to some other value. This implies that the marginal rate of substitution between apples and bananas depends

---

[6] In Savage's model, the mixing of two conditional acts with a common alternative is performed by mapping a subset of the states to a common set of consequences rather than by assigning objective probabilities to the alternative consequences, and in these terms the independence axiom is stated as his postulate P2. This setup does not lend itself to taking differences or linear combinations of acts so as to apply the separating hyperplane theorem, but it is equivalent to Anscombe and Aumann's, which does.

[7] Here and elsewhere, italic symbols such as $x_1, x_2, \ldots$ denote scalars (e.g., quantities of money or goods) while boldface symbols such as $x_1, x_2, \ldots$ denote vectors or more general objects of choice (e.g., state-contingent payoffs or probability distributions over a set of consequences).

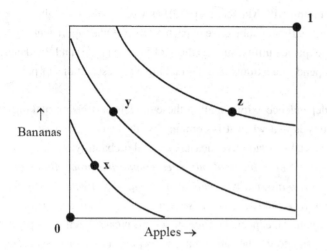

FIGURE 2.1: An ordinal utility function allows the construction of indifference curves but not comparisons of differences in utility.

only on how many apples and bananas the consumer possesses, not how many cantaloupes she has too. A fully general proof of this result and its connection with the theory of choice under uncertainty was given by Debreu in 1960. We will return to the multi-commodity version of the independence condition when we get to state-preference theory, which is Arrow and Debreu's theory of choice under uncertainty that is obtained by applying consumer theory to state-contingent commodities. But the punch line in all these results is the same: *if* preferences are independent of common consequences that are experienced in other events or received in units of other commodities, *then* they are representable by an additive utility function that is cardinally measurable.

To illustrate how the independence conditions allow the construction of cardinal utility functions, consider a consumer of apples and bananas. Suppose that there is a total of one apple and one banana in the market and they are continuously divisible, so a bundle of them is representable by a point x in the unit square, where x_1 is the fraction of an apple consumed and x_2 is the fraction of a banana consumed. Let $\mathbf{1} := (1,1)$ denote the best bundle and let $\mathbf{0} := (0,0)$ denote the worst. If the consumer has preferences that are complete, transitive, and convex,[8] her indifference curves might look something like those shown in Figure 2.1. All bundles on

---

[8]Convexity of preferences means that if the individual is indifferent between x and y, then she at least weakly prefers a convex combination of them to either of them alone, i.e., $x \sim y$ implies $\lambda x + (1-\lambda)y \succsim x$ and $\lambda x + (1-\lambda)y \succsim y$ for $0 < \lambda < 1$. This property captures the intuition of diminishing marginal utility without assuming the existence of a differentiable utility function.

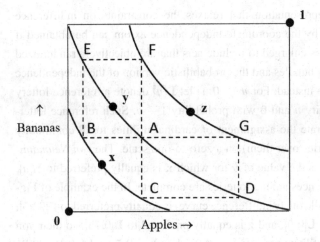

FIGURE 2.2: For an additive utility function, equally spaced indifference curves satisfy the hexagon condition.

the same curve have the same utility by definition, but absent any other restrictions the utility function is only quasi-concave, and there is no meaningful way to compare, say, the difference in utility between **x** and **y** against the difference in utility between **y** and **z**. If the coordinate independence assumption is added, it becomes possible to construct a cardinal utility scale that makes such comparisons possible. However, the utility function must be additive across coordinates, which restricts the shapes that apple-banana indifference curves are allowed to take. In two dimensions, the standard version of the independence condition can't be used, because it refers to bundles of three or more commodities, but if the utility function is additive, indifference curves that are equally spaced in utility must satisfy the "hexagon condition," illustrated by Figure 2.2.[9] This property can be used to triangulate the plane and construct the entire utility function, as shown by Wakker (1989). If there is a third dimension for cantaloupes, then the apple-banana indifference curves must look the same on any slice through the cantaloupe axis, and a more general variant of the independence condition known as "generalized triple cancellation" can be used to construct the utility function.

---

[9]Starting from a point A on the indifference curve for **y**, draw horizontal and vertical lines, respectively, that intersect the indifference curve for some other point **x**. From those two points of intersection B and C, draw vertical and horizontal lines, respectively, to get back to the **y** difference curve again at points E and D. From those two (new) points on the **y** indifference curve, draw horizontal and vertical lines, respectively, that intersect the original vertical and horizontal lines that were drawn through the starting point. If the utility function is additive, the latter two intersection points F and G lie on the same indifference curve, which is the indifference curve for **z** in the figure, and the three indifference curves are equally spaced in utility.

A cardinal utility representation that relaxes the constraints on indifference curves that are imposed by the coordinate independence axiom can be obtained if the space of alternatives is enlarged to include acts that are objectively randomized lotteries over commodity bundles and the probabilistic version of the independence axiom is applied to them instead. For $u \in [0,1]$ let $\mathbf{L}[u]$ denote a reference lottery yielding 1 with probability u and 0 with probability $1 - u$. Such reference lotteries can be used to calibrate the assignment of cardinal utilities to all commodity bundles (as well as lotteries over them) on a zero-to-one scale. The *von Neumann-Morgenstern utility* of $\mathbf{x}$ is the value of u for which $\mathbf{x}$ is equally preferred to $\mathbf{L}[u]$, which is unique if preferences among lotteries are complete. In the example of Figure 2.3, $\mathbf{x}$ (and every bundle on its indifference curve) is equally preferred to $\mathbf{L}[0.25]$, $\mathbf{y}$ is equally preferred to $\mathbf{L}[0.5]$, and $\mathbf{z}$ is equally preferred to $\mathbf{L}[0.75]$, so their von Neumann-Morgenstern utilities are $U(\mathbf{x}) = 0.25$, $U(\mathbf{y}) = 0.5$, and $U(\mathbf{z}) = 0.75$ when the scale is anchored by the endpoints $U(\mathbf{0}) := 0$ and $U(\mathbf{1}) := 1$.

The von Neumann-Morgenstern utilities of lotteries are linear functions of the probabilities they assign to different commodity bundles (which means that compound lotteries can be reduced), but they need not be additive across commodities unless the coordinate independence axiom is imposed separately, so the indifference curves in the subspace of determinate commodity bundles (which are the projections of the curves in Figure 2.3 onto a single horizontal plane) can have very general shapes.

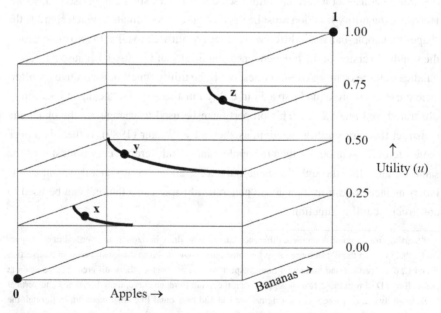

FIGURE 2.3: Von Neumann-Morgenstern utilities are determined by comparison with reference lotteries yielding 1 with probability u and 0 with probability $1 - u$.

2.3 The difficulty of measuring utility

The method of construction of a cardinal von Neumann-Morgenstern utility function for consequences, as illustrated in the last example above, raises some big issues. First, you need to perform rich and novel psychological measurements: assessments of preferences among probability distributions over consequences, not just preferences among the consequences themselves. The decision-maker must now contemplate randomized acts as well as determinate acts. However, randomized acts are easier to evaluate when their consequences are amounts of money rather than bundles of apples and bananas, because in that case they have fewer dimensions of taste to integrate with probabilities and they are somewhat familiar objects, exactly what you might get by visiting a casino or buying lottery tickets or insurance. Preferences are better quantifiable and more consistent and more communicable when they refer to acts that have a monetary dimension rather than acts whose only consequences are non-monetary goods or personal experiences, and this is especially true under conditions of uncertainty. And even where money *is* involved, it is not easy or natural for unaided individuals to assess complete and consistent preferences among *arbitrary* lotteries with objective probabilities. Most of us do not actually gamble very often, so we are not skilled and level-headed in making these sorts of comparisons, which is why we have casinos and state lotteries and overpriced warranties and why it is often a good idea to get advice from others when spending money on uncertain prospects. A good financial decision generally has an interpersonal dimension.

Second, these richer preferences need to satisfy all the axioms, of which the most subtle is the independence axiom that allows the construction of a cardinal utility scale. In the example of Figure 2.3, the utility of **y** is indirectly revealed to be exactly halfway between those of **x** and **z** based on separate comparisons of these acts with reference lotteries yielding only the outcomes **0** and **1**. If utility measurements are to be consistent, a direct comparison should reveal that **y** is equally preferred to a 50-50 lottery between **x** and **z**. Von Neumann and Morgenstern simply assumed that equivalences such as this must hold because compound lotteries can be reduced by multiplying out the probabilities. Here, $U(\mathbf{x}) = 0.25$ means that **x** is preferentially equivalent to a lottery yielding a 25% chance of **1** and an 75% chance of **0**, and $U(\mathbf{z}) = 0.75$ means that **z** is equivalent to a lottery yielding a 75% chance of **1** and a 25% chance of **0**. The 50-50 lottery between **x** and **z** is therefore equivalent to a lottery that yields a 50% chance of {a 25% chance of **1** and an 75% chance of **0**} and a 50% chance of {a 75% chance of **1** and a 25% chance of **0**}. This reduces to a lottery that yields a 50% chance of 1 and a 50% chance of 0, which is the definition of $L[0.5]$, which is equally preferred to y. Malinvaud (1952) showed that the application of these mixing operations to equivalence classes of lotteries is justified only if the independence axiom holds.

For the first couple of decades after it was introduced in various forms by de Finetti, von Neumann and Morgenstern, and Savage, the probabilistic version of the independence axiom was widely regarded as a self-evident property of rational preferences under risk and uncertainty. Disjoint events are alternative universes, and consequences experienced in one universe cannot serve as complements or substitutes for consequences that might be experienced in some other universe. However, the paradoxical thought-experiments devised by Allais (1953) and Ellsberg (1961) showed that many persons do not always think in this fashion, and later experiments and field studies by behavioral decision theorists revealed a menagerie of similar anomalies. These findings eventually led, in the late 1970s, to an explosion of interest in economic models in which the independence axiom is relaxed in one way or another, giving rise to theories of non-expected-utility, one of which will be discussed in Chapters 6 and 7.

Third, once you have introduced probabilities into the definitions of acts and the measurement of utilities in this fashion, you have made a bargain with the devil, and you can't get rid of them again. You cannot attach any cardinal meaning to the utility values *except* in a context that involves risky choices.[10] If the utility of y is found to lie exactly halfway in between the utilities of x and z, then you cannot conclude that the increase in personal well-being (however defined) that results from the exchange of x for y is exactly the same as the increase that would result from the exchange of y for z. All you can say is that y is equally preferred to a coin flip between x and z, a choice that almost certainly will never come up. So, the von Neumann-Morgenstern utility function technically is not the same utility function that had been sought by earlier generations of utilitarians. It is an index to be used in evaluating choices under risk, not a measure of the intrinsic values of consequences to the decision-maker. As such, it conflates the intuitively distinct psychological phenomena of risk aversion, comparative tastes, and satiation of needs and wants.

---

[10]Savage did not introduce objective probabilities directly into his axiom system, but it is as if he did, because the state space includes whatever events the decision-maker cares to think about, and she might as well contemplate random number generators along with everything else. The alternative, and simpler, axiomatization of subjective expected utility by Anscombe and Aumann (1963) does explicitly use objective probability distributions to calibrate the subjective ones.

2.4 The fixed point theorem

Another pivotal step in the development of modern rational choice theory was the emergence of a new paradigm of designing *solution concepts* to characterize the interplay between two or more agents who satisfy the principles of individual rationality outlined above. Such concepts may take the form of axioms of group rationality or conditions of equilibrium that are expected to be satisfied in environments such as games and markets. Precursors were developed in the 19th century by Cournot, Walras, Pareto, and others, but the modern paradigm shift occurred when Nash published his elegant solution methods for noncooperative games and bargaining problems. In his solution concept for games, the key tool is a "fixed point theorem."

There are many different kinds of fixed-point theorems in mathematics, and the topic has a long history, but the one that is best known and perhaps most widely applied is due to Brouwer (1912). It was first used in economics by von Neumann (1928) to prove the existence of a minimax solution in a 2–person zero-sum game and again (1937) to prove the existence of a growth equilibrium in an expanding economy. The theorem states that every continuous mapping of a convex compact set into itself has a fixed point, i.e., at least one point that is mapped to itself. For example, if you briefly stir the water in a cup so as to cause it to swirl around while the surface remains level, then at least one molecule in the cup must be in its original position at any given time, although not necessarily the same one: the fixed point itself need not be fixed while the water is still moving.[11]

A more general fixed-point theorem due to Kakutani (1941), which applies to set-valued functions, was used by Nash (1950a, 1951) to prove that any noncooperative game (not merely any zero-sum game) has an equilibrium in pure or independently randomized strategies.[12] The method of proof is to consider the mapping from the set of all independently randomized strategies into itself that is determined by the best-reply correspondence, in which each player's strategy is mapped to her best reply(s) to the other players' strategies. Roughly speaking, it goes like this: each player's strategy set is convex because of the possibility of randomization, and the set of independent joint strategies of all players is convex for the same

---

[11]Brouwer supposedly noticed this phenomenon while stirring particles of sugar into a coffee cup.

[12]Kakutani 's (1941) version of the theorem is a set-valued generalization of Brouwer's theorem and was explicitly designed for simplifying von Neumann's proof of the minimax theorem. It states that any upper hemicontinuous set-valued mapping of a convex compact set into itself has a fixed point. It is better suited for application to the existence proof of Nash equilibria because in general the best reply to a given strategy may not be unique.

reason, and the expected payoff of any one of a player's strategies is a continuous function of those of the other players, and her expected payoff is also a continuous function of her own strategy holding those of the others fixed, and therefore the strategy(s) of her own that maximizes her expected payoff is a continuous function, so the best-reply correspondence for the joint strategies is continuous and hence must have a fixed point, which is an equilibrium in independently randomized strategies.

Over several decades this beautiful and powerful discovery launched a paradigm shift in economic theory,[13] but the normative and descriptive validity of the Nash concept is open to many questions that have been widely studied and which do not have good answers in most situations. How and why does convergence to an equilibrium occur, especially in a game that is not repeated? Can the players independently reason their way to a Nash equilibrium and get to it in one leap? How much information about each others' utility functions do they need, and does it need to be common knowledge? What would it mean for utility functions to be common knowledge, anyway? Do the players need to believe, like the game theorist, in the concept of Nash equilibrium? What if it is inefficient or otherwise undesirable, as in the prisoner's dilemma? What if there is more than one equilibrium? If the players randomize, must they always do it independently—why not flip a coin together, as they might wish to do in a battle-of-sexes game? *And if the players don't deliberately select a Nash equilibrium, or if they otherwise choose strategies that are not supported by a Nash equilibrium, so what?* In exactly what sense would an observer say that they are irrational? Nash's proof was quickly adapted by Arrow and Debreu (1954) to solve the long-standing problem of the existence of a general competitive (Walrasian) equilibrium of an economy, and questions of convergence and stability and uniqueness and authority arise there too.[14]

---

[13]Myerson (1999) puts it this way: "Nash's theory of noncooperative games should now be recognized as one of the outstanding intellectual advances of the twentieth century. The formulation of Nash equilibrium has had a fundamental and pervasive impact in economics and the social sciences which is comparable to that of the discovery of the DNA double helix in the biological sciences." Nash received a somewhat different reception when he visited von Neumann in the summer of 1949 to describe his new equilibrium concept and its proof: "[B]efore he had gotten out more than a few disjointed sentences, von Neumann interrupted, jumped ahead to the yet unstated conclusion of Nash's argument, and said abruptly, 'That's trivial, you know. That's just a fixed point theorem.'" (Nasar 1998, p. 94) Von Neumann never took much interest in Nash's noncooperative approach, especially as it applied to games with 3 or more players. He continued to be more interested in issues of coalition formation and the "stable set" solution concept that had been central in his work with Morgenstern. (Leonard 2010). See footnote 8 in the preceding chapter for more perspective from Myerson.

[14]Arrow (1986) observes: "Even if we make all the structural assumptions needed for perfect competition (whatever is needed by way of knowledge, concavity in production, absence of sufficient size to create market power, etc.), a question remains. How can equilibrium be established? The attainment of equilibrium requires a disequilibrium process. What does rational behavior mean in the presence of disequilibrium? Do individuals speculate on the equilibrating process? If they do, can the disequilibrium

2.5 The separating hyperplane theorem

The most important tool in the mathematical toolkit of rational choice theory (and much of microeconomics in general) is the separating hyperplane theorem of convex analysis. In most of the fundamental representation theorems, *it is where the numbers come from*. Its centrality in the field began to be recognized in the 1940s, when techniques of mathematical programming were first developed.[15] Farkas' lemma (1902) is a special case of it that applies to systems of linear inequalities that arise in linear programming applications. The lemma has many variations, one of which appears in von Neumann and Morgenstern's (1944) book, where it is called the "theorem of the alternative" and is used for a simplified proof of the minimax theorem. Another variation, discussed below, is the basis of the various fundamental theorems in which no-arbitrage conditions are applied.

Theorem 2.1: Separation of convex sets by a hyperplane

Any two disjoint convex sets, at least one of which is open, can be separated by a non-trivial hyperplane.[16] That is, if a convex set X is disjoint from an open convex set Y, there exists a vector π and a constant δ such that X and Y are strictly separated by the hyperplane whose equation is $x \cdot \pi = \delta$, i.e., $x \cdot \pi \geq \delta$ for all $x \in X$ and $y \cdot \pi < \delta$ for all $y \in Y$.[17]

---

be regarded as, in some sense, a higher-order equilibrium process? Since no one has market power, no one sets prices; yet they are set and changed. There are no good answers to these questions, and I do not pursue them. But they do illustrate the conceptual difficulties of rationality in a multiperson world."

[15] Arrow learned about convexity and separating hyperplanes at RAND in 1948 and applied them in his 1951 paper "An Extension of the Basic Theorems of Classical Welfare Economics."

[16] A set is convex if a convex combination of any two of its points (i.e., every point on the line segment connecting them) is also in the set. This means it has no holes nor any indentations on its surface. It is intuitively plausible that two such disjoint (non-intersecting) sets can always be separated by a straight line in two dimensions or by a plane in three dimensions or, inductively, by a hyperplane in higher dimensions. (A hyperplane is an $(n\text{-}1)$-dimensional plane that divides an n-dimensional space into two halves.) Two disjoint sets that are not convex need not be separable in this way. For example, a ring is not convex, and you cannot pass a flat sheet of paper between your finger and the ring when you are wearing it.

[17] The following notational conventions will be used throughout the book: lower-case italic Roman or Greek letters (x, π, f, ...) denote scalars or scalar functions. Lower-case boldface Roman or Greek letters ($\mathbf{x}$, $\mathbf{f}$, $\boldsymbol{\pi}$, ...) denote finite-dimensional vectors. $f(\mathbf{x})$ denotes the vector $(f(x_1), ..., f(x_N))$ when f is a scalar function of a single variable that is applied to an N-vector $\mathbf{x}$. $\mathbf{f}(\mathbf{x})$ denotes a more general vector-valued function of a vector argument, such as a gradient function. Italic upper-case letters denote scalar functions of a vector argument ($U(\mathbf{x})$). The boldface upper-case letters $\mathbf{A}$, $\mathbf{B}$, $\mathbf{C}$, and $\mathbf{D}$ will be used to denote events and their corresponding indicator vectors or matrices. All other boldface upper-case

Figure 2.4 provides a 2-dimensional illustration. The vector π is called the *normal vector* of the hyperplane (the vector "at right angles" to it), and the constant δ depends linearly on the distance of the hyperplane from the origin of coordinates. π and δ are unique only up to a common positive scale factor, and if the hyperplane passes through the origin, then $\delta = 0$. The sets $\mathcal{X}$ and $\mathcal{Y}$ can be infinite sets, or even infinite-dimensional sets (e.g., sets of functions), although for the purposes of this book it will usually suffice to consider them to be subsets of $\mathbb{R}^n$. In this picture and the others in this chapter, the boundaries of the two sets have a point in common, but in general they could be separated by some distance.

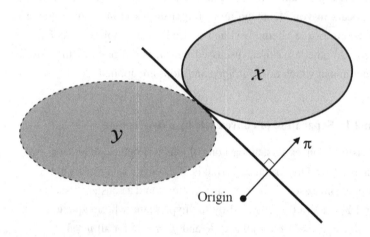

FIGURE 2.4: Separating hyperplane theorem

Another way to state the separating hyperplane theorem is as follows. Given two sets $\mathcal{X}$ and $\mathcal{Y}$, of which both are convex and $\mathcal{Y}$ is open, consider the following two problems:

1. Find a point **x** that belongs to both sets.

2. Find a vector π and a constant δ such that the hyperplane whose equation is $\mathbf{x} \cdot \pi = \delta$ strictly separates them.

---

symbols will denote matrices ($\mathbf{X}$, $\mathbf{F}$, $\mathbf{M}$, $\mathbf{\Pi}$, ...). Upper-case calligraphic letters ($\mathcal{X}$, $\mathcal{Y}$,...) denote sets. $\mathbf{x} \cdot \pi$ denotes the inner product (a.k.a. dot product or sumproduct) of the vectors $\mathbf{x}$ and π, which is equal to $\sum_m x_m \pi_m$, i.e., the sum of the products of the respective elements of $\mathbf{x}$ and π. If α is a K-vector and $\mathbf{W}$ is an $K \times M$ matrix whose km^{th} element is w_{km}, then $\mathbf{W} \cdot \alpha = \mathbf{W}^T \alpha$, where $\mathbf{W}^T$ denotes $\mathbf{W}$-transpose. This is the (column) M-vector whose m^{th} element is $\sum_{k=1}^{K} \alpha_k w_{km}$. ($\alpha \cdot \mathbf{W}$ is the same vector in row format.) If π is an M-vector, then $\mathbf{W}\pi$ is the matrix product of $\mathbf{W}$ and π, i.e., the K-vector whose k^{th} element is $\sum_{m=1}^{M} w_{km} \pi_m$. And finally, $\alpha \cdot \mathbf{W}\pi$ is the scalar whose value is $\sum_{k=1}^{K} \sum_{m=1}^{M} \alpha_k w_{km} \pi_m$.

Then problem 1 does *not* have a solution if and only if problem 2 *does* have a solution. These two problems are said to be *dual* to each other.

An example of a convex set is a *cone* formed by taking all non-negative linear combinations of some finite set of vectors. A cone extends infinitely in some directions, but it still meets the requirements of convexity, so it can be separated from other convex sets (e.g., other cones) from which it is disjoint. An important special case of the separating hyperplane theorem is the one in which $\mathcal{X}$ is the convex cone that is generated by non-negative linear combinations of the rows of some matrix $\mathbf{W}$ and $\mathcal{Y}$ is the open negative orthant, i.e., the set of all strictly negative vectors.

Lemma 2.1: Separation from the open negative orthant

For any matrix W, either there exists $\alpha \geq 0$ such that $W \cdot \alpha < 0$ or else there exists $\pi \geq 0$, with $\pi \neq 0$, such that $W\pi \geq 0$.

Proof: let $\mathcal{X} = \{\mathbf{W} \cdot \alpha \mid \alpha \geq 0\}$ denote the closed convex cone generated by the row vectors of the matrix $\mathbf{W}$ and let $\mathcal{Y}$ denote the open negative orthant. If one of the vectors in $\mathcal{X}$ is strictly negative (i.e., if there is a non-negative α such that $\mathbf{W} \cdot \alpha < \mathbf{0}$), then $\mathcal{X}$ and $\mathcal{Y}$ intersect. Otherwise they are disjoint, in which case they are separated by a non-trivial hyperplane. Let π denote the normal vector of that hyperplane. Then $\mathbf{W}\pi \geq \mathbf{0}$—i.e., all the row vectors of $\mathbf{W}$ lie "on or above" the hyperplane—while $\mathbf{y} \cdot \pi < 0$ for all vectors $\mathbf{y}$ in the open negative orthant—i.e., all the strictly negative vectors are "below" the hyperplane. The latter inequality implies that π is non-negative and strictly positive in at least one component.

In two dimensions, the environment of Lemma 2.1 looks like Figure 2.5, where $\mathbf{x}_1$ and $\mathbf{x}_2$ denote two row vectors of $\mathbf{W}$. The closed cone $\mathcal{X}$ is formed by non-negative linear combinations of $\mathbf{x}_1$ and $\mathbf{x}_2$ and it recedes from the origin toward the upper right. The open cone $\mathcal{Y}$ is the open negative orthant, which also recedes from the origin (toward the lower left) but does not include it. These two convex sets are disjoint, so they must be separated by some hyperplane, which is not unique in this case. The separating hyperplane passes through the origin (because the origin is a limit point of both sets), so its equation is $\mathbf{x} \cdot \pi = 0$, where π is a non-negative vector.

Figure 2.6 illustrates the special case in which $\mathcal{X}$ is an entire half-space (all points on or above the sloping line) that is separated from the open negative orthant, which describes the situation in which preferences are complete. In this case, the normal vector of the separating hyperplane is uniquely determined up to a positive scale factor.

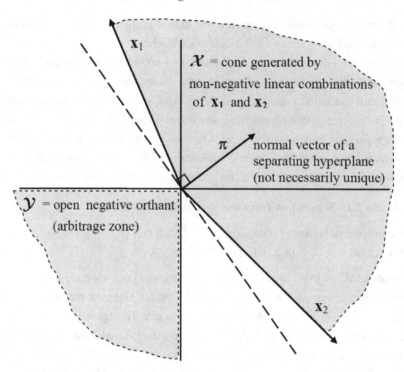

FIGURE 2.5: Separation of a convex cone from the open negative orthant

Another very important special case is that in which some but not all elements of $\mathbf{W} \cdot \alpha$ are strictly negative, which will turn out to be central to the modeling of "ex-post-coherent" subjective probabilities and equilibria of noncooperative games. It is described by the following more nuanced version of Lemma 2.1:

Lemma 2.2: Separation from the semi-open negative orthant

For any matrix $\mathbf{W}$, either there exists $\alpha \geq 0$ such that $\mathbf{W} \cdot \alpha \leq 0$ and $[\mathbf{W} \cdot \alpha]_m < 0$ or else there exists $\pi \geq 0$, with $\pi_m > 0$, such that $\mathbf{W}\pi \geq 0$.

Proof: The following stronger theorem by Gale (1960),[18] a variant of Farkas' lemma, will be used here: *For any matrix* $\mathbf{W}$ *and vector* β, *either there exists* $\alpha \geq 0$ *such that* $\mathbf{W} \cdot \alpha \leq \beta$ *or else there exists* $\pi \geq 0$ *such that* $\beta \cdot \pi < 0$ *and* $\mathbf{W}\pi \geq \beta$. Apply this result with $\beta = (0, ..., 0, -1, 0...0)$ where the -1 is in position m. It also provides a proof of Lemma 2.1 by using $\beta = (-1, -1, ..., -1)$.

---

[18]Here $\mathbf{M} = \mathbf{A}^T$, $\alpha = \mathbf{x}$, $\beta = b$, $\pi = \mathbf{y}$ in terms of Gale's version in Perng (2017, p. 2177, Theorem 6)

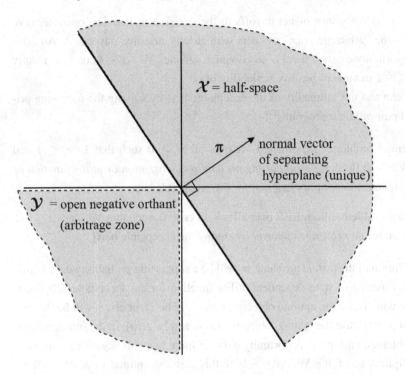

FIGURE 2.6: Separation of a half-space from the open negative orthant

2.6 Primal/dual linear programs to search for arbitrage opportunities

In the applications to be discussed later in the book, the objects of communication and exchange are bets whose payoffs are either in units of money attached to states of the world or in units of probability attached to consequences (experiences of some kind) whose receipt could be state-dependent. The rows of the matrix $\mathbf{W}$ are payoff vectors of acceptable bets, and the columns correspond to states or consequences or state-consequence pairs in which those bets yield their payoffs. π is a vector of probabilities or utilities or state-dependent utilities, and the condition $\mathbf{W}\pi \geq 0$ means that the acceptable bets all have non-negative expected monetary value or expected utility with respect to π.[19] α is a non-negative vector of coefficients for the rows of $\mathbf{W}$, representing quantities of the bets taken by an opponent,

---

[19]The symbols $\hat{u}$ and $\hat{v}$ will be used instead of π in the settings of EU and SEU theory, respectively.

and $\mathbf{W} \cdot \alpha$ is the vector of net payoffs to the decision-maker. The open negative orthant is the "arbitrage zone" of bets with strictly negative payoffs.[20] An arbitrage opportunity exists if there is an α which satisfies $\mathbf{W} \cdot \alpha < 0$, or equivalently $\mathbf{W} \cdot \alpha \leq -1$ in view of positive scalability of α.

You can test the rationality of the acceptable bets by solving the following primal/dual pair of linear programs[21]:

- Primal problem: maximize $1 \cdot \pi$ over all $\pi \geq 0$ such that $1 \cdot \pi \leq 1$ and $\mathbf{W}\pi \geq 0$ [look for a *supporting probability distribution or utility function or state-dependent utility function* for the acceptable bets]

- Dual problem: minimize z over all $z \geq 0$, $\alpha \geq 0$, such that $\mathbf{W} \cdot \alpha - z \leq -1$ [look for an *arbitrage opportunity* among the acceptable bets]

In the solution of the primal problem, π will be a supporting probability distribution or utility function or state-dependent utility function for the acceptable bets if one exists (in which case the optimal objective value will be 1), or else it will be the zero vector (in which case the optimal objective value will be zero). In the solution of the dual problem, an arbitrage opportunity exists if there is a non-negative vector α of bet multipliers satisfying $\mathbf{W} \cdot \alpha \leq -1$. In this case the optimal value of z will be its lower bound of zero. Otherwise α will the zero vector and the optimal value of z will be 1. As with any primal-dual pair, the optimal objective value is the same in both problems and the optimal values of the decision variables in one problem are the optimal values of the shadow prices of constraints in the other problem, so you just need to solve one or the other. You can use Solver in Excel to solve either one of these problems and get the solution to the other on the "sensitivity report." The various special cases of this model that arise in different branches of rational choice theory are presented and compared in Chapter 11.

The conventional parameterization of generic LP's that is given in most textbooks is:

- Primal problem: Maximize $\mathbf{c} \cdot \mathbf{x}$ over all $\mathbf{x} \geq 0$ subject to $\mathbf{Ax} \leq \mathbf{b}$.

- Dual problem: Minimize $\mathbf{b} \cdot \mathbf{y}$ over all $\mathbf{y} \geq 0$ subject to $\mathbf{A} \cdot \mathbf{y} \geq \mathbf{c}$.

---

[20]This setup can also be used in a consumer-theory setting in which bets are vectors of commodity trades and π is a vector of prices.

[21]There are many other variations on these LP's that could be used to obtain the same results or branch off in other directions. This particular variation has the convenient property that the solutions to both problems are always feasible and bounded. The choice of which problem to label as "primal" and which one to label as "dual" is also arbitrary. There are merely dual to each other. The problem of searching for π has been labeled as primal here because it is one that is normally graphed and solved in practice. Also it will be convenient to have payoff vectors of bets be the rows in the coefficient matrix.

where the last constraint can also be written as $-\mathbf{A} \cdot \mathbf{y} \leq -\mathbf{c}$. Letting K and M denote the numbers of rows and columns of $\mathbf{W}$ (number of bets and number of states or consequences or state-consequence pairs respectively), these correspond to the primal and dual problems given above with the following mapping of parameters:

- $\mathbf{x} = \boldsymbol{\pi}$ [the M-vector of primal variables, which are probabilities or utilities or state-dependent utilities]

- $\mathbf{y} = (z, \boldsymbol{\alpha})$ [the $(K+1)$-vector of dual variables, whose 1st element is a measure of no-arbitrage and whose 2nd through $(K+1)^{st}$ elements are multipliers of the acceptable bets]

- $\mathbf{c} = (1, 1, ..., 1)$ [an M-vector of 1's for summing the probabilities or utilities in the primal problem and adding the no-arbitrage measure z to the left-hand-side of constraints in the dual problem]

- $\mathbf{b} = (1, 0, ..., 0)$ [a $(K + 1)$-vector of right-hand-side constants to enforce $1 \cdot \boldsymbol{\pi} \leq 1$ and $\mathbf{W}\boldsymbol{\pi} \geq \mathbf{0}$ in the primal problem and to set z as the objective to minimize in the dual problem]

- $\mathbf{A} = \begin{bmatrix} 1,...,1 \\ -\mathbf{W} \end{bmatrix}$ [a $(K+1) \times M$ matrix whose 1st row consists of 1's and whose 2nd through $(K + 1)^{st}$ rows are payoff vectors of acceptable bets multiplied by -1]

The semi-negative version of the separating hyperplane theorem (Lemma 2.2) can be implemented by a generalization of these linear programs in which the primal objective and dual constraints are "maximize $\mathbf{c} \cdot \boldsymbol{\pi}$" and "$\mathbf{W} \cdot \boldsymbol{\alpha} - z \leq -\mathbf{c}$," where $\mathbf{c}$ is a non-negative vector. In particular, if $\mathbf{c} = (0, ..., 1, ...0)$ where the 1 is in the mth coordinate (only), then the primal LP seeks a solution that assigns positive probability to (at least) state m. The dual LP seeks a solution that yields an "ex-post-arbitrage" profit in (at least) state m, i.e., a positive payoff to the opponent in that state and non-negative payoffs in other states. The latter situation will be discussed further in Section 3.7 and in the chapter on game theory.

2.7   No-arbitrage and the fundamental theorems of rational choice

The separating hyperplane theorem is the basis of the fundamental theorem of linear programming, as shown above, and it is equivalent to the minimax theorem for two-player zero-sum games (von Neumann (1928)),[22] which in turn is known as the "fundamental theorem of game theory" (although limited to such games). Separating hyperplane arguments provide the basis for *all* of the fundamental theorems in different branches of rational choice theory, some of which were originally proved by less transparent methods. In a typical such theorem, $\mathcal{X}$ is the set of acceptable moves away from the status quo (e.g., acceptable bets or trades). The primal rationality condition is the existence of a system of probabilities or utilities or prices or weights that rationalizes the acceptable moves in the sense that they have a non-negative inner product with respect to it, i.e, they are all in the direction of non-decreasing expected value or expected utility or market value or social value. The equivalent dual rationality condition can be framed as a no-arbitrage requirement, namely that a move whose elements are all strictly undesirable is not acceptable.

In the "fundamental theorem of [subjective] probability" (de Finetti (1931, 1937, 1974)), the rates at which an individual is willing to bet money on events are *coherent* (do not create an arbitrage opportunity) if and only if there exists a probability distribution on states of the world under which all the bets she accepts have non-negative expectation. As such it is interpretable as a subjective probability distribution with respect to which she rationally seeks to maximize the expected value of her wealth.[23] The mirror image of de Finetti's theorem, in which the roles of probabilities and payoffs are reversed, is the fundamental theorem of expected utility, so it also follows from a separating hyperplane argument, as shown by Fishburn (1975b) and Gilboa (2009). In this setting, arbitrage can be defined as a transfer of positive probability to the worst consequence from all of the better ones, although that is not

---

[22]The duality theorem of linear programming is actually due to von Neumann, who introduced it to George Dantzig, the developer of the first practical algorithm for solving linear programs, who is remembered as "the father of linear programming." Dantzig tells the following story (Albers et al. (1986)): "On October 3, 1947, I visited [von Neumann] for the first time at the Institute for Advanced Study. I began by explaining the formulation of the linear programming model in terms of activities and items and so forth. I described it to him as I would describe it to an ordinary mortal. He responded in a way which I believe was uncharacteristic of him. 'Get to the point,' he snapped. ... In less than a minute I slapped the geometric and the algebraic versions of my problem on the blackboard. He stood up and said 'Oh, that.' For the next hour and a half he proceeded to give me a lecture on the mathematical theory of linear programs ... That was the way I learned for the first time about Farkas' lemma and the duality theorem." A proof of the duality theorem of linear programming and its equivalence to von Neumann's minimax theorem was first publicly presented at the 1949 Cowles Foundation conference by Gale, Kuhn, and Tucker (and later published in 1951), who were unaware that it had been worked out earlier (though not published) by von Neumann himself.

[23]De Finetti gave separate proofs of the additive and multiplicative laws of probabilities using arguments based on determinants of matrices. The separating hyperplane theorem was explicitly used to characterize coherent betting rates and Bayesian statistical inference in various settings by Blackwell and Girshick (1954, Theorem 4.3.1), Smith (1961), Cornfield (1969), Freedman and Purves (1969), Heath and Sudderth (1972), Pierce (1973), and Buehler (1976), among others.

the way in which they framed the problem.[24] The generalization to a fundamental theorem of subjective expected utility (a state-dependent version of Anscombe and Aumann's result) can be obtained in the same way by merely redefining the alternatives to be mappings from events to lotteries over consequences (Karni (1993), Nau (2006b)). In all of these models, it is trivial to relax the completeness assumption, merely getting convex sets rather than point values for the probabilities, utilities, and state-dependent utilities.

The "fundamental theorem of asset pricing" (Merton (1973), Black and Scholes (1973), Ross (1978)) is the same as de Finetti's theorem, except that its setting is a contingent claims market in which bets on different events may be placed by different individuals and payoffs of bets may be received at a later date with some amount of discounting. The theorem states that there is no arbitrage opportunity if and only if there exists a probability distribution on states under which the purchase of any asset at its market price has non-negative expected net present value, as if there is a "risk-neutral representative investor" who rationally prices all the assets according to their expected net present values.[25,26]

In an exchange economy, the relevant no-arbitrage condition is Pareto efficiency, because a violation of Pareto efficiency would present an arbitrage opportunity to an alert trader.[27] Arrow (1951a) used a separating hyperplane argument to prove the "[second] fundamental theorem of welfare economics," which states that an allocation of goods among consumers is Pareto efficient if and only if there is a system of prices such that no one can be made better off by cashing in her current allocation to buy some other allocation at the same prices.[28] In Harsanyi's (1955) social

---

[24]In Gilboa's proof a hyperplane is used to separate a convex set of weakly preferred differences between pairs of lotteries (analogous to bets) from the convex set of strictly dispreferred differences, as opposed to separation from the set of strictly negative differences.

[25]The representative investor is actually risk averse rather than risk neutral, and the probability distribution with respect to which asset prices are equal to expected net present values is not her true subjective probability distribution. Rather, it is interpretable as the product of her true probabilities and state-dependent marginal utilities for money, where the latter depend on her risk tolerance and status quo portfolio of assets. This principle will play an important role throughout the book, as noted earlier.

[26]Varian (1987) observes: "One of the major advances in financial economics in the past two decades has been to clarify and formalize the exact meaning of 'no arbitrage' and apply this idea systematically to uncover hidden relationships in asset prices. Many important results of financial economics are based squarely on the hypothesis of no arbitrage, and it serves as one of the most basic unifying principles of the study of financial markets." See Garman (1980) for a rigorous synthesis. The course notes by Dahl (2012) make explicit use of linear programming in this context.

[27]An inefficient allocation can arise from a failure of communication or from market friction or incomplete markets, in which case the exact parameters of the agents' preferences are not commonly known in a way that can be verified by public offers to trade, even if they are individually rational. So, it is more accurate to say that no-arbitrage means that Pareto inefficiency cannot be common knowledge.

[28]In the context of this result, Arrow (1951a) and Debreu (1951) cited the 1948 edition of Bonnesen and Fenchel's book on the theory of convex bodies, which originally had been published in 1934.

aggregation (cardinal utilitarian welfare) theorem, a fundamental theorem of social choice theory, the preferences of society satisfy a Pareto condition with respect to those of its individuals if and only if there is a system of weights with respect to which society's utility function is a weighted sum of the individual utility functions. Proofs of this result that use a separating hyperplane argument have been given by Border (1985), Weymark (1994), and De Meyer and Mongin (1995).

In the context of a decision under uncertainty—a "one-player game"—with finite sets of states and acts, under an as-if assumption of subjective expected utility preferences with or without state-independence of utility, it is possible to partially (but perhaps not completely) separate the decision-maker's probabilities from her utilities by publicly eliciting two types of small gambles she is willing to accept: "belief" gambles and "preference" gambles. These do not involve any objective randomization or counterfactualism. Belief gambles are ordinary bets on events, as in de Finetti's model of subjective probabilities. Preference gambles are of a special form: they are bets on events that are *conditioned on acts chosen by the decision-maker,* i.e., conditioned on the "moves" that she may make in the game. Preference gambles reveal information about the beliefs that the decision-maker *would need to hold* in order to choose one particular act over the others available. They do not directly reveal any information about her intentions nor her unconditional beliefs with regard to her own actions. As such they do not change her incentives. The separating hyperplane theorem can then be used to prove a fundamental theorem of decision analysis in which the decision-maker avoids arbitrage if and only if she chooses an alternative that maximizes her subjective expected utility (Nau (1995a), Smith and Nau (1995)). This means the decision-maker's problem can be solved by asking her to contemplate small gambles that focus separately on her beliefs and values in a decomposed fashion and then checking to see which alternatives are ruled out on the basis of arbitrage.

The decision-analysis-by-arbitrage model extends *without modification* to a setting in which there are two or more decision-makers with interlocking decision problems, i.e., a *noncooperative game*, in which acceptable preference gambles are the medium through which the players' payoff functions are commonly known. This yields a fundamental theorem of n-player noncooperative games (Nau and McCardle (1990)) in which the players as a group avoid arbitrage (they are *jointly* coherent) if and only if the game's outcome is one that has positive probability in a correlated equilibrium, a generalization of Nash equilibrium in which randomized strategies are permitted to be correlated. In other words, the game's outcome is rational in the no-arbitrage sense, from the perspective of an observer with the same public information as the players, if and only if it is one that could have arisen from the implementation of a correlated equilibrium. The same logic leads to a correlated generalization of Harsanyi's Bayesian equilibrium concept when it is applied to a game of incomplete information and the players' conditional beliefs about each others' types are also backed up by side bets (Nau (1992a)). This approach to modeling games fills in the discontinuities in the spectrum that are usually perceived to lie in between the theory of a single agent, the theory of "2, 3, 4... agents," and the theory of a large number of agents, insofar as it does not invoke a separate fixed-point condition to explain what happens when there are strategic interactions. It also directly addresses the issue of what it would mean for the rules of the game to be common knowledge and where a common prior might come from. The existence of a common prior is one of the conditions for no-arbitrage. In general the solution of the game may not be unique. From an observer's perspective it is the whole set of correlated equilibria, which is a possibly-imprecise probability distribution, absent further information.

The decision-analytic and game-theoretic models of no-arbitrage are easily generalized to settings in which decision-makers or game players may have state-dependent utility for money and/or unobservable prior stakes in events, in which case their subjective probabilities are not observable. In those situation the units of analysis merely become risk-neutral probabilies rather than probabilities that are measures of pure belief. Taken together, these results emphasize that:

> The separating hyperplane theorem is the tool of mathematics that
> quite literally puts the "rational" in rational choice theory by giving the
> decision-maker's inner self a numerical representation across the whole
> spectrum of personal and strategic and competitive and social decisions.

Of course, separation theorems are among the most basic tools of mathematical economics, but in the context of decision theory their role often goes unmentioned in presentations that emphasize original sources.[29]

Every one of the fundamental theorems mentioned above can be framed in such a way that the dual test of quantitative rationality is a no-arbitrage condition, and when they are formulated in this manner they are robust against relaxations of many of their less-compelling assumptions and they are also tempered by considerations of what is interpersonally observable and what isn't. There is more than one way to axiomatize a given pattern of behavior, but they may not be equivalent starting points. The discussion is channeled by the way in which the primitives of a model are chosen, which has implications for parameter measurement, for assumptions about personal and interpersonal knowledge, for empirical testing, and for determining directions along which the model can be or should be generalized or refined.

---

[29]For example, separating hyperplanes do not appear anywhere in Kreps' *Notes on the Theory of Choice* (1988), a popular reference.

If the focus is on properties of a preference relation defined on some set of real and unreal alternatives, then it is necessary to imagine that an individual's direction of preference between any pair of alternatives in that set can be measured. If agents are assumed to know (or even commonly know) each other's preferences, then corresponding reciprocal measurements must be possible. If the predictions of the model are falsified in experiments, then evidently one or more of the axioms is suspect, which suggests that further experiments ought to be designed to sharply test one axiom independently of the others. If a particular axiom is found to be violated in such an experiment, this suggests that a better model might be constructed by relaxing it in some way, holding the other axioms fixed. However, the most problematic axiom—completeness of preferences—often cannot be weakened at all without destroying the method of proof or trivializing the experimental design. If a fixed-point condition is used to determine an equilibrium solution of a noncooperative game, but it is found to be insufficiently strong to yield a unique one, this suggests that different fixed points should be compared to determine whether some are more plausible or desirable than others, so that refinements can be designed expressly to eliminate the inferior ones. And even if a unique equilibrium can be selected, the questions of how convergence occurs and whether the agents will wish to go along with it may still be problematic. Equilibrium solutions of noncooperative games need not be efficient or attractive precisely because their computation generally ignores those parameters of the game that measure the benefits of cooperation.

In the chapters that follow, the approach to modeling rationality that invokes the separating hyperplane theorem in the context of no-arbitrage for monetary outcomes will be shown to cast a different light on many of these issues.

Chapter 3

Subjective probability

3.1   Elicitation of beliefs

The subjective theory of probability defines beliefs in a way that makes them cardinally measurable and financially relevant. The theory was sketched by Ramsey (1926) and more clearly articulated by de Finetti (1931, 1937) prior to its integration with expected utility theory by Savage (1954). There are two distinct approaches to its construction, one that is *qualitative* (in which a binary relation of comparative belief or preference is a primitive concept[1]) and one that is *quantitative* (in which beliefs or preferences are revealed by willingness to bet or trade or be scored). De Finetti presented both and showed their equivalence. The same two approaches can be applied to expected utility and subjective expected utility, as will be shown in the next two chapters.

The qualitative approach to defining subjective probability is described by de Finetti (1937) as follows:

---

[1]Koopman (1940) also presented a theory of qualitative probability: "The intuitive thesis in probability holds that both in its meaning and in the laws which it obeys, probability derives directly from the intuition, and is prior to objective experience; it holds that it is for experience to be interpreted in terms of probability and not for probability to be interpreted in terms of experience; and it holds that all the so-called objective definitions of probability depend for their effective application to concrete cases upon their translation into the terms of intuitive probability."

DOI: 10.1201/9781003527145-3

"If we acknowledge only, first, that one uncertain event can only appear to us as (a) equally probable, (b) more probable, or (c) less probable than another; second, that an uncertain event always seems to us more probable than an impossible event, and less probable than a necessary event; and finally, third, that when we judge an event E' more probable than an event E, which is itself more probable than an event E'', the event E' can only appear more probable than E'' (transitive property), it will suffice to add to these three evidently trivial axioms a fourth, itself of a purely qualitative nature, in order to construct rigorously the whole theory of probability. This fourth axiom tells us that inequalities are preserved in logical sums: if E is incompatible with E_1 and also with E_2, then $E_1 \cup E$ will be more or less probable than $E_2 \cup E$, or they will be equally probable, according to whether E_1 is more or less probable than E_2, or they are equally probable."

The fourth axiom, the only one that is "non-trivial," is the independence axiom, and it is stated quite plainly. Arguably, de Finetti ought to get credit for being first to clearly invoke the independence axiom in representing preferences with regard to acts that yield uncertain payoffs. The subjective probability model and the expected utility model have the same axioms and same representation, theorem, merely with a bit of relabeling of variables, as will be discussed in the next chapter.[2]

The quantitative approach, which uses monetary bets, is described by him in this way:

"One can, however, also give a direct, quantitative, numerical defini-tion of the degree of probability attributed by a given individual to a given event, in such a fashion that the whole theory of probability can be deduced immediately from a very natural condition having an ob-vious meaning. It is a question of making mathematically precise the trivial and obvious idea that the degree of probability attributed by an individual to a given event is revealed by the conditions under which he would be disposed to bet on that event. ...[O]nce it has been shown that one can overcome the distrust that is born of the somewhat too concrete and perhaps artificial nature of the definition based on bets, the second procedure is preferable, above all for its clarity... Once an individual has evaluated the probabilities of certain events, two cases can present themselves: either it is possible to bet with him in such as way as to be

---

[2]Just interchange money with probability as the objectively given variable and interchange events with consequences as the objects to be subjectively evaluated. In both cases the two kinds of variables, probabilities and measures of outcomes, multiply each other in the additive representation.

assured of gaining, or else this possibility does not exist. In the first case one clearly should say that the evaluation of the probabilities given by the individual contains an incoherence, an intrinsic contradiction; in the other case we shall say that the individual is coherent. It is precisely this condition of coherence which constitutes the sole principle from which one can deduce the whole calculus of probability."

The betting approach allows the laws of probability to be derived from a no-arbitrage condition, a result that is commonly known as the Dutch Book Argument.[3] This is easily proved via the separating hyperplane theorem (although de Finetti did not do it that way), and it also provides a foundation for modeling common knowledge of beliefs and for integrating subjective probability with game theory, as will be shown. A superficially different but variationally equivalent method of measuring subjective probabilities via monetary transactions (either real money or "play money"), which was also used by both de Finetti and Savage, is that of *scoring rules*, in which the agent is penalized according to some measure of distance (such as squared or logged difference) between her assessed probabilities of the events and the 0-1 indicators of

---

[3] The origin of the term itself is obscure, even in Holland. (http://people.few.eur.nl/wakker/miscella/dutchbk.htm) The following speculation is offered by Andrew Colman in *A Dictionary of Psychology* (2001): "**Dutch book n.** A money pump in which a person holds intransitive or cyclic preferences, for example, preferring x to y, y to z, and z to x. Such a person is therefore willing to pay to have any of them exchanged for one that is preferred to it, and is then willing to pay to have that one exchanged for another, and so on indefinitely. The property that differentiates a Dutch book from any other money pump is that the price for each exchange is lowered until the person is willing to pay for it, although the term is often used as a synonym for a money pump. [Probably named after a Dutch auction in which the price of a lot is reduced by steps until a buyer comes forward]." (https://www.oxfordreference.com/view/10.1093/oi/authority.20110803095736804)

The term is mentioned in the 1955 article "Confirmation and Rational Betting" by R.S. Lehman, which in turn cites its appearance in the 1942 edition of *The American Thesaurus of Slang*: "dutch book, round book: a book with no house percentage," where the term "round book" as a synonym for a sure-win strategy is traced to an 1864 book *The Thesaurus of Slang*.

An intriguing possibility is that the 1940s version of the slang derives from the term "Dutching" that refers to a racetrack betting strategy in which last-minute odds manipulation is used to obtain an almost-sure win of a constant amount, which was employed by the infamous gangster Arthur Flegenheimer, a.k.a. "Dutch Schultz," who was gunned down in 1935. (http://en.wikipedia.org/wiki/Dutch_Schultz)

However, the online "Dutching guide" contains the following:

"Dutch Schultz aka Arthur Flegenheimer allegedly originated the dutching bet strategy. Alternatively, there is a train of thought that dutching originates from the term going dutch, where people share the price of a meal or a round of drinks.

'Dutching' in betting terms, is when a punter places more than one selection in a race, or any fixed odds betting medium, to glean the same amount of profit from any winner.

Not to be confused with a 'Dutch Book', which is the description used when a bookmaker bets over-broke in any race or betting event. An overbroke book means no matter what wins, the bookie will lose on the event." http://www.sandracer.com/2007/03/dutching-guide-how-to-dutch-bet.html

their occurrence. The latter method has long been used for evaluating professional forecasters and it is related to cross-entropy measures in information theory (Jose et al. (2008), Nau et al. (2007, 2009)), another connection between physics and decision theory. Another variation on the betting approach is to compare bets on events with subjective probabilities against bets on events with objective probabilities, a technique which is used in applied decision analysis and which links up with the use of objective probabilities to calibrate assessments of utility (Raiffa (1968)).

In a footnote to the 1937 paper, de Finetti mentioned two properties of the betting approach which he admitted might be viewed as shortcomings:

> "Such a formulation could better, like Ramsey's, deal with expected *utilities*. I did not know of Ramsey's work before 1937, but I was aware of the difficulty of money bets. I preferred to get around it by considering sufficiently small stakes, rather than build up a complex theory to deal with it... Another shortcoming of the definition—or the device for making it operational—is the possibility that people accepting bets against our individual have better information than he has (or know the outcome of the event considered). This would bring us to game theoretic considerations."

De Finetti's footnote is laden with significance. Both observations are important qualifications of the subjective probability model of beliefs, but they are not shortcomings of the use of monetary bets. Rather, the opposite is true. The issues raised by monetary bets with regard to how measurements of belief may be distorted or made less precise by nonlinear utility for money and by game theoretic considerations are fundamental to questions of what different agents are able to know about each others' preferences and how they should interact. These issues cannot be finessed away by introducing higher levels of abstraction in the definitions of consequences and the scope of preferences, and they will be directly addressed in later chapters of this book.[4]

---

[4]Surveys and discussions of Dutch Book arguments in general can be found in Hájek (2009), Pettigrew (2020), and Vineberg (2022). A wide range of critiques have been leveled at them, particularly from within the field of philosophy (e.g., see Kyburg (1978), Kennedy and Chihara (1979), Schick (1986), Hájek (2005)). These critiques typically start from a primitive assumption that a rational decision-maker has degrees of belief ("credence") that are representable by numbers. The issues explored are, inter alia, (i) whether and why and under what conditions those numbers should behave like probabilities, (ii) whether those probabilities are revealed by the decision-maker's preferences among monetary bets, and (iii) whether money is all that important anyway. In the narrative of this book, by comparison, the acceptance of a monetary bet is a primitive measurement (significantly, a potentially-public measurement), and the goal is to apply no-arbitrage requirements to obtain representations of mental states that support the acceptable bets under varying assumptions about their properties (e.g., whether they are complete, ad-

3.2   A 3-state example of probability assessment

To fix ideas, consider a very simple example of eliciting beliefs about the outcome of the next soccer World Cup. Suppose that you and your friends are debating
who will win, and your interest centers on the rivalry between Argentina and Brazil,
who are in different brackets of the tournament and can play each other only in the final game. So, the question of interest might be framed as that of whether the outcome
will be a victory by Argentina over Brazil in the final game, or a victory by Brazil
over Argentina in the final game, or a less exciting matchup involving one or more
other countries. For the purposes of modeling this debate, the events "Argentina-
beats-Brazil (A)," "Brazil-beats-Argentina (B)," and "other matchup (C)" may be
considered as *states of the world* of which exactly one will turn out to be true. In
this example they are states of a future world, but that need not always be the case.
Perhaps the winner has already been decided but you and your friends are marooned
on a desert island and have not yet heard the news: you are merely speculating about
what happened in your absence. A "state" is a state of *information* in which you
may reside at some future date and which might or might not be shared with others.
In this case it is state of information of you and your friends with respect to the
outcome of a soccer tournament.

You could discuss your opinions in informal, unscientific terms such as "It's
likely that Argentina and Brazil will meet in the final game" or "I'm confident
that Brazil can beat Argentina," but you can try to communicate with each other
more precisely by directly or indirectly assigning subjective probabilities to the
three states. There are a number of ways in which you might do this. One way is
to make direct intuitive judgments of your relative degrees of belief, as proposed by
Koopman.[5] For example, you might say "I believe that in a head-to-head matchup

---

ditive, linear, state-independent, intrapersonal vs. interpersonal, etc.). The representations of those mental
states may or may not involve precise numbers that have the interpretation of probabilities that are measures of unadulterated beliefs. In this setup the presence of money is critical for performing numerical
reasoning and numerical measurement at all. Another issue that has been raised in the Dutch Book Argument literature is whether the argument ought to be applied to the characterization of rational beliefs when
there are two or more decision-makers in the scene, and it has been claimed (e.g., by Christensen (1991))
that it should not. In this book, by comparison, application of the Dutch Book Argument to two or more
agents in the same scene ("joint" coherence) provides a foundation for modeling of rational behavior in
noncooperative games (Chapter 8) and linking up with asset pricing theory (Chapter 9).

[5]In Koopman's theory the primitive judgments are one-sided comparisons of conditional probabilities
with possibly-different conditioning events, e.g., "A_1 given B_1 is equally or less probable than A_2 given
B_2," which also allows for incompleteness of beliefs.

Argentina and Brazil are equally likely to win, but I believe that a different matchup is just as likely as a matchup between those two teams." This could be interpreted to mean that you assign probabilities of 1/4, 1/4, and 1/2 to the events $\mathbf{A}$, $\mathbf{B}$, and $\mathbf{C}$, respectively. Things work out nicely in this example because you just happen to feel that some combinations of these three states are equally likely. More precise statements could be made in terms of comparisons with a finer partition of events. For example, if there is some other set of 100 mutually exclusive and collectively exhaustive events $\{\mathbf{D}_1, \mathbf{D}_2, ..., \mathbf{D}_{100}\}$ that you (and perhaps you alone) can discriminate, you could say "I believe $\mathbf{D}_1, \mathbf{D}_2, ..., \mathbf{D}_{100}$ to be equally probable and I believe that $\mathbf{A}$ is exactly as probable as the union of $\mathbf{D}_1, \mathbf{D}_2, ..., \mathbf{D}_{23}$. This would suggest that you regard the probability of $\mathbf{A}$ to be exactly 0.23, although it would remain a purely verbal expression of belief. Someone else might carry out a similar exercise using his own set of personal reference events and come up with, say, a value of 0.27. This would suggest that he thinks $\mathbf{A}$ is more probable than you do, although you would not be thinking in literally the same terms and (absent further assumptions) your difference of opinion would have no practical implications.

A second approach is to calibrate your subjective probabilities against publicly observable reference events that have agreed-upon objective probabilities, which enables you to express your beliefs in a language of theoretically-enforceable monetary bets. This approach provides financial incentives for accuracy and consistency and it also makes your probability judgments common knowledge insofar as it provides opportunities for reciprocal action by others. For example, you could say: "A lottery ticket that pays $100 if $\mathbf{A}$ occurs is exactly as desirable to me as one that pays $100 if a RAND( ) function on this Excel spreadsheet returns a value less than 0.23 on the next recalculation, and I am willing to trade either one for the other at the discretion of someone else."[6] Letting the symbols $\mathbf{A}$ and $\mathbf{D}_{0.23}$ also stand for indicator variables for the events ($\mathbf{A} = 1$ if $\mathbf{A}$ occurs, $\mathbf{A} = 0$ otherwise, $\mathbf{D}_{0.23} = 1$ if RAND( )<0.23, $\mathbf{D}_{0.23} = 0$ otherwise), the payoff functions of the two lottery tickets can be written as $\$100\mathbf{A}$ and $\$100\mathbf{D}_{0.23}$. If you are willing to trade either one for the other, this means that you are willing to accept a bet whose payoff function is $\$100(\mathbf{D}_{0.23} - \mathbf{A})$ and also willing to accept one whose payoff function is $\$100(\mathbf{A} - \mathbf{D}_{0.23})$. Now suppose, hypothetically, that you assign probabilities p and q to $\mathbf{D}_{0.23}$ and $\mathbf{A}$ respectively and that you are someone who only accepts bets that have non-negative expected value for you. The expected values for the two bets are $\$100(p - q)$ and $\$100(q - p)$, and if you find both of them equally acceptable, then

---

[6]The text string "RAND( )," if entered in a cell or embedded in a formula, yields a uniform [0,1] random number when any action is taken that results in a recalculation of all the formulas, such as hitting the F9 key or making an edit somewhere..

evidently you feel that $p = q$, and everyone knows that $p = 0.23$, so everyone now knows that your belief concerning the event that Argentina will defeat Brazil in the final game can be represented by the probability 0.23.

There are many implicit assumptions behind this method of quantifying your beliefs, besides belief in the uniform randomness of RAND( ). They involve psychological issues such as facility with numbers and value for money and enjoyment of gambling and strategic incentives and distinctions between risk and uncertainty and reliance on heuristics and social norms. $100 needs to be a large enough sum for you to take seriously but not so large as to raise questions of diminishing marginal value or aversion to uncertainty, you should value it the same regardless of which teams win and lose, you should attach no value to gambling on soccer per se, you should not have qualitatively different attitudes toward objective and subjective un-certainties, you should be able to quantify your beliefs on a 0-to-1 scale without bias and with a precision of at least 0.01, you should not be trying to exploit your friends based on speculation about how and why their beliefs might differ from yours, and you should regard this whole exercise as a familiar and natural mode of thinking.[7] All of these issues are relevant to the modeling of rational beliefs and (with the exception of love of gambling) will be discussed in more depth later on.

A third approach (de Finetti's), which also involves bets for money but does not rely on objective randomization, is to give *odds* at which you are willing to bet on events for arbitrary finite stakes (within reason), or equivalently, to name *prices* at which you are willing to buy or sell lottery tickets (or more exotic financial assets) whose payoffs depend on the events. For example, you might say "I'll give four-to-one odds against **A**," which means that you are willing to accept a bet in which you pay out 4α if **A** occurs and you get paid α if it doesn't occur, where α is a positive number chosen at the discretion of the person on the other side of the bet, within reason. This conveys that you believe the probability of **A** is no greater than 0.2. If at the same time you say "I'll also *take* four-to-one odds against **A**," this means you are also willing to take the opposite bet, which altogether conveys that you believe the probability of **A** is *exactly* 0.2.

---

[7]Random devices such as adjustable spinners ("probability wheels") *are* used to calibrate probability assessments (and also utility assessments) in applied decision analysis, under the conventional but un-verifiable assumption of state-independence of utility, but this is another exception that proves the rule. Most persons don't think to do it except at the urging of a consultant or instructor.

Another way to say the same thing is that for a price of $0.20 each you are willing to buy or sell lottery tickets that yield a prize of $1 if **A** occurs and $0 otherwise,[8] giving you the following net payoffs:

TABLE 3.1: Payoffs for purchase of a unit lottery ticket

	A	B	C
Payoff of **A** ticket	$1.00	$0.00	$0.00
Less price paid	−$0.20	−$0.20	−$0.20
Net payoff to buyer	$0.80	−$0.20	−$0.20

Furthermore, suppose your pricing policy is linear, i.e., the price at which you are willing to buy or sell β such tickets is equal to β times the price of one ticket. If you buy β of these lottery tickets at this price, thus paying 0.20β in hopes of getting back β if **A** occurs, you will enjoy a gain of 0.80β if **A** occurs and suffer a loss of 0.20β otherwise.

The price p at which you would indifferently buy or sell an **A** lottery ticket appears to be your subjective probability for **A**, insofar as the ticket has an expected value of zero for you in that case. However, to justify the interpretation of lottery ticket prices as subjective probabilities, they must obey all the relevant laws of probability. One of these laws is the additive law which states that the probability of the union of two disjoint events should equal the sum of their probabilities. For example, your probability for the event that the outcome will be either a victory for Argentina over Brazil or a victory of Brazil over Argentina, which is the union of **A** and **B**, should be the sum of your probabilities for the two separate events.

In order to force your lottery ticket prices to satisfy the additive law, we need to make two more assumptions. First, your prices for bundles of different lottery tickets should be linear and additive. If you are willing to buy or sell lottery tickets on **A** and **B** for $0.20 and $0.30 apiece, respectively, then you ought to be willing to buy or sell a bundle consisting of one of each for $0.50. You should also be willing to buy two **A** tickets and sell one **B** ticket for a total of $0.10 ($= 2 \times \$0.20 - \$0.30$), and so on, even in fractional amounts. The second assumption is that you should not expose yourself to arbitrage (a Dutch book) by offering to buy or sell the same lottery for two different prices. To continue the example, if you also offer to buy or sell a lottery ticket on the event **A**∪**B**, then your price for it must be $0.50, otherwise someone can sell you the separate **A** and **B** lottery tickets for a total of $0.50 and buy back the **A**∪**B** ticket at the other price, or vice versa, and walk off with your

---

[8]In the modeling of financial markets, a lottery ticket that yields a $1 prize in a given state is called an "Arrow security."

money. For instance, if you sell the A∪B ticket for $0.49 while buying the **A** and **B** tickets for $0.20 and $0.30, you will suffer a sure loss of $0.01:

TABLE 3.2: Arbitrage due to violating the additive law of probability

	A	B	C
Net payoff for buying **A** ticket @ $0.20	$0.80	−$0.20	−$0.20
Net payoff for buying **B** ticket @ $0.30	−$0.30	$0.70	−$0.30
Net payoff for selling A∪B ticket @ $0.49	−$0.51	−$0.51	$0.49
Total (a sure loss)	−$0.01	−$0.01	−$0.01

The assumptions that pricing is linear and additive and that there is no arbitrage are sufficient to guarantee that your prices for lottery tickets that yield $1 prizes obey the additive law of probability.

Conditional probabilities can be elicited in the same way by introducing the option to "call off" a bet on one event if some other event does not occur. For example, you might offer to buy or sell an **A** ticket for $0.40 conditional on the event of Argentina and Brazil meeting in the final game, i.e., conditional on the event A∪B. If someone sells β of these lottery tickets to you at this price, then you enjoy a gain of 0.60\beta$ if **A** occurs and suffer a loss of 0.40\beta$ if **B** occurs and the deal is called off if **C** occurs. Your offer of this bet suggests that your conditional probability for A|A∪B is 0.40.[9]

When conditional probabilities are elicited through called-off bets in this manner, the assumptions made above (linearity, additivity, no-arbitrage) are also sufficient to enforce the multiplicative law of probability and Bayes' rule. Suppose that an opponent buys a 0.4 fraction of an A∪B lottery ticket from you at the per-unit price of $0.50, which gives her a net gain of $0.20 if **A** or **B** occurs and a net loss of $0.20 otherwise. Suppose that at the same time she buys one lottery ticket on A|A∪B for $0.40, which yields her a net gain of $0.60 if **A** occurs, a net loss of $0.40 if **B** occurs, and zero otherwise. The sum of these two transactions yields her a net gain of $0.80 if **A** occurs and net loss of $0.20 if either **B** or **C** occurs, which is equivalent to paying $0.20 for the lottery ticket on **A**:

TABLE 3.3: Illustration of the multiplicative law of probability

	A	B	C
Net payoff for buying 0.4 A∪B ticket @ $0.50	$0.20	$0.20	−$0.20
Net payoff for buying 1 A|A∪B ticket @ $0.40	$0.60	-$0.40	$0.00
Total (equivalent to buying **A** ticket @ $0.20)	$0.80	−$0.20	−$0.20

---

[9]The notation A|A ∪ B means "A given {A ∪ B}," i.e., the occurrence of **A** conditional on the event that either **A** or **B** occurs.

Quoting any other price for the **A** ticket would expose you to arbitrage. It is no accident that $0.20 = 0.40 \times 0.50$, i.e., the price of the **A** ticket that is indirectly constructed out of the **A|A∪B** ticket and the **A∪B** ticket is the product of their prices, which is consistent with the multiplicative law of probability.

To prove that the multiplicative law holds generally, let the prices offered for lottery tickets on events **A**, **A∪B** and **A|A∪B** be denoted by $P(\mathbf{A})$, $P(\mathbf{A}\cup\mathbf{B})$ and $P(\mathbf{A}|\mathbf{A}\cup\mathbf{B})$, and similarly for other events. Let **x** and **y** denote the payoff vectors associated with the purchase of the respective lottery tickets, which are:

TABLE 3.4: Payoff vectors for separate bets on **A∪B** and **A|A∪B**

	A	B	C
x	$1 - P(\mathbf{A}\cup\mathbf{B})$	$1 - P(\mathbf{A}\cup\mathbf{B})$	$-P(\mathbf{A}\cup\mathbf{B})$
y	$1 - P(\mathbf{A}|\mathbf{A}\cup\mathbf{B})$	$-P(\mathbf{A}|\mathbf{A}\cup\mathbf{B})$	0

Suppose that an opponent buys quantities α and β of the two lottery tickets, where $\alpha = P(\mathbf{A}|\mathbf{A}\cup\mathbf{B})$ and $\beta = 1$. (She could also choose the negatives of the same amounts.) This yields her the following vector of payoffs:

TABLE 3.5: Payoffs of $\alpha\mathbf{x}+\beta\mathbf{y}$

A	B	C
$1 - P(\mathbf{A}|\mathbf{A}\cup\mathbf{B})P(\mathbf{A}\cup\mathbf{B})$	$-P(\mathbf{A}|\mathbf{A}\cup\mathbf{B})P(\mathbf{A}\cup\mathbf{B})$	$-P(\mathbf{A}|\mathbf{A}\cup\mathbf{B})P(\mathbf{A}\cup\mathbf{B})$

which is the same as that of a lottery ticket on **A** sold at a price of $P(\mathbf{A}|\mathbf{A}\cup\mathbf{B})P(\mathbf{A}\cup\mathbf{B})$. No-arbitrage then requires $P(\mathbf{A}) = P(\mathbf{A}|\mathbf{A}\cup\mathbf{B})P(\mathbf{A}\cup\mathbf{B})$, which is the multiplicative law.

The same sort of construction can be used to demonstrate Bayes' rule, which applies when there is a two-fold partition of the states into a set of hypotheses and a set of observations. In the simplest (2×2) case, the state space consists of $\{\mathbf{AB}, \mathbf{A\overline{B}}, \mathbf{\overline{A}B}, \mathbf{\overline{A}\,\overline{B}}\}$ for some events **A** and **B**,[10] where, say, **A** is a hypothesis that might or might not be true and **B** is an observation of a signal that might or might not be positive Suppose that prices $P(\mathbf{A})$, $P(\mathbf{B})$, and $P(\mathbf{B}|\mathbf{A})$ are offered for lottery tickets on the events **A**, **B**, and **B|A**,[11] and let **x**, **y**, and **z** denote the payoff vectors of bets corresponding to their purchase at these prices:

---

[10]$\mathbf{\overline{B}}$ denotes the complement of B, i.e., not-B, whose indicator function is $1 - \mathbf{B}$.

[11]If the data consists of prices of lottery tickets on **A**, **B|A**, and **B|$\overline{\mathbf{A}}$** (rather than B directly), then the additive and multiplicative laws can be applied first to obtain $P(\mathbf{B}) = P(\mathbf{A})P(\mathbf{B}|\mathbf{A}) + (1 - P(\mathbf{A}))P(\mathbf{B}|\overline{\mathbf{A}})$.

TABLE 3.6: Payoff vectors for separate bets on **A**, **B**, and **B**|**A**

	AB	**A$\overline{\text{B}}$**	**$\overline{\text{A}}$B**	**$\overline{\text{AB}}$**
x	$1 - P(\mathbf{A})$	$1 - P(\mathbf{A})$	$-P(\mathbf{A})$	$-P(\mathbf{A})$
y	$1 - P(\mathbf{B})$	$-P(\mathbf{B})$	$1 - P(\mathbf{B})$	$-P(\mathbf{B})$
z	$1 - P(\mathbf{B}|\mathbf{A})$	$-P(\mathbf{B}|\mathbf{A})$	0	0

Suppose that an opponent buys quantities α, β, and γ of these three lottery tickets, where $\alpha = P(\mathbf{B}|\mathbf{A})$, $\beta = -P(\mathbf{A})P(\mathbf{B}|\mathbf{A})/P(\mathbf{B})$, and $\gamma = 1$. This yields the following payoffs:

TABLE 3.7: Payoffs of $\alpha\mathbf{x}+\beta\mathbf{y}+\gamma\mathbf{z}$

AB	**A$\overline{\text{B}}$**	**$\overline{\text{A}}$B**	**$\overline{\text{AB}}$**
$1 - P(\mathbf{A})P(\mathbf{B}|\mathbf{A})/P(\mathbf{B})$	0	$-P(\mathbf{A})P(\mathbf{B}|\mathbf{A})/P(\mathbf{B})$	0

which is the same as the payoff vector of a lottery ticket on the conditional event **A**|**B** that is purchased at a price of $P(\mathbf{A})P(\mathbf{B}|\mathbf{A})/P(\mathbf{B})$. No-arbitrage therefore requires that $P(\mathbf{A}|\mathbf{B}) = P(\mathbf{A})P(\mathbf{B}|\mathbf{A})/P(\mathbf{B})$, which is Bayes' rule.

The betting-odds/lottery-ticket method of eliciting subjective probabilities is subject to many of the same qualifications as the one that uses the RAND( ) function, with the following important exception. Deals like this really do take place and have gigantic institutions devoted to making them possible. Casinos and gambling web sites offer real-time odds and money lines at which they are willing to take bets on football games or horse races or tennis matches.[12] Securities markets make it

---

[12]Online sports betting has become a huge industry and is the setting in which precise quantification of uncertainty is most widely practiced by non-specialists. Significantly the standard language for expressing the terms of sports bets is that of "money lines," not probabilities or odds directly. These are the numbers that are typically displayed on web sites and TV broadcasts, and they focus your attention on concrete amounts of money. A money line of $+x$ on the underdog in a game, where $x > 100$, means that you must risk a loss of \$100 for a chance to gain \$$x$ if the underdog wins, so the implied upper probability of that event is $100/(100 + x)$. A money line of $-x$ on the favorite, where $x > 100$, means that you must risk a loss of \$$x$ for a chance to gain \$100 if the favorite wins, for which the implied upper probability is $x/(100+x)$. The sign on x focuses your attention on whether it is an amount to be won or lost in a gamble whose alternative outcome is ± 100. In economic terms, the implied upper probabilities are market prices of Arrow securities. Their sum, minus 1, is the so-called "overround", which in online betting is often in a range from 0.03 to 0.10 depending on the depth of the market. The overround divided by the sum of implied probabilities is the "juice" or "vigorish", the casino's riskless profit margin if there is equal betting on both sides. The sum of implied upper probabilities can be normalized to yield a predictive distribution. This system applies equally well to games with 3 or more outcomes. In betting on soccer games there are money lines for win, lose, and draw outcomes. For example, in a game between teams A

possible to buy and sell complicated forms of lottery tickets whose payoffs depend
on future stock prices and commodity prices and interest rates and events such as
credit defaults. Insurance companies invite you to bet large amounts of money on
the loss of your house or your car or your life. The deals you can get from casinos or
stock exchanges[13] don't necessarily reflect the beliefs of anyone in particular—they
aggregate the beliefs of everyone in the market and (in the case of the stock market)
may include adjustments for risk—but the principle is the same for individuals and
markets: prices of lottery tickets reveal information about subjective beliefs that can
be expressed in terms of numbers, and no-arbitrage forces those numbers to obey all
the relevant laws of probability when transactions may be linearly combined.

3.3   The fundamental theorem of subjective probability

In this section the analysis sketched above will be formalized to derive the fundamental theorem of subjective probability. In the next chapter it will be shown that the
fundamental theorem of expected utility (von Neumann and Morgenstern's result) is
the same theorem with the parts merely interchanged.

Let $S := \{s_1, ..., s_M\}$ denote a finite set of mutually exclusive and exhaustive
states of the world about whose realization the individual is uncertain. Let $\mathbf{A}$ and $\mathbf{B}$
denote *events* (subsets of S) as well as their respective indicator vectors, i.e., $\mathbf{A} =
(a_1, ..., a_M)$, where $a_m = 1$ if event $\mathbf{A}$ contains state s_m and $a_m = 0$ otherwise.
For example, the vector $\mathbf{A} = (1, 1, 0, 0, ..., 0)$ represents an event consisting of the
union of s_1 and s_2. Let $\mathcal{F}$ denote an M-dimensional convex subset of $\mathbb{R}^M$ whose
elements are "SP-acts" that assign levels of monetary wealth to states of the world,
measured relative to some personally meaningful zero-point. (For the remainder of
this chapter, these will just be referred to as "acts.") $\mathcal{F}$ could consist of all of $\mathbb{R}^M$, but
it would be more realistic to assume that it is a bounded set of financial assets among
which the statewise variations in wealth are not so large as to raise questions of
nonlinear utility for money. A vector such as $\mathbf{f} = (f_1, ..., f_M) \in \mathcal{F}$ represents an act

---

and B, the money lines might be 340 for a win by team A, 210 for a win by team B, and -105 for a draw.
These translate into implied upper probabilities of 0.227, 0.323, and 0.512, whose sum is 1.062, which
has a "vig" of $0.062/1.062 = 5.8\%$. Normalized, they yield a predictive point probability distribution
of $(0.214, 0.304, 0.482)$. Another advantage of money lines is that they help to visualize the impact of
low-probability events. For example, a money line of $+8000$ (a quantity of dollars to be won) translates
into an implied upper probability of 1.23% for the event of winning. The implications of the former
number are easier to grasp for most individuals.

[13] The deals you get from insurance companies are based to a large extent on actuarial data analysis
and forecasting models, but there is still some room for subjectivity and for portfolio effects.

in which f_m is wealth enjoyed in state m. This vector notation allows probabilities and expected values to be written as inner products. If $\boldsymbol{\pi} = (\pi_1, ..., \pi_M)$ is a vector of state probabilities, then

$$P_{\boldsymbol{\pi}}(\mathbf{A}) := \boldsymbol{\pi} \cdot \mathbf{A} = \sum_{m=1}^{M} \pi_m a_m \qquad (3.1)$$

is the probability of $\mathbf{A}$ and $E_{\boldsymbol{\pi}}(\mathbf{f}) := \boldsymbol{\pi} \cdot \mathbf{f}$ is the expected value of $\mathbf{f}$.[14]

Rational beliefs can be characterized in terms of axioms that should be satisfied by an individual's preferences among acts. The primitive concept of preference that will be used here and throughout the book is *weak* preference: "$\mathbf{f} \succsim \mathbf{g}$" means that "$\mathbf{f}$ is preferred or indifferent to $\mathbf{g}$," in terms of which strict preference ("$\mathbf{f} \succ \mathbf{g}$") is defined to mean "$\mathbf{f} \succsim \mathbf{g}$ and not $\mathbf{g} \succsim \mathbf{f}$" and indifference ("$\mathbf{f} \sim \mathbf{g}$") is defined to mean "both $\mathbf{f} \succsim \mathbf{g}$ and $\mathbf{g} \succsim \mathbf{f}$." Some authors (e.g., Fishburn (1970), Seidenfeld et al. (1995), Karni (2011c)) prefer to use strict preference as the primitive. One reason for starting from weak preference is that an assertion of weak preference translates directly into a weak inequality on the set of probabilities that represents the individual's beliefs, which is therefore closed, and weak inequalities are more natural to use in constrained optimization models. Another reason is that weak preference supports a direct distinction between indifference ("both $\mathbf{f} \succsim \mathbf{g}$ and $\mathbf{g} \succsim \mathbf{f}$") and incomparability ("neither $\mathbf{f} \succsim \mathbf{g}$ or $\mathbf{g} \succsim \mathbf{f}$"), which is useful for the modeling of incomplete preferences. The argument that completeness is the first of the standard assumptions that ought to be given up will recur throughout the book.

As a starting point, suppose that the individual's preferences satisfy the following axioms[15] for any acts $\mathbf{f}$, $\mathbf{g}$, and $\mathbf{h}$:

Axiom SP1^a (completeness): $\mathbf{f} \succsim \mathbf{g}$ or $\mathbf{g} \succsim \mathbf{f}$ or both.

Axiom SP2^a (continuity): the sets $\{\mathbf{g} : \mathbf{f} \succsim \mathbf{g}\}$ and $\{\mathbf{g} : \mathbf{g} \succsim \mathbf{f}\}$ are closed.

Axiom SP3^a (transitivity): $\mathbf{f} \succsim \mathbf{g}$ and $\mathbf{g} \succsim \mathbf{h} \Longrightarrow \mathbf{f} \succsim \mathbf{h}$.

Axiom SP4^a (independence): $\mathbf{f} \succsim \mathbf{g} \Longleftrightarrow$
$\alpha\mathbf{f} + (1 - \alpha)\mathbf{h} \succsim \alpha\mathbf{g} + (1 - \alpha)\mathbf{h}$ for all $\mathbf{h}$ and all $\alpha \in (0, 1)$.

---

[14]For any vector $\mathbf{x}$, $P_{\boldsymbol{\pi}}(\mathbf{x})$ and $E_{\boldsymbol{\pi}}(\mathbf{x})$ are actually the same quantity, namely $\boldsymbol{\pi} \cdot \mathbf{x}$. However, it will be written as $P_{\boldsymbol{\pi}}(\mathbf{x})$ when $\mathbf{x}$ is the indicator of an event and as $E_{\boldsymbol{\pi}}(\mathbf{x})$ when $\mathbf{x}$ is the payoff vector of an act or bet. When the terms $P(\mathbf{x})$ and $E(\mathbf{x})$ are not subscripted by a probability distribution, this indicates that they are subjectively assessed probabilities and expectations. Indicator vectors of events will usually be represented by upper-case symbols such as $\mathbf{A}$, consistent with the conventional use of upper-case letters as names for events.

[15]The labels of these axioms have the superscript "a" to indicate that they refer to acts, while the labels of the alternative axioms given later have the superscript "b" to indicate that they refer to bets.

Definition (strict dominance): An act **f** *strictly dominates* another act **g**, written **f** $\gg$ **g**, if $f_m > g_m$ for every m.

Axiom SP5^a (strict monotonicity): **f** $\gg$ **g** $\Longrightarrow$ **f** $\succ$ **g**.

The first three are standard ordering axioms, and the fifth is the preferential equivalent of no-arbitrage. The psychologically interesting axiom is the independence axiom, and it accomplishes several things. First, it has a transparent linearity property: it requires that if you prefer act **f** to act **g** and you consider moving from each of them in a straight line toward *or away from* some other act **h** by the same proportional amount α, your direction of preference between them is maintained along the way. Second, it interacts with the transitivity axiom to imply that two different preferences may be added. These properties are summarized in the following two lemmas:

Lemma 3.1: f $\succsim$ **g** and **f**$^* - $**g**$^* \propto$ **f** $-$ **g** $\Longrightarrow$ **f**$^* \succsim$ **g***.

Proof: **f**$^* - $**g**$^* \propto$ **f** $-$ **g** implies that **f**$^* - $**g**$^* = \frac{1}{\beta}($**f** $-$ **g**$)$ for some $\beta > 0$, in terms of which β**f**$^* - $**f**$ = \beta$**g**$^* - $**g** Let **h** $= (\frac{1}{1-\alpha})((1 - \alpha\beta)$**e**$ + \alpha(\beta$**f**$^* - $**f**$))$, where **e** is an arbitrary vector and α is a positive number satisfying $\alpha < 1$ and $\alpha\beta < 1$.[16] Then α**f**$ + (1 - \alpha)$**h**$ = \alpha\beta$**f**$^* + (1 - \alpha\beta)$**e**, and similarly α**g**$ + (1 - \alpha)$**h**$ = \alpha\beta$**g**$^* + (1 - \alpha\beta)$**e**. A double application of the independence axiom yields:

$$\mathbf{f} \succsim \mathbf{g} \iff \alpha\mathbf{f} + (1 - \alpha)\mathbf{h} \succsim \alpha\mathbf{g} + (1 - \alpha)\mathbf{h}$$
$$\iff \alpha\beta\mathbf{f}^* + (1 - \alpha\beta)\mathbf{e} \succsim \alpha\beta\mathbf{g}^* + (1 - \alpha\beta)\mathbf{e}$$
$$\iff \mathbf{f}^* \succsim \mathbf{g}^*$$

Lemma 3.2: f $\succsim$ **g** and **f**$^* \succsim$ **g**$^* \Longrightarrow$
    α**f**$ + (1 - \alpha)$**f**$^* \succsim \alpha$**g**$ + (1 - \alpha)$**g***

Proof: By the independence axiom, **f** $\succsim$ **g** implies α**f**$ + (1 - \alpha)$**f**$^* \succsim \alpha$**g**$ + (1 - \alpha)$**f*** and **f**$^* \succsim$ **g*** implies α**g**$ + (1 - \alpha)$**f**$^* \succsim +\alpha$**g**$ + (1 - \alpha)$**g*** (note that **f*** is the common element in the first case and **g** is the common element in the second case), from which α**f**$ + (1 - \alpha)$**f**$^* \succsim \alpha$**g**$ + (1 - \alpha)$**g*** follows by transitivity

Lemma 3.1 establishes that the direction of preference between **f** and **g** depends only on the direction of the vector **f** $-$ **g**, not on the separate elements of **f** and **g** nor

---

[16]If **f** and **g** are probability distributions, **e** can be taken to be a uniform distribution, in which case **h** is also a probability distribution if λ is sufficiently small. Hence the same lemma will be applicable later on when the objects of choice are lotteries or lottery acts.

on the absolute magnitude of their differences. This means that the individual effectively has linear utility for money (although not necessarily the *same* linear utility for money in each state) and her preferences are represented by a set of *spatial* directions in which she would like to move her distribution of wealth. Lemma 3.2 implies that this set of spatial directions is convex. In fact, the properties of preferences derived in Lemmas 3.1 and 3.2 are jointly equivalent to the properties assumed in axioms SP3^a and SP4^a, so they could have just as well been taken as the 3rd and 4th axioms. They seem somewhat reasonable on-the-margin where changes in wealth are not too dramatic.

Rational beliefs can also be characterized in terms of axioms that should be satisfied by bets or trades that an individual would be willing to make, whose payoff vectors are directions in which she would like to move. One who *prefers* **f** to **g** should privately be willing to *exchange* **g** for **f** if the opportunity should arise, which would result in a change of **f** − **g** in her wealth distribution. And it follows from Lemma 3.1 that she should also be willing to exchange some non-negative multiple of **g** for the same multiple of **f** regardless of whether she starts with **g**. For example, if she is currently in possession of act **h**, she should be willing to exchange **h** for **h**+α(**f** − **g**) for any α > 0. This is equivalent to saying that, regardless of her current wealth distribution, she is privately willing to take a bet whose payoff vector is any non-negative multiple of **f** − **g**. For example, the Argentina-wins lottery ticket has payoff vector **f** = $(1, 0, 0)$, and a lump sum of twenty cents has the payoff vector **g** = $(1/5, 1/5, 1/5)$. An individual who is willing to exchange the latter for the former is accepting a "\$-bet" whose payoff vector is **f** − **g** = $(4/5, -1/5, -1/5)$, which is a bet *on* Argentina of a stake of size \$0.20 at odds of 4-to-1-against.

The set of bets that are acceptable in the above sense can be defined in terms of preferences as follows:

Definition (feasible \$-bets): **x** is a *feasible* \$-bet if and only if **x** = **f** − **g** for some **f** ∈ $\mathcal{F}$ and **g** ∈ $\mathcal{F}$. That is, a feasible \$-bet is an exchange of one act for another. The set of feasible \$-bets is denoted by $\mathcal{X}^{\mathcal{F}}$, and it is a convex set that is symmetric around the origin, i.e., **x** ∈ $\mathcal{X}^{\mathcal{F}}$ if and only if −**x** ∈ $\mathcal{X}^{\mathcal{F}}$.

Definition (preferable and acceptable \$-bets) **x** is a *preferable* \$-bet if and only if **x** ∈ $\mathcal{X}^{\mathcal{F}}$ and **f**+α**x** $\succsim$ **f** for any α > 0 such that α**x** ∈ $\mathcal{X}^{\mathcal{F}}$ and the set of such **f** is non-empty. That is, a \$-bet is preferable if it is a trade that is weakly desirable, privately, regardless of the status quo. It is *acceptable* if it is preferable *and* the individual publicly asserts that she is willing to accept the trade at the discretion of an observer who chooses α. This means that the preference is not merely private: it is

common knowledge in a material, specular, contractual sense. The set
of acceptable \$-bets is denoted by $\mathcal{X}$.

Assumption CK (common knowledge): every preferable bet is also
acceptable.

Henceforth it will be tacitly assumed that common knowledge applies, but it is to
be understood that, epistemically, "preferable" and "acceptable" are not necessarily
the same thing. What goes on in the imagination is not always made public. Also
note that in the SP model the common knowledge assumption leverages the fact that
payoffs are in units of money, which is a conserved quantity that is visible and understood without definition and physically exchangeable and universally desirable, ceteris paribus. Thus, an observer is really capable of executing the trade and implictly
has opposite incentives to those of the individual. These knowledge assumptions will
become more farfetched in the context of expected utility and subjective expected
utility, where acts are mostly counterfactual and have non-monetary consequences
that may be entirely personal in nature.

    With the definitions and assumptions above, it follows that $\mathcal{X}$ satisfies a corre-
sponding set of axioms:

**Theorem 3.1: Equivalence of SP axioms for preferences and accept-
able \$-bets**

The preference relation $\succsim$ over SP-acts satisfies SP1^a-SP5^a if and only
if the set $\mathcal{X}$ of acceptable \$-bets satisfies:

Axiom SP1^b (completeness): for any $\mathbf{x} \in \mathcal{X}^{\mathcal{F}}$, either $\mathbf{x} \in \mathcal{X}$ or
   $-\mathbf{x} \in \mathcal{X}$ or both.

Axiom SP2^b (continuity): $\mathcal{X}$ is closed.

Axiom SP3^b (convexity): $\mathbf{x} \in \mathcal{X}$ and $\mathbf{y} \in \mathcal{X} \Longrightarrow$
   $\alpha\mathbf{x} + (1-\alpha)\mathbf{y} \in \mathcal{X}$ for any $\alpha \in (0,1)$.

Axiom SP4^b (linear extrapolation): $\mathbf{x} \in \mathcal{X} \Longrightarrow \alpha\mathbf{x} \in \mathcal{X}$
   for any $\alpha > 1$ such that $\alpha\mathbf{x} \in \mathcal{X}^{\mathcal{F}}$.

Axiom SP5^b (no arbitrage): $\mathbf{x} < \mathbf{0} \Longrightarrow \mathbf{x} \notin \mathcal{X}$.

Proof: If the "a" axioms hold, then by choosing $\mathbf{x} = \mathbf{f} - \mathbf{g}$, it follows
from Lemma 3.1 and the definition of acceptable bets that $\mathbf{f} - \mathbf{g} \in \mathcal{X}$
if and only if $\mathbf{f} \succsim \mathbf{g}$. In this case, SP1^b, SP2^b, SP4^b, and SP5^b are
directly implied by SP1^a, SP2^a, SP4^a and SP5^a, respectively, and SP3^b
is directly implied by Lemma 3.2. Conversely, if the "b" axioms hold,

then SP1^b implies SP1^a, and $\mathbf{f}^* \succsim \mathbf{g}^*$ is implied by $\mathbf{f} \succsim \mathbf{g}$ and $\mathbf{f}^* - \mathbf{g}^* = \mathbf{f} - \mathbf{g}$, which is a weak form of Lemma 3.1. SP2^a, SP4^a and SP5^a then follow directly from SP2^b, SP4^b, and SP5^b, respectively. To establish SP3^a, plug $\mathbf{x} = \mathbf{f} - \mathbf{g}$ and $\mathbf{y} = \mathbf{g} - \mathbf{h}$ and $\alpha = \frac{1}{2}$ into SP3^b to show that $\frac{1}{2}\mathbf{f} \succsim \frac{1}{2}\mathbf{h}$ is implied by $\mathbf{f} \succsim \mathbf{g}$ and $\mathbf{g} \succsim \mathbf{h}$, then apply SP4^b with $\alpha = 2$ to scale up the inference to $\mathbf{f} \succsim \mathbf{h}$.

So, either the preference relation $\succsim$ or the set of acceptable bets $\mathcal{X}$ can be viewed as the psychological primitive, with the other derived from it. We could just as well start with a set $\mathcal{X}$ of acceptable bets that satisfy the SP^b axioms and then define a corresponding preference relation $\succsim$ by the requirement that $\mathbf{f} \succsim \mathbf{g}$ if and only if $\mathbf{f} - \mathbf{g} \in \mathcal{X}$, from which it would follow that $\succsim$ satisfies the SP^a axioms. However, the two axiom systems package the linearity and additivity assumptions in slightly different ways. The "b" system refers directly to the geometry of the decision-maker's behavior, which (literally) makes its assumptions easier to visualize, and it clearly distinguishes between the operations of interpolating between preferences and extrapolating from them. More importantly, the fundamental measurement in the latter system (the acceptance of a bet) is one that lends itself to a symmetric transaction with another agent, the parameters of which are commonly known with precision. The significance of this will be emphasized throughout the book.

The "b" axioms are essentially those used by de Finetti in the quantitative version of his theory of subjective probability, in which he refers to the no-arbitrage condition as *coherence*. These axioms set the stage for applying the separating hyperplane theorem to obtain a representation of $\mathcal{X}$ and $\succsim$.

Theorem 3.2: Fundamental theorem of subjective probability

The preference relation $\succsim$ over SP-acts satisfies SP1^a–SP5^a, and equivalently the set $\mathcal{X}$ of acceptable \$-bets satisfies SP1^b–SP5^b, if and only if there exists a unique probability distribution π such that $\mathbf{f} \succsim \mathbf{g}$ if and only if $E_\pi(\mathbf{f}) \geq E_\pi(\mathbf{g})$, and equivalently $\mathbf{x} \in \mathcal{X}$ if and only if $E_\pi(\mathbf{x}) \geq 0$.

Proof: SP1^b–SP3^b imply that $\mathcal{X}$ is a closed convex subset of $\mathcal{X}^{\mathcal{F}}$ that includes the origin. SP4^b further implies that it is the intersection of $\mathcal{X}^{\mathcal{F}}$ with a convex cone that recedes from the origin. SP1^b carries the further implication that the latter cone must be at least a half-space, if not the whole space. SP5^b requires that $\mathcal{X}$ not contain any points in the open negative orthant, which is itself a convex set. Therefore (by Lemma 2.1) a non-trivial hyperplane separates $\mathcal{X}$ from the open negative orthant and it must be the same hyperplane that is the boundary of $\mathcal{X}$ in the vicinity of the origin, because the origin is a limit point of both sets. Let π denote the normal vector of this separating hyperplane, which is unique up to positive scaling. Then $\pi \cdot \mathbf{x} \geq 0$ for all $\mathbf{x} \in \mathcal{X}$ and $\pi \cdot \mathbf{x} < 0$ for all $\mathbf{x} < 0$. The latter condition means that π is non-negative with at least one positive element, and without loss of generality it can be normalized to be a probability distribution. Conversely, if $\mathcal{X}$ is the set of $\mathbf{x} \in \mathcal{X}^{\mathcal{F}}$ satisfying $\pi \cdot \mathbf{x} \geq 0$ for some probability distribution π, then it is the intersection of $\mathcal{X}^{\mathcal{F}}$ with a half-space that includes the origin but does not contain any strictly negative vectors, so it satisfies SP1^b–SP5^b.

Hence, the rational bettor or investor behaves as if her objective is to maximize the expected value of her wealth with respect to a unique subjective probability distribution π that *appears* to represent her beliefs. However, insofar as her marginal utility for money or the purchasing power of money could be state-dependent, the elicited probabilities could be distorted. In those kinds of situations, the elicited parameters of beliefs are *risk-neutral probabilities*, which are interpretable as products of subjective probabilities and relative marginal utilities for money with reference to a particular currency as discussed in Section 1.4.[17] Risk-neutral probabilities provide a foundation for modeling aversion to uncertainty and ambiguity in operational terms that do not require a separation of probability and utility, as well as for pricing of assets in financial markets, and they will be discussed in detail in Chapters 6–9.

The picture looks like the one shown in Figure 2.6 in the 2-state case with $\mathcal{F}$ and $\mathcal{X}^{\mathcal{F}}$ both consisting of all of $\mathbb{R}^2$. The set of acceptable bets is a half-space, and no-arbitrage requires it to be disjoint from the open negative orthant. The normal vector of the separating hyperplane must be non-negative with at least one strictly positive element, so it can be normalized to be a probability distribution that assigns non-negative expected value to all acceptable bets.

All the usual laws of subjective probabilities and expectations are implied by the fundamental theorem, as summarized in the following corollaries.[18]

Corollary 3.1 (existence of unconditional probabilities and expectations): For any event $\mathbf{A}$ there is a unique number $P(\mathbf{A})$, called the individual's probability of $\mathbf{A}$, for which both $\mathbf{A} - P(\mathbf{A})$ and $P(\mathbf{A}) - \mathbf{A}$ are

---

[17]Seidenfeld et al. (1990b) and Schervish et al. (2013) point out that in a situation where the exchange rate between two currencies is state-dependent, the values of the elicited probabilities will depend on which currency is used for the bets. That is not a problem if the analysis is carried out in terms of risk neutral probabilities, which are by definition currency-dependent.

[18] The qualifier "subjective" is dropped for the rest of this section because only subjective probabilities and expectations are considered here. De Finetti used the nice term "previsions" to refer to both subjective probabilities and subjective expectations, although the more conventional terms are used here.

acceptable \$-bets. In particular, $P(\mathbf{A}) = P_\pi(\mathbf{A})$, where π is the unique distribution whose existence is established by the fundamental theorem. [Proof: both bets are acceptable if and only if $P_\pi(\mathbf{A}) - P(\mathbf{A}) = 0$.] Similarly, for any act $\mathbf{f}$ there is a unique number $E(\mathbf{f})$, henceforth called the *expectation* of $\mathbf{f}$, such that both $\mathbf{f} - E(\mathbf{f})$ and $E(\mathbf{f}) - \mathbf{f}$ are acceptable \$-bets. In particular, $E(\mathbf{f}) = E_\pi(\mathbf{f})$.

Definition (conditionally acceptable \$-bets): $\mathbf{x}$ is a conditionally acceptable \$-bet given the occurrence of event $\mathbf{B}$ if $\mathbf{x_B}$ is an acceptable bet, where $\mathbf{x_B}$ denotes $\mathbf{x}$ with its elements zeroed-out in states where $\mathbf{B}$ is not true.[19]

Corollary 3.2 (existence of conditional probabilities and expectations): For any events $\mathbf{A}$ and $\mathbf{B}$ such that $P(\mathbf{B}) > 0$, there is a unique number $P(\mathbf{A}|\mathbf{B})$, henceforth called the individual's *conditional probability of $\mathbf{A}$ given $\mathbf{B}$*, for which both $\mathbf{A} - P(\mathbf{A}|\mathbf{B})$ and $P(\mathbf{A}|\mathbf{B}) - \mathbf{A}$ are conditionally acceptable \$-bets given the occurrence of event $\mathbf{B}$. In particular, $P(\mathbf{A}|\mathbf{B}) = P_\pi(\mathbf{AB})/P_\pi(\mathbf{A})$, where π is the unique distribution whose existence is established by the fundamental theorem. [Proof: both bets are acceptable if and only if $\pi \cdot ((\mathbf{A} - P(\mathbf{A}|\mathbf{B}))_\mathbf{B}) = 0$, which requires $P_\pi(\mathbf{AB}) = P(\mathbf{A}|\mathbf{B})P_\pi(\mathbf{B})$. This is trivially true if $P_\pi(\mathbf{B}) = 0$, because that implies $P_\pi(\mathbf{AB}) = 0$. If $P_\pi(\mathbf{B}) > 0$, both sides can be divided by it to obtain $P(\mathbf{A}|\mathbf{B}) = P_\pi(\mathbf{AB})/P_\pi(\mathbf{A})$.] Similarly, for any act $\mathbf{f}$ and event $\mathbf{B}$ such that $P(\mathbf{B}) > 0$, there is a unique number $E(\mathbf{f}|\mathbf{B})$, henceforth called the *conditional expectation of $\mathbf{f}$ given $\mathbf{B}$*, such that both $\mathbf{f} - E(\mathbf{f}|\mathbf{B})$ and $E(\mathbf{f}|\mathbf{B}) - \mathbf{f}$ are conditionally acceptable \$-bets given the occurrence of event $\mathbf{B}$. In particular, $E(\mathbf{f}|\mathbf{B}) = E_\pi(\mathbf{f}_\mathbf{B})/P_\pi(\mathbf{B})$.

Corollary 3.3 (additive law of probability): $P(\mathbf{A} \cup \mathbf{B}) = P(\mathbf{A}) + P(\mathbf{B})$ if $\mathbf{A} \cap \mathbf{B} = \varnothing$. [Proof: follows from Corollary 3.1.]

Corollary 3.4 (multiplicative law of probability): $P(\mathbf{A}|\mathbf{B}) = P(\mathbf{AB})/P(\mathbf{B})$ if $P(\mathbf{B}) > 0$. [Proof: follows from Corollary 3.2.]

---

[19] For example, if $\mathbf{x} = (1, 2, 3)$ and $\mathbf{B} = (1, 1, 0)$, i.e., $\mathbf{B}$ is the union of states s_1 and s_2, then $\mathbf{x_B} = (1, 2, 0)$. The same quantity could also be denoted by the pointwise product $\mathbf{xB}$ in this case, but subscript notation will be used for multiplication by an indicator vector to avoid confusion with matrix multiplication in what comes later, when some upper-case bold letters (e.g., $\mathbf{F}$) may denote matrices instead of vectors. Also, using subscripts to denote conditioning events for bets will synch up with the use of subscripts to denote conditioning events for preferences, in which $\succsim_\mathbf{A}$ denotes conditional preference given the occurrence of event $\mathbf{A}$.

Corollary 3.5 (Bayes' theorem): $P(A|B) = (P(A)P(B|A))/P(B)$ if $P(B) > 0$. [Proof: multiply both sides of $P(A|B) = (P(A)P(B|A))/P(B)$ by $P(B)$ and then apply the multiplicative law on both sides, obtaining the identify $P(AB) = P(AB)$. $P(B|A)$ is technically undefined if $P(A) = 0$, but in that case the product $P(A)P(B|A)$ must be zero for any finite value of $P(B|A)$, so $P(A) = 0 \Rightarrow P(A|B) = 0$.]

The additive and multiplicative laws have a nice geometric interpretation when there are 3 states, in which case probability distributions can be plotted in triangular coordinates. Each vertex of the probability triangle is a distribution assigning probability 1 to one of the states (**A**, **B**, **C**), and a line drawn drawn parallel to its opposite side is an iso-probability line for that state. For example, a line drawn parallel to the side connecting the vertices for states **B** and **C** is an iso-probability line for state **A**.

In Figure 3.1, the two dark line segments (parallel to the sides opposite the **B** and **C** vertices) are the sets of distributions satisfying $P(B) = 0.2$ and $P(C) = 0.5$, and they intersect at a single point. No-arbitrage requires that the probability assigned to state **A** should determine an iso-probability line passing through the same point, which is the dashed line consisting of the distributions satisfying $P(A) = 0.3$, fulfilling the additive-law requirement that $P(A) = P(B) + P(C)$. Thus, the implied

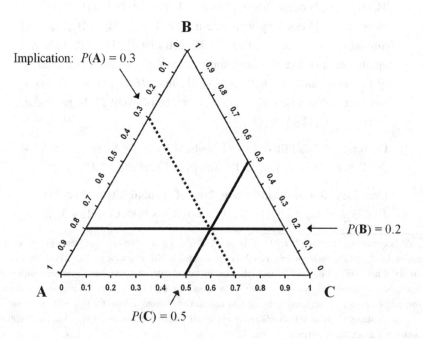

FIGURE 3.1: Illustration of the additive law of probability

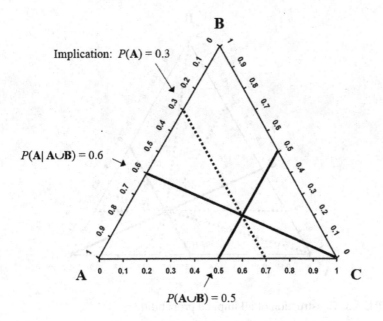

FIGURE 3.2: Illustration of the multiplicative law of probability

probability of **A** can be determined either by applying the additive-law formula or by using a straight-edge to find the iso-probability line for **A** that passes through the intersection of the iso-probability lines for **B** and **C**.

In Figure 3.2, the two dark line segments are the sets of distributions satisfying $P(A \cup B) = 0.5$ and $P(A|A \cup B) = 0.6$, which also intersect at a single point. The set of distributions satisfying a conditional-probability constraint is a straight line drawn through the vertex at which the conditioning event has probability zero. In this case the conditioning event $A \cup B$ has zero probability at the vertex where **C** has a probability of 1. No-arbitrage requires that the probability assigned to state **A** should determine an iso-probability line passing through the intersection point, which is the dashed line consisting of the distributions satisfying $P(A) = 0.3$, fulfilling the multiplicative-law requirement that $P(A) = P(A|A \cup B)P(A \cup B)$. Again, the implied probability for the target event can be computed either by applying the relevant probability law or by using a straight-edge.

In Figure 3.3, additional lines have been drawn through the same intersection point, from which the remaining conditional and unconditional probabilities can be determined: $P(B) = 0.2$, $P(B|B \cup C) = 0.286$, and $P(C|A \cup C) = 0.625$.

Arbitrage and Rational Decisions

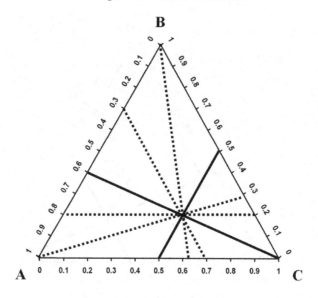

FIGURE 3.3: Construction of all implied probabilities

3.4 Bayes' theorem and (not) learning over time

A common misconception about Bayes' theorem is that it is a model of learning over time. The terminology of "prior" and "posterior" probabilities is suggestive of this view, yet it means no such thing in general.

> Hacking (1967): "Well known properties of [Bayes'] theorem lead us to a model of learning from experience.... The idea of the model of learning is that P(H|E) represents one's personal probability after one learns E. But formally the conditional probability represents no such thing. If, as in all of Savage's work, conditional probability is a defined notion, then P(H|E) stands merely for the quotient of two probabilities. It in no way represents what I have learned after I take E as a new datum point.... A man knowing E would be incoherent if the [betting] rates offered on H unconditionally differed from his rates on H conditional on E. But no incoherence obtains when we shift from the point before E is known to the point after it is known. The man's P(H|E) before learning E differs from his P(H|E) after learning E. Why not, he says: the change represents how I have learned from E!"

Goldstein (1985): "The question as to what information is conveyed by the expression P(H|E) is central to the Bayesian argument. Most developments of Bayesian theory proceed by defining conditional probabilities and deriving their properties in terms of called off penalties or bets, determined before the actual conditioning event is revealed. This definition is then taken to represent your actual beliefs about H having seen E. This transition is not justified and is rarely even pointed out. However, it is clearly a false transition on both theoretical and practical grounds. As no coherence principles are used to justify the equivalence of conditional and a posteriori probabilities, this assumption is an arbitrary imposition on the subjective theory. As Bayesians rarely make a simple updating of actual prior probabilities to the corresponding conditional probabilities, this assumption misrepresents Bayesian practice."

Coherence (no-arbitrage) requires only that conditional and unconditional probabilities that are assessed *at the same moment in time*, in terms of bets simultaneously offered to an observer, must obey all of the relevant probability laws, including Bayes' theorem. Thus, Bayes' theorem describes *expected* learning before the fact, not realized learning after the fact. The passage of time raises other philosophical and practical issues.[20]

Let $P_0(.)$ and $P_1(.)$ denote your subjective probabilities or expectations (previsions) held at times 0 and 1. It is currently time 0 and an event E will be observed at time 1 (say, the result of some experiment that you are planning to perform) and a second event H will be observed later (say, an event on which some payoff depends). Can $P_1(H|E)$ be determined in advance by applying Bayes' theorem to the beliefs that you hold right now, i.e., $P_1(H|E) = P_0(H|E) := P_0(H)P_0(E|H)/P_0(E)$? No, *temporal* coherence requires the following condition to be satisfied:

$$P_0(P_1(H|E)|E) = P_0(H|E) \tag{3.2}$$

That is, at time 0 your subjective conditional probability for H given E, which is the quantity $P_0(H|E)$, must be equal to your *conditional expectation* of what the value of $P_1(H|E)$ will be at time 1, after you have observed E. (Goldstein (1985), Theorem 1) From the vantage point of time 0, the quantity $P_1(H|E)$ is itself uncertain due to unmodeled changes in your mental state that may arise with the passage of some amount of time. Perhaps other information will trickle in, or perhaps you will just give the whole matter more thought. Bayes' theorem is a model of how your beliefs will be updated over time *only in expectation*. From an observer's

---

[20]This point is also made by Binmore (2009) in regard to learning within the "small world" of a Bayesian model versus learning in the "large world" of real experience.

perspective, only the bets based on P_0 are available at time 0. They need to be internally consistent but they do not have to predict the future.

Of course, you can try to build an explicit model of the extra information that might arrive between time 0 and time 1. For example, you could add an extra layer of detail to model a range of possible values for $P_1(H|E)$ that might be in effect at time 1, with higher-order probabilities attached to them. But no matter how much detail you add, the passage of real time will add a continuing element of surprise. If nothing that you didn't precisely foresee has happened to you lately, then time has stopped for you.

A special case of this result is that in which there is no conditioning event, just an unconditional probability of H to be assessed and updated over time. (Goldstein (1983)) The relation that must hold there is:

$$P_0(P_1(H)) = P_0(H) \tag{3.3}$$

Thus, temporal coherence requires that subjective probabilities should evolve over time as a *martingale* process: today's probability ought to be equal to the expected value of tomorrow's probability, as is the case for (discounted) stock prices.[21]

These facts are problematic for traditional decision tree analysis, in which tree diagrams are constructed to show sequences of events and decisions that will happen in the future, and optimal sequences of decisions are determined by backward induction ("rollback"). Sometimes the tree diagram (the so-called "extensive form" of the model) is converted to a table (the so-called "normal form") in which every complete sequence of conditional decisions is treated as single option that can be chosen right now from a list, eliminating the time dimension altogether and allowing the decision-maker to go on vacation. In principle the two modeling approaches are same (Raiffa (1968)), but only if there are no surprises. The use of either type of model in an intertemporal setting ignores the fact that there may be "option value" in delaying a decision so that adjustments can be made if unexpected things happen or new decisions are imagined.[22] One possible proxy for the effects of unexpected

---

[21]Another desideratum that has been proposed for the updating of beliefs from one time period to another is the "reflection principle" (van Fraassen (1984)), which applies to a situation in which a decision-maker bets on whether her future subjective probability for some event will have a particular value. (See Christensen (1991) and Vineberg (1997) for some discussion.) It requires that the decision-maker's present (time 0) subjective probability for H conditional on the event that its future (time 1) subjective probability will be q must be equal to q itself. That is, $P_0(H|P_1(H) = q) = q$. As in Goldstein's (1985) theorem on the proper use of Bayes' theorem, this is really a static requirement of conditional beliefs held at time 0, not a dynamic requirement of the evolution of beliefs in real time. The updating rule is deterministic and the bets are agreed upon in advance.

[22]"[We] speak and think as though decisions were creative. Our spontaneous and intuitive habits of mind treat a decision as introducing a new strand into the tapestry as it is woven, as injecting into the dance and play of events something essentially new, something not implicit in what had gone before." (Shackle (1959))

events and limits of imagination would be to make direct subjective judgments of option value without trying decompose them. Higher option values would indicate more anticipated "volatility" in model parameters or plans for further soul-searching.

3.5 Incomplete preferences and imprecise probabilities

The idea that subjective probabilities may be imprecise or indeterminate to varying degrees, whether due to the quantity and quality of information on which they are based or merely due to the fact that they have been incompletely revealed, is an old one. Koopman's (1940) theory of intuitive probability is cached in terms of lower and upper probabilities, and de Finetti's fundamental theorem of probability, as given in his 1974 book, is stated in terms of lower and upper bounds on the probability of a target event that are determined from probabilities of other events by the requirement of no-arbitrage. (De Finetti himself did not actually endorse the idea that subjective probabilities are imprecise, though, as discussed by Vicig and Seidenfeld (2012).) The completeness assumption plays no important role in the proof of this theorem, so the concept that beliefs should be represented by intervals of subjective probabilities rather than by point values is a natural one to explore in this framework. It also resonates with the use of bid-ask spreads in asset pricing. Many other authors have explored relaxations of the completeness axiom in the context of modeling beliefs, including Smith (1961), Good (1962), Suppes (1974), Kyburg (1974), Williams (1976), Levi (1974, 1980), and Seidenfeld et al. (1989, 1990a). Research on this topic accelerated with the publication of Walley's (1991) seminal book, leading to the establishment of the Society for Imprecise Probabilities: Theory and Applications (sipta.org) and an expansive literature.[23]

There are (at least) four good reasons why it might to desirable to represent subjective probabilities in terms of sets of values rather than unique values.

1. The problem may be one of Bayesian statistical inference, and there may be a lack of information or consensus on which to base the selection of a prior distribution. By using a set of priors rather than a unique prior, a more robust analysis can be carried out.

2. The decision-maker may feel that she has insufficient information or expertise to determine a single number that represents her degree of belief with respect

---

[23] See Augustin, Coolen, de Cooman, Matthias, and Troffaes, eds. (2014), *Introduction to Imprecise Probabilities*, for overviews of major topics in the field.

to a given event, but she is comfortable in giving a range of values. To the extent that there is any amount of subjectivity in the decision-maker's evaluation of a probability (or any other quantity, for that matter), it is logical to expect that some will be assessed more precisely than others.

3. The decision-maker may feel that she can assign a unique probability to any event given sufficient time for analysis and reflection, but time and effort are limited. If there are M states of the world and the decision-maker has assessed unique probabilities for fewer than $M - 1$ logically independent events, then probabilities for some events will not be uniquely determined (yet).

4. The situation may be one that really involves betting (or some other form of interaction with an opponent), and the decision-maker may have strategic or competitive reasons for not wanting to disclose her probabilities in precise terms. In this case, there is an *intersubjective* element in the public expression of her beliefs. By asserting distinct lower and upper probabilities for an event, she is effectively able to adjust her own probability estimate up or down in response to the action of her betting opponent, which she might want to do if she regards his behavior as informative or if she wishes to have a positive expected gain from any transaction.

The generalization of the fundamental theorem of subjective probability to allow beliefs to be represented by convex sets of probabilities rather than unique values is mathematically trivial. It merely involves weakening the completeness axioms to:

Axiom SP1^{a*} (weak monotonicity): $\mathbf{f} \geq \mathbf{g} \Longrightarrow \mathbf{f} \succsim \mathbf{g}$

Axiom SP1^{b*} (weak acceptability): $\mathbf{x} \geq \mathbf{0} \Longrightarrow \mathbf{x} \in \mathcal{X}$

With these substitutions, Theorem 3.2 generalizes to:

Theorem 3.3: Fundamental theorem of lower and upper probabilities

The preference relation $\succsim$ satisfies SP1^{a*} and SP2^a–SP5^a, and equivalently the set $\mathcal{X}$ of acceptable \$-bets satisfies SP1^{b*} and SP2^b–SP5^b, if and only if there exists a non-empty convex set $\mathbf{\Pi}$ of probability distributions such that $\mathbf{f} \succsim \mathbf{g}$ if and only if $E_\pi(\mathbf{f}) \geq E_\pi(\mathbf{g})$ for all $\pi \in \mathbf{\Pi}$, and equivalently $\mathbf{x} \in \mathcal{X}$ if and only if $E_\pi(\mathbf{x}) \geq 0$ for all $\pi \in \mathbf{\Pi}$.

The proof is identical to the previous one except that the set $\mathcal{X}$ of acceptable bets needs to be only a convex cone, not a half-space. Therefore the hyperplane which

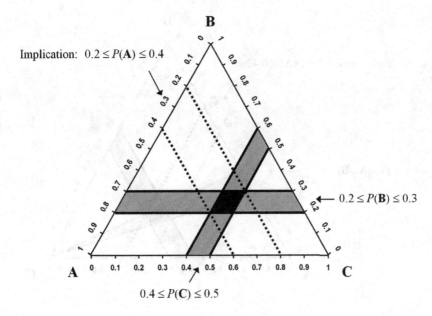

FIGURE 3.4: Addition of lower and upper probabilities

separates it from the negative orthant may not be unique. The picture now looks like Figure 2.5 rather than Figure 2.6. The convex set of (normalized) normal vectors of the separating hyperplanes is the set Π of probability distributions representing the decision-maker's imprecise beliefs. Figures 3.4 and 3.5 are generalizations of Figures 3.1 and 3.2 which illustrate how the additive and multiplicative laws of probability apply to lower and upper probabilities.

Once it is acknowledged that some personal probabilities may be imprecise, it is hard to avoid the implication that they are all imprecise, only to varying degrees. Some are based on better information than others, and the act of assessing a subjective probability does not merely consist of reading a number off an internal scale: it is a process of mental construction. Even in the physical sciences, every measurement perturbs the thing being measured, and in the limit of very fundamental measurements this leads to indeterminacies. It might seem paradoxical, though, that an interval of probabilities representing an imprecise degree of belief should itself have uniquely determined endpoints. On what basis should the individual decide where to draw the lines?

One possible canonical answer to this question is that the decision-maker may believe that her opponent has private information that will be partially revealed by which side of the bet he chooses to take, and consequently she may wish to build in an adjustment for it in advance. For example, suppose she feels that her best estimate

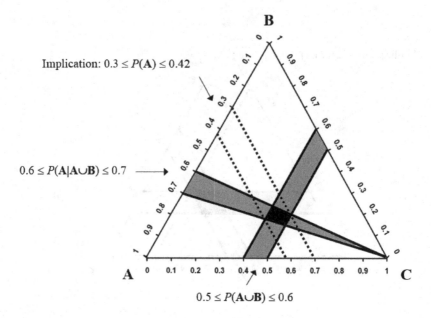

FIGURE 3.5: Multiplication of lower and upper probabilities

of a probability is 0.4 in the sense that it is the medial value she would choose if she were forced to name a unique price at which she would be willing to either buy or sell $1 lottery tickets at the discretion of her opponent. But if he turns out to be eager to buy tickets from her at that price, it reveals that his own probability for the event is greater than 0.4. Upon learning that, the decision-maker may want to revise her revise her own probability estimate upward.[24] She might therefore ask herself: what is the lowest price for which she would be willing to sell the ticket given the knowledge that her opponent wants to buy it at that price, and similarly, what is the highest price for which she would be willing to buy it given the knowledge that her opponent wants to sell it to her at that price?

If she is able to answer these questions, they will determine an interval that represents her degree of belief in a way that is both subjective and intersubjective. To the extent that she is or is not confident about the quality of her own information vis-a-vis that of others, the interval will be narrower or wider, and this is useful information to have. It reveals more about her state of mind than would obtained by asking for a single number, and it is more relevant for understanding what she

[24]She might also want to profit from a difference of opinion, particularly if she feels that the opponent is *less* well informed than she is, but for the present purposes it will be assumed that she is just trying to honestly reveal as much information about her beliefs as she can without harming herself in the process, in the spirit of probability assessment methods in general.

will do or ought to do in situations where she interacts with others. If the result of this exercise is a very wide interval, she should ask better-informed people for their opinions!

A further generalization of the fundamental theorem can be obtained by dropping the linear extrapolation axiom. This allows the decision-maker to quote several lower or upper probabilities for an event, each associated with a "confidence weight" that establishes the maximum loss that she would be willing to risk when taking $-bets on the event. In this way it is possible to place "soft" or "fuzzy" constraints on beliefs, while otherwise remaining in the Bayesian paradigm. Details of this model can be found in Nau (1992b, 2002).

3.6 Continuous probability distributions

In many applications the sets of states are infinite, ranging over a continuum of values, in which case the separating hyperplane theorem for Euclidean space (Lemma 2.1) is not applicable. The situation in which the consequences are amounts of money measured on a continuous scale is of particular interest in the modeling of attitudes toward risk, insofar as risk aversion may be described by nonlinear utility for money. In this book, risk aversion and more general phenomena such as ambiguity aversion will be studied mainly in the context of the *state-preference* model of choice under uncertainty, which allows for subjectivity in both beliefs and tastes without using objective randomization or counterfactual mappings of states to consequences, and which does not emphasize riskless reference points or the unique separation of probabilities from utilities. However, a brief treatment of the topic of continuous sets of states or consequences in the context of the SP and EU models will be presented here and in the next chapter.

The fundamental theorems of subjective probability and expected utility can be generalized to continuous distributions by using an infinite-dimensional version of the separating hyperplane theorem. There are many such theorems tailored to different spaces and topologies, but one that has been used in this context (Lemma 3 in Buehler (1976)) is the following:

Lemma 3.3: Let S and T be sets, and let $X = \{x_t(s) : t \in T\}$ denote a family of bounded, real-valued functions defined on S. Exactly one of the following alternatives holds:

(i) There exist $t_1, ..., t_K \in \mathcal{T}$ and positive scalars $\alpha_1, ..., \alpha_K$ such that $\sup_{s \in S} \sum_{k=1}^{K} \alpha_k x_{t_k}(s) < 0$ (i.e., a positive weighted sum of finitely many elements of $\mathcal{X}$ is strictly negative and bounded away from the origin).

(ii) There exists a finitely additive probability measure p on S such that $E_p(x) := \int x(s) dp(s) \geq 0$ for all $x \in \mathcal{X}$ (i.e., there exists a hyperplane separating $\mathcal{X}$ from the negative orthant, and its normal vector is a finitely additive probability measure).

$\mathcal{T}$ and $\mathcal{X}$ can be the same set (elements of $\mathcal{X}$ can just be indexed by themselves), so condition (i) can be rewritten to begin "There exist $x_1..., x_K \in \mathcal{X}$..." The generalization of the SP theorem then proceeds as follows. Let $f, g, ...$ denote acts that are elements of a set $\mathcal{F}$ of bounded scalar functions defined on S, whose values represent monetary payoffs received by the decision-maker. (Without loss of generality the unit of money could be scaled so that the minimum and maximum possible payoffs are 0 and 1.) As before, a \$-bet is a difference between two acts, e.g., $x = f - g$, and x is an *acceptable* \$-bet if $f + x \succsim f$ wherever $f \in \mathcal{F}$ and $f + x \in \mathcal{F}$. The set of acceptable \$-bets is again denoted by $\mathcal{X}$. To fit the terms of Lemma 4.1, the definitions of strict dominance and strictly negative bets are first amended to require a finite amount of separation from a dominated act or from the origin:

Definition (strict dominance): f *strictly dominates* g, denoted $f \gg g$, if $\inf_{s \in S}(f(s) - g(s)) > 0$.

Definition (strictly negative \$-bets): x is a *strictly negative* \$-bet, denoted $x \ll 0$, if $\sup_{s \in S} x(s) < 0$.

The strict monotonicity and no-arbitrage axioms are then amended to require that no finite combination of preferences or acceptable bets should be strictly dominated or strictly negative:

Axiom SP5^{a*} (strict monotonicity): There do not exist $f_1, ..., f_K$, and $g_1..., g_K \in \mathcal{F}$ such that $f_k \succsim g_k$ for all k and $\sum_{k=1}^{K} g_k \gg \sum_{k=1}^{K} f_k$

Axiom SP5^{b*} (no arbitrage): There do not exist $x_1..., x_K \in \mathcal{X}$ such that $\sum_{k=1}^{K} x_k \ll 0$

These definitions hold the would-be arbitrageur to the same standard that is inherent in the finite-dimensional version of the model: he must achieve a finite arbitrage profit with a finite number of trades, not merely come infinitesimally close to an arbitrage profit with an infinite sequence of trades. In these terms, the fundamental theorem of subjective probability is stated exactly the same as before, and the proof is also the same except with Lemma 3.3 replacing Lemma 2.1 as the separating hyperplane tool:[25]

Theorem 3.5: Fundamental theorem of subjective probability for infinite sets of states

$\succsim$ satisfies SP1^a–SP4^a and SP5^{a*}, and equivalently $\mathcal{X}$ satisfies SP1^b–SP4^b and SP5^{b*}, if and only if there exists a finitely additive probability measure p such that $f \succsim g$ if and only if $E_p(f) \geq E_p(g)$, and equivalently $x \in \mathcal{X}$ if and only if $E_p(x) \geq 0$

In general p is not unique because of things that may happen on sets of measure zero, but in light of the completeness axiom the expected value of any act or bet is uniquely determined.

3.7 Prelude to game theory: no-ex-post-arbitrage and zero probabilities

Subjective probabilities have been defined to be coherent if bets through which they are revealed do not admit an arbitrage opportunity, i.e., a sure loss for the individual who assesses them. It is OK for the individual to explicitly or implicitly assign zero probability to one or more states, just not all of them at once. The asserted or implied state probabilities merely need to be non-negative and sum to 1. If an observer places a bet on a state that was assigned zero probability and it does not occur, no harm is done. Presumably the individual who offered the bet knows the state to be impossible. She might even be in control of it. Who knows? But what if it *does* occur? The observer then walks away with a riskless profit, and the individual's behavior may be rightfully judged to be irrational ex post. This observation gives rise to the following:

Definitions (ex-post-coherence, no-ex-post-arbitrage): An *ex-post-arbitrage opportunity* in state m is an acceptable bet that yields a negative payoff in state m and non-positive payoffs in all other states. A collection of acceptable bets is *ex-post-coherent* in state m if it does not contain an ex-post-arbitrage opportunity in state m.

---

[25]This result is equivalent to the one given on p. 1057 of the article by Buehler (1976). There, non-preference for strictly dominated acts and non-acceptance of strictly negative bets are called "PR-coherence" and "w-coherence" respectively.

This more nuanced criterion of no-arbitrage requires that the state that occurs should not be one that was assigned zero probability:

Theorem 3.6: Ex-post fundamental theorem of subjective probability

(a) For unique probabilities (generalization of Theorem 3.2): the set $\mathcal{X}$ of acceptable \$-bets satisfies SP1^b–SP5^b and is ex-post-coherent in state m if and only if there exists a unique probability distribution π with $\pi_m > 0$ such that $\mathbf{x} \in \mathcal{X}$ if and only if $E_\pi(\mathbf{x}) \geq 0$.

(b) For lower and upper probabilities (generalization of Theorem 3.3): $\mathcal{X}$ satisfies SP1^{b*} and SP2^b–SP5^b and is ex-post-coherent in state m if and only if there exists a non-empty convex set Π of probability distributions, *at least one of which* satisfies $\pi_m > 0$, such that $\mathbf{x} \in \mathcal{X}$ if and only if $E_\pi(\mathbf{x}) \geq 0$ for all $\pi \in \Pi$. This means that zero cannot be an upper probability for state m.

Proof: Apply Lemma 2.2 to the matrix $\mathbf{W}$ whose rows are the payoff vectors of the acceptable bets that generate $\mathcal{X}$.via convex combinations and positive linear extrapolation.

If the individual has some amount of control over the occurrence of events, then ex-post-coherence is a constraint on both her beliefs and her actions. And insofar as an observer doesn't care if he is betting against one person or more than one person at the same time in the same scene, this rationality criteron applies equally well to a group of agents who are accepting bets and manipulating events in each others' presence. In particular, it applies to situations in which the states are outcomes of a noncooperative game that is played out in public and whose "rules" are revealed by players offering bets on the strategies played by others, conditional on their own choices of strategies. The public acceptance of bets of this nature reveals precisely the information that is normally assumed to be common knowledge in a noncooperative game. If fact, it operationalizes the idea of common knowledge of a game's numerical parameters, which is otherwise rather tricky to do. It will be shown in Chapter 8 that the no-ex-post-arbitrage criterion applied to such bets ("joint" coherence) is all that is needed to characterize jointly rational behavior of multiple players in a noncooperative.game, and by linear programming duality it yields the result that rational outcomes of a game are those which have positive probability in a correlated equilibrium. (Nau and McCardle (1990)) The fundamental theorem of noncooperative game theory thus turns out to be a simple twist on the ex-post fundamental theorem of subjective probability, and it leads (only) to correlated equilibrium rather than Nash equilibrium as the elementary solution concept.

Chapter 4

Expected utility

4.1 Elicitation of tastes

The preceding chapter analyzed the situation in which events do not have objective probabilities but decisions have objective consequences (money). It was shown that under a strong but appealing set of rationality postulates, an individual behaves as if her beliefs about events are represented by subjective probabilities and she chooses among wealth distributions or accepts bets on the basis of their expected values, which seems somewhat reasonable if the differences among the wealth distributions or the sizes of the bets are not too large. Under these conditions an objective scale of monetary payoffs can be used to calibrate a subjective scale of beliefs.

An individual who makes decisions on the basis of maximizing the expected value of her wealth is said to be "risk-neutral." Such behavior may be inappropriate or even impossible to describe in situations where stakes are large or consequences are not monetary or the presence of risk arouses anxiety or excitement. The theory of expected utility addresses the problem of how to quantify rational *tastes* among alternatives that have exactly these sorts of properties. The approach is to assume that probabilities are objective, and the objective probabilities are then used to calibrate a subjective scale of utility that captures all the intangibles. Of course, important decisions almost never involve objective probabilities, but never mind that for now.

To fix ideas, suppose you are a lucky contestant in a game show in which you will appear on live television tomorrow night and be given the opportunity to choose between a sure prize and a risky gamble. The sure prize is $20K$ in cash plus a 7-day Caribbean cruise. The risky gamble is a lottery that will yield one of the following three prizes: (i) $1M$ plus two tickets to the Academy Award ceremony and a celebrity ball, with a his-and-hers shopping spree for the formal attire, (ii) $100K$ in cash plus a two-week all-expenses-paid trip to a 5-star resort, or (iii) $0 plus a ride back to your home town on a bus that departs at midnight. The non-monetary components of the prizes are valid for your consumption only: you are not allowed to sell or trade them. Let's denote the four prizes in descending order of monetary value by $1000K^+$, $100K^+$, $20K^+$, and $0K^+$. (The "+" indicates that the prize is something more than the indicated mount of cash. Note that the sure prize is third in this ordering.)

If you take the risky gamble and end up with $0K^+$ you may suffer embarrassment and remorse, and if you end up with $1000K^+$ you will be famous for a couple of weeks and may even get an appearance on a TV talk show or a photo of yourself in a tabloid magazine. The outcome of the lottery will be determined by drawing a ball from an urn containing 1000 marbles of the following three colors: gold for the $1000K^+$ prize, silver for the $100K^+$prize, and black for the $0K^+$ prize. You do not yet know the proportions of the different colors, which are the objective probabilities of the three prizes. That will be revealed to you on the show, and you will then have 5 minutes to make your choice.

Tonight you are reflecting on your preferences in order to be prepared to make a level-headed decision tomorrow in the face of a manipulative master of ceremonies and a frenzied crowd and a national TV audience. The objects of your preferences in this case are the sure thing and a set of of risky gambles that potentially includes all probability distributions on the prize set $\{1000K^+, 100K^+, 0K^+\}$. One line of thinking would be to try to price out the prizes as objectively as possible, converting them into equivalent amounts of cash based on your own estimates of their market worth, and then assess your subjective values for the cash amounts to adjust for diminishing marginal utility for money. However, this wouldn't take into account the uniquely personal pleasure or displeasure you might also experience. You might or might not be thrilled by the prospect of attending a celebrity ball or having your picture on a magazine cover. You might or might not enjoy going on cruises[1] or staying in fancy resorts or wearing formal clothes. You might or might not care about how you will look in social media. So, you will need to make some additional adjustments for these other attributes of the prizes. The extra psychological dimensions could in principle be represented by some kind of multi-attribute value function. However, such a function would not necessarily represent your attitude toward risk per se. In the moment of choice you might or might not be anxious and afraid "Utility," in

---

[1] Your ultimate enjoyment of the cruise might depend on subjectively uncertain events like bad weather or an engine failure on the ship, as the author can attest, but never mind that either.

the sense of expected utility theory, may depend on all of these subjective elements of your tastes.

Suppose that you are undaunted by the many issues and you set about to construct a quantitative decision model. What you will need by morning is a formula into which you can plug a set of probabilities in order to determine which alternative to choose. Here another issue rears its head: how should the probabilities be used in this calculation? Even if you could satisfactorily price out all the outcomes in terms of equivalent amounts of money, it is not clear whether you would be justified in pricing out the lottery by using the probabilities to compute the expected value of its equivalent monetary outcomes. But then you hit upon another way in which to frame the problem in order to integrate probabilities into the analysis, namely, you can contemplate the use of randomized strategies. That is, you could use a random device such as a coin flip or the roll of a die to select between the sure prize and the risky gamble when the time comes. There doesn't appear to be any tangible benefit in doing this, but conceptually it enlarges your set of alternatives to include probability distributions over all four prizes, not just three. For example, if the risky gamble is the probability distribution $(0.1, 0.6, 0.3)$ on the 3-prize set $\{1000K^{+}, 100K^{+}, 0K^{+}\}$, then a coin flip between the sure prize and the risky gamble is a lottery that gives you a probability distribution of $(0.05, 0.3, 0.5, 0.15)$ on the 4-prize set $\{1000K^{+}, 100K^{+}, 20K^{+}, 0K^{+}\}$. Your full set of alternatives is therefore the set of lotteries that assign objective probabilities to all four prizes. A generic lottery is of the form $\mathbf{f} = (f_1, ..., f_4)$, where f_n is the objective probability of receiving prize n that is determined by the proportions of balls in the urn and by the randomizing strategy you choose to employ, if any.

Let us go on to suppose that your preferences among 4-prize lotteries are transitive and complete, where completeness means that you are able to determine a direction of weak preference between any two lotteries *given* sufficient time for analysis and reflection. You will not have that luxury tomorrow, which is why you must do the analysis and reflection tonight. Now comes the important assumption: suppose that your preferences among lotteries also satisfy the same independence assumption that was used in the theory of subjective probability. This assumption states that $\mathbf{f} \succsim \mathbf{g}$ if and only if $\alpha\mathbf{f} + (1-\alpha)\mathbf{h} \succsim \alpha\mathbf{g} + (1-\alpha)\mathbf{h}$ for all $\mathbf{h}$ and all $\alpha \in (0,1)$, except that here $\mathbf{f}, \mathbf{g}$, and $\mathbf{h}$ are distributions of probability over a set of prizes rather than distributions of payoffs over a set of states, and $\alpha\mathbf{f} + (1-\alpha)\mathbf{h}$ is a convex combination of probability vectors rather than payoff vectors. The intuition behind this assumption in the present setting is that it is suggestive of contingent planning. In the comparison between $\alpha\mathbf{f} + (1-\alpha)\mathbf{h}$ and $\alpha\mathbf{g} + (1-\alpha)\mathbf{h}$ it is as if with probability α you will get to make a choice tomorrow between $\mathbf{f}$ and $\mathbf{g}$, and otherwise you will end up with some other lottery $\mathbf{h}$. If you have an intrinsic preference for $\mathbf{f}$ over $\mathbf{g}$, then you should plan to choose $\mathbf{f}$ over $\mathbf{g}$ tomorrow if you get the chance, which from today's perspective is equivalent to choosing $\alpha\mathbf{f} + (1-\alpha)\mathbf{h}$ over $\alpha\mathbf{g} + (1-\alpha)\mathbf{h}$, and (very importantly) this should be true for any $\mathbf{h}$ and α. The resolution of uncertainty does not really need to be sequential in order for this argument to work. Rather, this is just a way of framing the problem that you might or might not find compelling in a situation where two lotteries you are comparing just happen to be decomposable as $\alpha\mathbf{f} + (1-\alpha)\mathbf{h}$ and $\alpha\mathbf{g} + (1-\alpha)\mathbf{h}$ for some $\mathbf{f}$, $\mathbf{g}$, $\mathbf{h}$ and α. Let's suppose you find it compelling.

The constraint imposed the independence assumption dramatically simplifies the rest of your analysis. It implies (in the presence of the other assumptions) that you can assign cardinal utilities to the four prizes in such a way that your preferences among lotteries are determined by the magnitudes of their expected utilities. And the assessment of the utilities is quite simple: all you need to do is compare each of the intermediate prizes against lotteries involving the best and worst prizes. It will help to introduce some additional notation at this point. For $n = 1, 2, 3, 4$, let $\mathbf{L}_n$ denote a reference lottery that yields prize n with certainty, and for $u \in [0, 1]$ let $\mathbf{L}[u]$ denote another kind of reference lottery that yields the best prize $(1000K^+)$ with probability u and the worst prize $(0K^+)$ with probability $1 - u$. That is, $\mathbf{L}_1 = (1, 0, 0, 0)$, etc., and $\mathbf{L}[u] = u\mathbf{L}_1 + (1 - u)\mathbf{L}_4$. Now let u_2 denote a hypothetical probability and ask yourself: for what value of u_2 do your preferences satisfy $\mathbf{L}_2 \sim \mathbf{L}[u_2]$? The question must have a unique answer if the ordering axioms are satisfied and $\mathbf{L}_1 \succsim \mathbf{L}_2 \succsim \mathbf{L}_4$ with at least one strict preference. Let us call this number your utility of $\mathbf{L}_2$ on a one-to-zero scale anchored by $\mathbf{L}_1$ and $\mathbf{L}_4$. Now do the same for prize 3 $(20K^+)$: for what hypothetical probability u_3 do your preferences satisfy $\mathbf{L}_3 \sim \mathbf{L}[u_3]$? These two assessments are sufficient to determine that your preferences are represented by the vector of utilities $\mathbf{u} = (1, u_2, u_3, 0)$ in the sense that $\mathbf{f} \succsim \mathbf{g}$ if and only if $E_\mathbf{u}(\mathbf{f}) \succsim E_\mathbf{u}(\mathbf{g})$, for all $\mathbf{f}$ and $\mathbf{g}$.[2] Suppose that in this case your assessments are $u_2 = 0.47$ and $u_3 = 0.28$.[3] If the urn that you are presented with tomorrow turns

---

[2] $E_\mathbf{u}(\mathbf{f})$ denotes the inner product $\mathbf{u} \cdot \mathbf{f}$, which is the expected utility yielded by the act $\mathbf{f}$ (an assignment of objective probabilities to outcomes) with respect to the utility function $\mathbf{u}$ (an assignment of subjective values to outcomes). This is, for deliberate reasons, the same notation that was used for expected values of acts in the preceding chapter. There $E_\pi(\mathbf{f})$ denoted the inner product $\pi \cdot \mathbf{f}$, which was the expected value of the payoff vector $\mathbf{f}$ with respect to the probability distribution π. The roles of probabilities and payoffs are merely reversed in the two models.

[3] This method of constructing a utility function is one that is actually used in applied decision analysis when risk aversion or non-monetary consequences are involved. A decision-maker is shown a "probability wheel," which is a spinner with (say) a blue background and a red wedge whose size is continuously adjustable and whose relative area represents the probability α of getting the best-case outcome in a lottery between best-case and worst-case alternatives. (This is the same sort of wheel that is used to calibrate the assessment of subjective probabilities.) For each intermediate outcome, the decision-maker is asked to adjust the size of the wedge until she is roughly indifferent between the intermediate outcome and a

out to contain 50 gold marbles, 550 silver marbles, and 400 black marbles, then the risky gamble's expected utility will be $1 \times 0.05 + 0.47 \times 0.55 + 0 \times 0.4 = 0.3085$, and after calculating this amount on the back of your hand and observing that it is greater than 0.28, you will choose to take the risk.

To prove that the method of analysis described above is really justified by the conditions that your preferences are assumed to satisfy, consider an arbitrary lottery $\mathbf{f}$ over these four prizes, and let $\mathbf{h}$ be the lottery obtained from it by zeroing-out the probability of prize 2 and then renormalizing so that $h_2 = 0$ and $h_n = f_n/(1 - f_2)$ for $n \neq 2$. Then by definition $(1 - f_2)\mathbf{h} = (f_1, 0, f_3, f_4)$ and $\mathbf{f} = f_2\mathbf{L}_2 + (1 - f_2)\mathbf{h}$. The independence axiom can now be applied to the indifference $\mathbf{L}_2 \sim \mathbf{L}[u_2]$ to obtain:

$$f_2\mathbf{L}_2 + (1 - f_2)\mathbf{h} \sim f_2\mathbf{L}[u_2] + (1 - f_2)\mathbf{h}. \tag{4.1}$$

The second term on the RHS is $(f_1, 0, f_3, f_4)$, which is $\mathbf{f}$ with all the probability removed from prize 2, and the first term on the RHS is $(u_2 f_2, 0, 0, (1 - u_2)f_2)$, which is a redistribution of the probability that was formerly on prize 2 onto prizes 1 and 4 in proportions of u_2 and $(1 - u_2)$, respectively. The LHS is equal to $\mathbf{f}$ as just shown. We therefore have:

$$\mathbf{f} \sim (f_1 + u_2 f_2, 0, f_3, (1 - u_2)f_2 + f_4). \tag{4.2}$$

The same trick can be played to shift the probability mass f_3 off of prize 3 and onto prizes 1 and 4 in proportions of u_3 and $1 - u_3$, respectively, yielding[4]:

$$\begin{aligned} \mathbf{f} \quad &\sim \quad (f_1 + u_2 f_2 + u_3 f_3, 0, 0, (1 - u_2)f_2 + (1 - u_3)f_3 + f_4) \tag{4.3} \\ &= \quad (\mathbf{f} \cdot \mathbf{u}, 0, 0, 1 - \mathbf{f} \cdot \mathbf{u}) \\ &= \quad (E_{\mathbf{u}}(\mathbf{f}), 0, 0, 1 - E_{\mathbf{u}}(\mathbf{f})). \end{aligned}$$

Now consider some other lottery $\mathbf{g}$ and perform the same substitutions to obtain:

$$\mathbf{g} \sim (E_{\mathbf{u}}(\mathbf{g}), 0, 0, 1 - E_{\mathbf{u}}(\mathbf{g})) \tag{4.4}$$

Finally, consider the question of whether $\mathbf{f} \succsim \mathbf{g}$. By applying the transitivity axiom to the indifferences just established, we get:

$$\mathbf{f} \succsim \mathbf{g} \Leftrightarrow (E_{\mathbf{u}}(\mathbf{f}), 0, 0, 1 - E_{\mathbf{u}}(\mathbf{f})) \succsim (E_{\mathbf{u}}(\mathbf{g}), 0, 0, 1 - E_{\mathbf{u}}(\mathbf{g})) \tag{4.5}$$

---

lottery in which the wheel will be spun and she will get the best-case outcome if the spinner lands on red and the worst-case outcome if it lands on blue. The wheel has a probability scale on the back that is used to read off the value of α, say, $\alpha = 0.28$ for the sure prize in this example. The advantage of this method is that it helps to avoid anchoring on round numbers ("50-50" or "1-in-10") and it also gives the reference lottery a concreteness that the decision maker can relate to her experience in playing games that involve randomization. It also yields utilities that are quantified to two decimal places of precision.

[4]The probabilities of the 4 consequences must sum to 1, so it must be the case that $f_1 + (1 - u_2)f_2 + (1 - u_3)f_3 = 1 - (u_2 f_2 + u_3 f_3 + f_4) = 1 - (\mathbf{f} \cdot \mathbf{u})$.

The two lotteries in the comparison on the right consist of one that places probability $E_\mathbf{u}(\mathbf{f})$ on the best prize and the remaining probability on the worst prize and one that places probability $E_\mathbf{u}(\mathbf{g})$ on the best prize and the remainder on the worst prize. It's better to have more probability on the best prize, from which it follows that $\mathbf{f} \succsim \mathbf{g}$ if and only if $E_\mathbf{u}(\mathbf{f}) \geq E_\mathbf{u}(\mathbf{g})$, i.e., the preferred act is the one that yields the greatest expected utility according to the utility function $\mathbf{u}$ that was elicited by simple comparisons among reference lotteries. By induction, the same method of analysis can be applied to situations involving any number of distinct prizes.

What happened to the many angles of the problem and the complex psychological issues that were mentioned at the outset? They are all finessed away by the appeal to the independence axiom. If you satisfy that axiom (along with the other less subtle ones), then all there is to know about your tastes is revealed by the u_2 and u_3 that you pull out of your head when asked to compare some simple lotteries Et voila!

4.2 The fundamental theorem of expected utility

As illustrated in the preceding section, the theory of expected utility is just the dual image of the theory of subjective probability. (Objective) probabilities are decision variables in one and (subjective) probabilities are dual variables in the other. (Objective) payoffs are decision variables in one and (subjective) payoffs are dual variables in the other. To obtain the expected utility model, we just need to reverse the roles of the variables in the subjective probability model and apply the separating hyperplane theorem again.[5] Let $\mathcal{C} := \{c_1, ..., c_N\}$ denote a finite set of mutually exclusive and exhaustive *consequences* that might be received or experienced by an individual. A consequence can be a physical or financial good or something uniquely personal. It need not have an ordinal or cardinal scale of measurement. It might be a pot of money or a day at the beach or a painful death. For convenience and without much loss of generality, let the consequences be labeled a priori so that, if they are to be received with certainty, c_1 is weakly most-preferred and c_N is weakly least-preferred, and c_1 is strictly preferred to c_N.[6]

---

[5]Gilboa (2009) gives separating-hyperplane derivations of both models that are similar but not identical as they are here.

[6]Actually it is not necessary to distinguish the dominating consequence a priori in order to label it as c_1. The important thing for one of the consequences to be weakly worst overall and strictly worse than at least one of the others. If preferences are incomplete, there is still not too much loss of generality insofar as it may be possible for the decision-maker to conceive of extreme positive and negative consequences for purposes of anchoring a scale of utility and determining what sort of transfer of probability among consequences is considered as a sure loss.

Suppose that the individual imagines herself in a casino in which the acts from which she can choose are roulette-wheel lotteries that yield known objective probability distributions over the set of consequences. Acts of this sort might be used to model risky decisions in situations where there is general agreement on the probabilities, or they might be used to model randomized strategies that are deliberately chosen in a game. Let a vector $\mathbf{f} = (f_1, ..., f_N)$ denote an "EU-act" that yields an objective probability f_n of receiving consequence c_n. (For the remainder of this chapter, these will just be referred to as "acts.") Acts can be objectively mixed by independent randomization: for any acts $\mathbf{f}$ and $\mathbf{h}$ and any $\alpha \in [0, 1]$, $\alpha \mathbf{f} + (1 - \alpha)\mathbf{h}$ is a lottery that yields consequence n with probability $\alpha f_n + (1 - \alpha)h_n$. The set of all acts is representable by $\mathcal{F} := \Delta^N$, the probability simplex in $\mathbb{R}^N$. Apart from the fact that acts are restricted to the simplex rather than arbitrary convex subsets of $\mathbb{R}^N$, which entails a corresponding redefinition of strict dominance, the axioms and theorems below are derived from the SP versions by merely replacing "state" with "consequence" and by replacing state-dependent amounts of money with consequence-dependent amounts of probability.

Rational tastes can be characterized in terms of consistency conditions that should be satisfied by an individual's preferences among acts. Suppose that they satisfy the following axioms for any acts $\mathbf{f}$, $\mathbf{g}$, and $\mathbf{h}$:

Axiom EU1^a (completeness): $\mathbf{f} \succsim \mathbf{g}$ or $\mathbf{g} \succsim \mathbf{f}$ or both.

Axiom EU2^a (continuity): the sets $\{\mathbf{g} : \mathbf{f} \succsim \mathbf{g}\}$ and $\{\mathbf{g} : \mathbf{g} \succsim \mathbf{f}\}$ are closed.

Axiom EU3^a (transitivity): $\mathbf{f} \succsim \mathbf{g}$ and $\mathbf{g} \succsim \mathbf{h} \Longrightarrow \mathbf{f} \succsim \mathbf{h}$.

Axiom EU4^a (independence): $\mathbf{f} \succsim \mathbf{g} \Longleftrightarrow$
$\alpha \mathbf{f} + (1 - \alpha)\mathbf{h} \succsim \alpha \mathbf{g} + (1 - \alpha)\mathbf{h}$ for all $\mathbf{h}$ and all $\alpha \in (0, 1)$.

Definitions (strict and weak dominance): $\mathbf{f}$ *strictly dominates* $\mathbf{g}$, written as $\mathbf{f} \gg \mathbf{g}$, if $f_n > g_n$ for $n = 1, ..., N - 1$, which means that $\mathbf{g}$ is obtained from $\mathbf{f}$ by shifting some positive probability mass to c_N from each of the other consequences, all of which are weakly preferred and at least one of which is strictly preferred, in every state. $\mathbf{f}$ *weakly dominates* $\mathbf{g}$, written as $\mathbf{f} \geqq \mathbf{g}$, if $f_n \geq g_n$ for $n = 1, ..., N - 1$.

Axiom EU5^a (strict monotonicity): $\mathbf{f} \gg \mathbf{g} \Longrightarrow \mathbf{f} \succ \mathbf{g}$

These are point-by-point identical to the "a" axioms of subjective probability. Axioms EU1–EU4 are standard in expected utility theory, and EU5 identifies one consequence as a priori "worst" for purposes of fixing the lower end of a utility scale.

Rational tastes can also be characterized in terms of consistency conditions that should be satisfied by bets or trades that an individual is willing to make. From the same lemmas proved in the preceding chapter, it follows that the direction of preference between any two acts depends only on the difference between them. The difference between two acts that are objective lotteries is the change in the individual's *risk profile* that she gets when she swaps one for the other. It is equivalent to a bet, as is a swap of two acts that are distributions of wealth over states of the world. The only difference is that swapping two objective lotteries changes the individual's risk profile by shifting probabilities attached to fixed consequences rather than by shifting amounts of money attached to fixed states. Let us call such bets *p-bets* to distinguish them from $-bets. A *p*-bet satisfies a couple of explicit constraints that a $-bet need not satisfy, because acts in this setting are probability distributions rather than arbitrary vectors. First of all, the elements of a *p*-bet must sum to zero. This is unimportant—it just means that *p*-bets live in an $(N-1)$-dimensional space rather than an N-dimensional space. Second, a *p*-bet is a feasible only if the sum of its positive elements is less than or equal to 1 in magnitude and the sum of its negative elements is less than or equal to 1 in magnitude, because it is not possible to shift more than 1 unit of total probability from one subset of consequences to another. This is also unimportant: what matters is that the set of *p*-bets should be convex.

The set of *p*-bets that are acceptable can be defined in terms of preferences as follows:

> **Definition (feasible *p*-bets):** $\mathbf{x}$ is a *feasible p-bet* if $\mathbf{x} = \mathbf{f} - \mathbf{g}$ for some $\mathbf{f} \in \mathcal{F}$ and $\mathbf{g} \in \mathcal{F}$. That is, a feasible *p*-bet is an exchange of one act for another. The set of feasible *p*-bets is denoted by $\mathcal{X}^{\mathcal{F}}$.

> **Definition (preferable and acceptable *p*-bets):** $\mathbf{x}$ is a preferable *p-bet* if and only if $\mathbf{x} \in \mathcal{X}^{\mathcal{F}}$ and $\mathbf{f} + \alpha\mathbf{x} \succsim \mathbf{f}$ for every $\mathbf{f}$ and positive α such that both $\mathbf{f} \in \mathcal{F}$ and $\mathbf{f} + \alpha\mathbf{x} \in \mathcal{F}$ and the set of such $\mathbf{f}$ is non-empty. That is, a preferable *p*-bet is a feasible change in probabilities assigned to consequences that is weakly desirable, regardless of the status quo. It is *acceptable* if it is preferable *and* the individual is publicly willing to accept the change at the discretion of an observer who chooses α. The set of acceptable *p*-bets is denoted by $\mathcal{X}$.

> **Definition (negative *p*-bets):** $\mathbf{x}$ is a *negative p*-bet if $\mathbf{x} \ll \mathbf{0}$. A negative *p*-bet transfers positive amounts of probability to the worst consequence from each of the other consequences, at least one of which is strictly better.

As in the SP model, we will henceforth adopt the common knowledge (CK) assumption and not distinguish between what is (privately) preferable and what is (publicly) acceptable, although that assumption does not make as much sense here insofar as there is not really a "common currency" for transactions. It follows that $\mathcal{X}$ satisfies a corresponding set of axioms:

Theorem 4.1: Equivalence of EU axioms for preferences and acceptable bets

$\succsim$ satisfies EU1^a-EU5^a if and only if $\mathcal{X}$ satisfies:

Axiom EU1^b (completeness): For any $x \in \mathcal{X}^{\mathcal{F}}$, either $x \in \mathcal{X}$ or $-x \in \mathcal{X}$ or both

Axiom EU2^b (continuity): $\mathcal{X}$ is closed.

Axiom EU3^b (convexity): $x \in \mathcal{X}$ and $y \in \mathcal{X} \Longrightarrow$
$\alpha x + (1 - \alpha)y \in \mathcal{X}$ for any $\alpha \in (0, 1)$.

Axiom EU4^b (linear extrapolation) $x \in \mathcal{X} \Longrightarrow \alpha x \in \mathcal{X}$ for any $\alpha > 1$ such that $\alpha x \in \mathcal{X}^{\mathcal{F}}$.

Axiom EU5^b (no arbitrage): $x \ll 0 \Longrightarrow x \notin \mathcal{X}$.

These are point-by-point identical to the "b" axioms of subjective probability except that in this context no-arbitrage means to not accept a bet that transfers positive probability to the worst consequence from all of the others, i.e., to not throw away "probability currency." The proof is identical to that of Theorem 3.1, because it is the same theorem apart from the interpretation of the symbols. The fundamental theorem of expected utility is obtained from the fundamental theorem of subjective probability in a similar way:

Theorem 4.2: Fundamental theorem of expected utility

The preference relation $\succsim$ on EU-acts satisfies EU1^a–EU5^a, and equivalently the set $\mathcal{X}$ of acceptable p-bets satisfies EU1^b–EU5^b, if and only if there exists a unique von Neumann-Morgenstern utility function **u** satisfying $1 = u_1 \geq u_n \geq u_N = 0$ for $n = 2, ..., N - 1$, such that $f \succsim g$ if and only if $E_{\mathbf{u}}(f) \geq E_{\mathbf{u}}(g)$, and equivalently $x \in \mathcal{X}$ if and only if $E_{\mathbf{u}}(x) \geq 0$.

Proof: This is the same as Theorem 3.2 except for the facts that elements of $\mathcal{X}$ sum to zero and that consequences 1 and N are distinguished a priori as being weakly best and worst, with 1 strictly better than N. Following the same line of reasoning as before, EU1^b implies that $0 \in \mathcal{X}$, and this together with EU2^b–EU4^b implies that $\mathcal{X}$ is the intersection of $\mathcal{X}^{\mathcal{F}}$ with a closed convex cone that recedes from the origin. EU1^b further guarantees that the cone is at least a half-space in the linear subspace of vectors whose elements sum to zero, while EU5^b prevents it from being the whole space, hence the bounding hyperplane is unique. Let $\mathbf{u}$ denote the normal vector of this hyperplane, and note that the elements of $\mathbf{u}$ sum to zero because it is in the same subspace. Then $\mathbf{u}$ has the interpretation of a utility function for consequences, because $\mathbf{f} \succsim \mathbf{g}$ if and only if $E_{\mathbf{u}}(\mathbf{f}) \geq E_{\mathbf{u}}(\mathbf{g})$. Any positive affine transformation of $\mathbf{u}$ represents the same preferences, so without loss of generality it can be renormalized to a range of $[0,1]$, which renders it unique, and in that form it must satisfy $u_1 = 1$, $u_N = 0$, and $0 \leq u_n \leq 1$ for $n < N$ to satisfy the assumptions about extremal consequences.

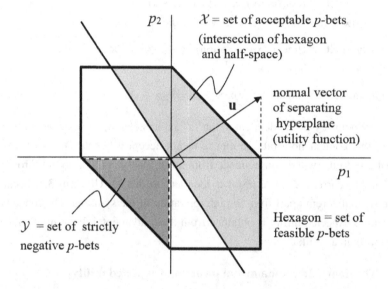

FIGURE 4.1: Acceptable bets that are swaps of probability distributions

The theorem is illustrated for the $N = 3$ case by Figure 4.1, which is the counterpart of Figure 2.6. Here a p-bet is represented by a 2-dimensional vector whose coordinates p_1 and p_2 are the amounts of probability transferred from the worst consequence (#3) to the other two consequences (#1 and #2). The set of feasible p-bets is the hexagon-shaped region determined by the constraints $|p_1| \leq 1$, $|p_2| \leq 1$, and $|p_1 + p_2| \leq 1$. The set of strictly negative p-bets (in which probability is transferred to the worst consequence from both of the others) is the shaded semi-open triangular region that is the intersection of the hexagon and the open negative orthant, and the set of acceptable p-bets is the intersection of the hexagon and a half-space. No-arbitrage requires the latter set to be disjoint from the former, hence they are separated by a hyperplane, and the normal vector of the hyperplane must be non-negative with at least one strictly positive component, as in the case of subjective probabilities. The elements of the normal vector $\mathbf{u}$ can be interpreted as utilities assigned to consequences 1 and 2, with a utility of zero assigned to consequence 3 by default, and it can be normalized so that its largest element is equal to 1, as shown here, in which case it is unique.

The EU model also generalizes to the case of *incomplete preferences* exactly like the SP model: when the completeness axiom is dropped, the set of acceptable p-bets is the intersection of a convex cone, rather than a half-space, with the set of feasible p-bets. The dual representation is a convex set of utility functions: a p-bet is acceptable if and only if it has non-negative expected utility with respect to every utility function in the set. The generalization of the fundamental theorem involves the same weakening of the completeness axioms to:

Axiom EU1^{a*} (weak monotonicity): $\mathbf{f} \geqq \mathbf{g} \Longrightarrow \mathbf{f} \succsim \mathbf{g}$

Axiom EU1^{b*} (weak acceptability): $\mathbf{x} \geqq 0 \Longrightarrow \mathbf{x} \in \mathcal{X}$

With these substitutions, Theorem 4.2 becomes:

Theorem 4.3: Fundamental theorem of lower and upper expected utilities

The preference relation $\succsim$ satisfies EU1^{a*} and EU2^a–EU5^a, and equivalently the set $\mathcal{X}$ of acceptable p-bets satisfies EU1^{b*} and EU2^b–EU5^b, if and only if there exists a non-empty convex set $\mathbf{U}$ of utility functions satisfying $1 = u_1 \geq u_n \geq u_N = 0$ for $n = 2, ..., N - 1$, such that $\mathbf{f} \succsim \mathbf{g}$ if and only if $E_\mathbf{u}(\mathbf{f}) \geq E_\mathbf{u}(\mathbf{g})$ for all $\mathbf{u} \in \mathbf{U}$, and equivalently $\mathbf{x} \in \mathcal{X}$ if and only if $E_\mathbf{u}(\mathbf{x}) \geq 0$ for all $\mathbf{u} \in \mathbf{U}$.

As in the SP case, the proof is identical to the complete-preference version except that the set $\mathcal{X}$ of acceptable bets needs to be only a convex cone, not a half-space. Therefore the hyperplane which separates it from the set of negative p-bets may not be unique, and $\mathbf{U}$ is the convex hull of the corresponding normal vectors This model was first introduced by Aumann (1962), and it is studied in much more depth and generality by Fishburn (1975a) and Dubra et al. (2004), also using separating hyperplane arguments in the proof. In Fishburn's version, the sets that are separated are $\mathcal{X}$ and the open convex hull of $-\mathcal{X}$.

Formally, then, the EU model has exactly the same structure as the SP model and places exactly the same consistency requirements on acceptable bets, but the nature of the bets to which it refers is much more abstract. In the SP model, the parameters of bets are measured in units of money, which is concrete and routinely handled by everyone. Markets in which money is wagered on commonly-observable events really exist and have important economic functions that include, among other things, the provision of a channel of communication through which beliefs about those events can be publicly revealed and aggregated. Whereas, in the EU model, the parameters of bets are measured in intangible units of probability and markets in which individuals swap objective probabilities of getting different commonly-observable consequences (say, markets in which state lottery tickets that offer different jackpots are traded) are neither very common nor very important. For this reason, the completeness axiom is even more problematic in the EU model than it was in the SP model. It also requires a bigger stretch of the imagination to conceive of situations in which there is *common knowledge* of von Neumann-Morgenstern utilities, except for the special case in which consequences are amounts of money and everyone's utility for money is linear.

4.3 Continuous payoff distributions and measurement of risk aversion

The fundamental theorem of expected utility can be generalized to the case of a continuum of payoffs (almost) exactly as the fundamental theorem of subjective probability was generalized to the case of a continuum of states. Let the set C of consequences be the unit interval, representing real-valued amounts of wealth ("cash") measured on a nominal scale of 0-to-1, with generic element c. Let the set $\mathcal{F}$ of acts consist of all probability distributions on C having a finite number of positive-mass points,[7] and let them be represented by their *decumulative* distribution functions, with the following notational convention: $f(z) := 1 - \Pr[c \leq z]$ for $z < 1$, and $f(1) = \lim_{z \to 1} \Pr[c > z]$. Thus, f is continuous at both 0 and 1 with $1 - f(0)$ being the point mass, if any, at $c = 0$ and $f(1)$ being the point mass, if any, at $c = 1$. The reason for adopting this convention is so that strict dominance of g over f, as defined below, requires $g(c) > f(c)$ by a finite margin for all c in the closed interval $[0, 1]$, which is needed to apply Lemma 4.1. (Note that if this condition holds, g must have positive mass at $c = 1$ and f must have positive mass at $c = 0$.) A p-bet is again defined as a difference between two acts, and $\mathcal{X}$ denotes the set of all acceptable p-bets, i.e., the set of all $f - g$ such that $f \succsim g$. Dominance and negativity are

---

[7]This condition, which is imposed for convenience without much loss of generality, allows piecewise-continuous distributions with finite numbers of pieces but rules out distributions having infinite numbers of discrete outcomes, such as the geometric distribution.

defined in terms of these decumulative representations of acts and bets, as illustrated by Figure 4.2:

> **Definitions (first-order stochastic dominance):** g *strictly dominates* f, denoted $g \gg f$, if $\inf_{c \in [0,1]}(g(c) - f(c)) > 0$.
>
> **Definitions (negative p-bets):** x is a *strictly negative p-bet*, denoted $x \ll 0$, if $\sup_{c \in [0,1]} x(c) < 0$.

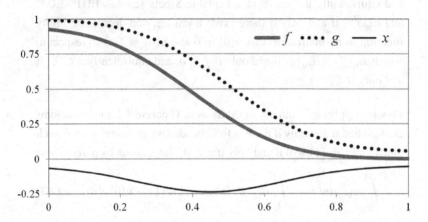

FIGURE 4.2: $x := f - g$ is the bet that is a swap of g for f. g strictly dominates f, as shown here, if the graph of its decumulative distribution is strictly greater than that of f by a finite margin, and correspondingly x is strictly negative by a finite margin.

The strict monotonicity and no-arbitrage axioms of the EU model are then amended to explicitly refer to finite linear combinations, as were those of the SP model:

> **Axiom EU5^{a*} (strict monotonicity):** There do not exist acts $f_1, ..., f_K$ and $g_1, ..., g_K$ and positive constants $\alpha_1, ..., \alpha_K$ summing to 1 such that $f_k \succsim g_k$ for all k and $\sum_{k=1}^{K} \alpha_k g_k \gg \sum_{k=1}^{K} \alpha_k f_k$.
>
> **Axiom EU5^{b*} (no arbitrage):** There do not exist p-bets $x_1..., x_K \in \mathcal{X}$ such that $\sum_{k=1}^{K} x_k \ll 0$.

Now let u denote a monotonic utility function on $[0, 1]$ scaled so that $u(0) = 0$ and $u(1) = 1$. The expected utility of f determined by u is:

$$E_u(f) := f(1) - \int_0^1 u(z)df(z), \tag{4.6}$$

in which the first term on the right assigns a utility of 1 to the possibly-finite probability mass at $c = 1$ and the second term has a negative sign because f is defined
in decumulative form. Correspondingly, the *change* in expected utility due to the
acceptance of a v-bet x is $E_u(x)$. In these terms we have:

Theorem 4.4: Fundamental theorem of EU for a continuum of consequences

The preference relation $\succsim$ on EU-acts satisfies EU1^a–EU4^a and EU5^{a*}
and equivalently the set $\mathcal{X}$ of acceptable \$-bets satisfies EU1^b–EU4^b
and EU5^{b*}, if and only if there exists a unique, non-decreasing utility
function u, normalized so that $u(0) = 0$ and $u(0) = 1$, with respect to
which $E_u(f) \geq E_u(g)$ if and only if $f \succsim g$, and equivalently $x \in X$ if
and only if $E_u(x) \geq 0$.

Proof: Applying Lemma 4.1 exactly as in Theorem 4.2, the EU axioms
are satisfied if and only if there is finitely additive measure[8] u on $\mathcal{S}$ such
that $\int_0^1 x(z)du(z) \geq 0$ if and only if $x \in \mathcal{X}$. Integration by parts yields:

$$\int_0^1 x(z)du(z) + \int_0^1 u(z)dx(z) = u(1)x(1) - u(0)x(0). \qquad (4.7)$$

The RHS is equal to $x(1)$ given the normalization of u, so this equation
can be rewritten as

$$\int_0^1 x(z)du(z) = x(1) - \int_0^1 u(z)dx(z) = E_u(x). \qquad (4.8)$$

Hence, the necessary and sufficient condition $\int_0^1 x(z)du(z) \geq 0$ for all
$x \in \mathcal{X}$ is equivalent to $E_u(x) \geq 0$ for all $x \in \mathcal{X}$, which in turn is
equivalent to $E_u(f) \geq E_u(g)$ for all f and g satisfying $f \succsim g$.

Risk aversion can be characterized as an aversion to taking a bet that is a *mean-
preserving spread (MPS)*, as illustrated in Figures 4.3–4.5. The mean of a payoff
distribution is the area under the graph of its decumulative distribution function, as
shown in Figure 4.3. A mean-preserving spread has a distribution of payoffs with
the same mean but more risk than the status quo in the sense of having more mass in
the tails on both sides. Specifically, if f is a mean-preserving spread of g, then the
bet x that is a swap of g for f has a decumulative distribution function that is non-
negative to the left of the mean and non-positive to the right with the same enclosed

---

[8]Strictly speaking, in the setting of this theorem u assigns a "utility mass" of $u(z)$ to the entire
monetary interval $[0, z]$, as if the individual has received a stream of marginal payoffs adding up to z,
which yield marginal utilities adding up to $u(z)$.

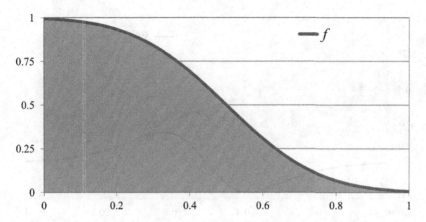

FIGURE 4.3: The expected value of an act is the area under the.graph of its decumulative distribution.

areas above and below zero, as shown in Figure 4.4. Using this construction, it is easy to see that risk aversion requires marginal utility to be decreasing. Suppose that the utility function u whose existence is established by Theorem 4.3 is differentiable, and let u' denote its derivative. Then the identity $E_u(x) = \int_0^1 x(z)du(z)$ that was used in the proof can be written as:

$$E_u(x) = \int_0^1 x(z)u'(z)dz, \qquad (4.9)$$

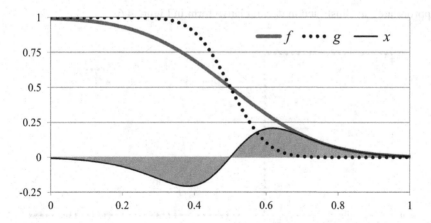

FIGURE 4.4: f is a mean-preserving spread of g if the graph of their difference, x, is non-positive to the left of the mean and non-negative to the right and the enclosed areas above and below are positive and equal in magnitude. Here the mean is 0.5.

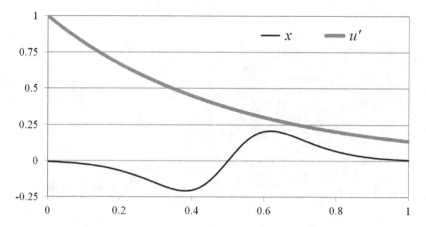

FIGURE 4.5: x is the decumulative representation of a bet, and u' is the marginal utility of money. The integral of xu' is the change in total expected utility yielded by x, which is negative for a mean-preserving spread is u' is decreasing.

from which it follows immediately that $E_u(x) < 0$ for every mean-preserving spread if and only if $u'(z)$ strictly decreasing, i.e., the utility function u must be strictly concave if the decision-maker is risk-averse, as illustrated in Figure 4.5.

The same construction can be used to derive the familiar formula for the Arrow-Pratt measure of risk aversion. Let g be a constant act that yields a certain payoff of α and let f be a binary mean-preserving spread of g that yields a gain of ε/p with probability p and a loss of $\varepsilon/(1-p)$ with probability $1-p$ for some small, finite number ε (large enough for a local second-order approximation to u to be appropriate), as illustrated in decumulative form in Figure 4.6.

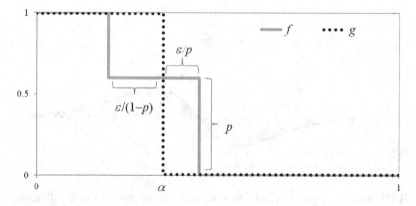

FIGURE 4.6: g is a constant act that yields α with certainty and f is a binary mean-preserving spread of g that yields a gain of ε/p with probability p and a loss of $\varepsilon/(1-p)$ with probability $(1-p)$.

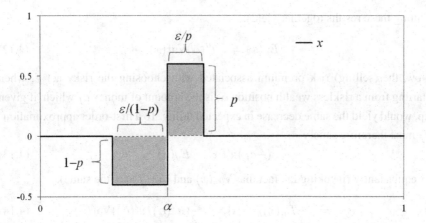

FIGURE 4.7: Cumulative payoff function of the gamble $x := f - g$. Its variance is $\varepsilon^2/(p(1-p))$, and its risk premium is approximately the variance multiplied by $-\frac{1}{2}(u''(\alpha)/u'(\alpha))$.

The corresponding bet $x = f - g$ is illustrated in Figure 4.7, in which the areas of the shaded rectangles above and below the horizontal axis, which are the positive and negative parts of the expected value calculation, are both equal to ε. The change in expected utility due to the acceptance of x (i.e., swapping g for f) is the integral of xu'. Now suppose that u is twice continuously differentiable and its second derivative u'' is approximately constant on the interval $[\alpha - \varepsilon/(1-p), \alpha + \varepsilon/p]$, so u' is approximately linear on that interval. Then the integral of xu' on each of the two subintervals $[\alpha - \varepsilon/(1-p), \alpha]$ and $[\alpha, \alpha + \varepsilon/p]$ is the area above or below graph of x (namely, ε) multiplied by the value of u' at the midpoint of the subinterval, which is $u'(\alpha) - \frac{1}{2}(\varepsilon/(1-p))u''(\alpha)$ on the left (where x is negative) and $u'(\alpha) + \frac{1}{2}(\varepsilon/p)u''(\alpha)$ on the right (where x is positive). This yields the following expression for the change in expected utility due to the acceptance of the bet x (which is negative because $u''(\alpha) < 0$):

$$
\begin{aligned}
E_u(x) &= -\varepsilon(u'(\alpha) - \frac{1}{2}(\varepsilon/(1-p))u''(\alpha)) + \varepsilon(u'(\alpha) + \frac{1}{2}(\varepsilon/p)u''(\alpha)) \\
&= \frac{1}{2}\varepsilon^2 u''(\alpha)(1/p + 1/(1-p)) \qquad (4.10) \\
&= \frac{1}{2}\varepsilon^2 u''(\alpha)/(p(1-p)).
\end{aligned}
$$

Meanwhile, the variance of x is:

$$
\begin{aligned}
\mathrm{Var}(x) &= p(\varepsilon/p)^2 + (1-p)(\varepsilon/(1-p))^2 \qquad (4.11) \\
&= \varepsilon^2/(p(1-p)).
\end{aligned}
$$

Putting these results together yields:

$$E_u(x) = \frac{1}{2}u''(\alpha)\mathrm{Var}(x) \qquad (4.12)$$

Now, the (selling) risk premium associated with choosing the risky act f, when starting from a riskless wealth position α, is the amount of money c_f which, if given up, would yield the same decrease in expected utility. To a first-order approximation[9] c_f must therefore satisfy

$$(-c_f)u'(\alpha) = E_u(x) \qquad (4.13)$$

or equivalently (invoking the fact that $\mathrm{Var}(x)$ and $\mathrm{Var}(f)$ are the same):

$$\begin{aligned} c_f &= -E_u(x)/u'(\alpha) = -\frac{1}{2}(u''(\alpha)/u'(\alpha))\mathrm{Var}(x) \qquad (4.14) \\ &= -\frac{1}{2}(u''(\alpha)/u'(\alpha))\mathrm{Var}(f), \end{aligned}$$

in which the term $-u''(\alpha)/u'(\alpha)$ is the local Arrow-Pratt measure of (absolute) risk aversion that determines the tradeoff between risk (as measured by variance of the payoff) and the amount of cash that the decision-maker is willing to pay to get rid of it. This formula, which is also valid for non-binary acts, depends sensitively on the use of a constant act as the reference point for evaluating risk and also on the assumption that the probability distribution of the risky act is observable. In Chapter 6 a more general approach to modeling and measuring risk aversion will be presented, in which these restrictions are not necessary. The unobservability of background uncertainty as well the unobservability of true probabilities (when the latter are subjective) will both be finessed by changing the units of measurement to risk-neutral probabilities.

The symmetry between the SP and EU models in infinite dimensions becomes exact when the state space in the SP model consists of the interval $[0, 1]$. The analog of decreasing marginal utility in this setting is a subjective probability distribution whose density decreases with s, such as the exponential distribution.

4.4   The fundamental theorem of utilitarianism (social aggregation)

The expected utility model is also used in applications to social choice, and in particular to the foundations of utilitarianism. A classic application is Harsanyi's

---

[9] A first-order approximation suffices here because c_f is of the same order of magnitude as ε^2, as is $\mathrm{Var}(x)$.

(1955) "social aggregation theorem" (also called the "utilitarian cardinal welfare theorem"), which addresses the following problem. Society must choose from among a set of alternatives (e.g., policies of some kind) that affect the welfare of a population of heterogeneous individuals. Under what conditions does society have preferences described by a cardinal utility function that is a sum or weighted sum of those of the individuals? This would mean that cardinal utility is interpersonally comparable, or more precisely, that *differences* in cardinal utility between alternatives are interpersonally comparable. Thus, it would possible to ask whether Alice's gain in utility if society switches from alternative **f** to alternative **g** will be greater than Bob's loss in utility—i.e., whether the total cardinal utility of Alice and Bob is increased or decreased by a move from **f** to **g**. The imposition of the vNM independence condition on society's preferences is the key to this result, as shown by Mongin (1994).

Harsanyi (1955) proved the following theorem: if society has expected-utility preferences with respect to imaginary lotteries, and so do the individuals, and society's preferences satisfy a Pareto indifference condition with respect to those of the individuals, then it follows that society's utility function must be a weighted sum of the individual utility functions, with weights that may be positive or negative. The theorem generalizes to other types of Pareto optimality conditions that apply to weak or strong preferences. This result might be appropriately called the "fundamental theorem of utilitarianism" because it can be proved by the same separating hyperplane argument[10] that proves all of the other fundamental theorems of rational choice theory, and a Pareto optimality condition is essentially a no-arbitrage requirement: a violation would create an opportunity for a paid negotiator to step in.[11]

Here is the setup. Suppose there are N individuals for whom society chooses among a set of M alternatives and (counterfactual) objective lotteries over them. Such lotteries are represented by M-vectors of probabilities, generically denoted by **f** or **g**. Suppose that all individuals have vNM utility functions for the lotteries, represented by M-vectors $\mathbf{u}_n, n = 1, ..., N$, and society also has a vNM utility function $\mathbf{u}_0$. The expected utility for lottery **f** can then be expressed as $\mathbf{f} \cdot \mathbf{u}_n$ for $n = 0, ...N$. U will denote the $M \times N$ matrix whose columns are the utility vectors of individuals. Let society's and the individuals' indifference and weak/strong preference relations be denoted by $\sim_n$, $\succsim_n$, and $\succ_n$ respectively. Thus, $\mathbf{f} \sim_n \mathbf{g}$ [resp. $\mathbf{f} \succsim_j \mathbf{g}$, $\mathbf{f} \succ_n \mathbf{g}$] means $\mathbf{f} \cdot \mathbf{u}_n = \mathbf{g} \cdot \mathbf{u}_n$ [resp. $\mathbf{f} \cdot \mathbf{u}_n \geq \mathbf{g} \cdot \mathbf{u}_n$, $\mathbf{f} \cdot \mathbf{u}_n > \mathbf{g} \cdot \mathbf{u}_n$].

Let **v** be an M-vector that is proportional to the difference between a pair of lotteries, i.e., $\mathbf{v} = \alpha(\mathbf{f} - \mathbf{g})$ for some acts **f** and **g** and constant $\alpha > 0$. Adapting the

---

[10] A separating hyperplane proof is given by Border (1985).

[11] More precisely, a violation of Pareto optimality is an arbitrage opportunity if it is common knowledge, as if revealed through public offers of material bets or trades or side payments, as discussed in Section 1.5.

terminology used earlier, $\mathbf{v}$ is a "social p-bet," a change in risk profile that results from the proportional swap of one lottery for another. Then $\mathbf{v} \cdot \mathbf{u}_n$ is individual n's gain in expected utility that follows from the acceptance of the social p-bet $\mathbf{v}$, $\mathbf{v} \cdot \mathbf{u}_0$ is society's own gain in expected utility, and $\mathbf{v} \cdot \mathbf{U}$ is the vector of expected utility gains for all individuals. A social p-bet is [*strictly, indifferently*] *acceptable* to an individual or to society if its gain in expected utility for them is non-negative [positive, zero]. Without loss of generality the elements of $\mathbf{u}_n$ and $\mathbf{u}_0$ can be assumed to sum to zero (because addition of a constant to a utility function does not matter), and the elements of $\mathbf{v}$ can also be adjusted by the addition of a constant so that they sum to zero (because an additive constant will drop out of the inner products $\mathbf{v} \cdot \mathbf{u}_n$ and $\mathbf{v} \cdot \mathbf{u}_0$). Also $\mathbf{v}$ can be arbitrarily scaled so that its elements are all less than or equal to 1 in magnitude, thus fulfilling the conditions of a p-bet.

Consider the following standard Pareto optimality principles:

Pareto indifference: if $\mathbf{f} \sim_n \mathbf{g}$ for all $n > 0$, then $\mathbf{f} \sim_0 \mathbf{g}$.

Semistrong Pareto: if $\mathbf{f} \succsim_n \mathbf{g}$ for all $n > 0$, then $\mathbf{f} \succsim_0 \mathbf{g}$.

Strong Pareto: if $\mathbf{f} \succsim_n \mathbf{g}$ for all $n > 0$ and $\mathbf{f} \succ_j \mathbf{g}$ for some j, then $\mathbf{f} \succ_0 \mathbf{g}$.

In terms of acceptable social p-bets, Pareto indifference means that if $\mathbf{v}$ is indifferently acceptable to every individual, then it is indifferently acceptable to society; semistrong Pareto means that if $\mathbf{v}$ is acceptable to every individual, then it is acceptable to society; and strong Pareto means that if $\mathbf{v}$ is acceptable to every individual and strictly acceptable to at least one, then it is strictly acceptable to society. The utilitarian implications of these principles are given by:

Theorem 4.5: Fundamental theorem of utilitarianism (social aggregation)

1. Pareto indifference is satisfied if and only if society's utility function is a weighted sum of the individual utility functions, with weights unrestricted in sign.

2. Semistrong Pareto is satisfied if and only if society's utility function is a non-negatively weighted sum of the individual utility functions.

3. Strong Pareto and semistrong Pareto are satisfied if and only if society's utility function is a positively weighted sum of the individual utility functions.

The Pareto indifference version is Harsanyi's (1955) original theorem.[12] A simple proof for all three versions in terms of linear programming is given by Mandler (2005).[13] In the proof below, a different parameterization will be used for the linear programs, one that is based on the formulation of the EU model used earlier in this chapter.

Proof: Let π be an N-vector of weights assigned to individuals, and consider the following canonical primal/dual pair of linear programs:

Primal LP 0: find $\mathbf{x}$ to maximize $\mathbf{c} \cdot \mathbf{x}$ subject to $\mathbf{Ax} = \mathbf{b}$ and $\mathbf{x} \geq \mathbf{0}$

Dual LP 0: find $\mathbf{y}$ to minimize $\mathbf{b} \cdot \mathbf{y}$ subject to $\mathbf{y} \cdot \mathbf{A} \geq \mathbf{c}$

By the duality theorem of linear programming, which rests on the separating hyperplane theorem, the primal LP is feasible if and only if the solution to the dual LP is bounded and has the same optimal objective value, otherwise the primal LP is infeasible and the dual LP solution is unbounded.[14] Satisfaction of each of the Pareto principles corresponds to the existence of a solution to an appropriate primal LP.

Case 1: Pareto indifference. Let $\mathbf{x} = (\pi^+, \pi^-)$ where π^+ and π^- are the positive and negative parts of π (whose sign is unconstrained here), let $\mathbf{A} = [\mathbf{U} | -\mathbf{U}]$ (so that $\mathbf{Ax} = \mathbf{U}\pi$), let $\mathbf{b} = \mathbf{u}_0$, let $\mathbf{c} = \mathbf{0}$, and let $\mathbf{y} = \mathbf{v}$ (note that $\mathbf{y} \cdot \mathbf{A} \geq \mathbf{0}$ means $\mathbf{v} \cdot \mathbf{U} = \mathbf{0}$). Then the LP's become:

Primal LP 1: find π to satisfy $\mathbf{U}\pi = \mathbf{u}_0$

Dual LP 1: find $\mathbf{v}$ to minimize $\mathbf{v} \cdot \mathbf{u}_0$ subject to $\mathbf{v} \cdot \mathbf{U} = \mathbf{0}$

The primal LP seeks a set of *positive or negative weights* for which society's utility function is a weighted sum of the individual utility functions. This is not feasible if and only if the dual LP solution is unbounded, i.e., there exists a social p-bet $\mathbf{v}$ such that $\mathbf{v} \cdot \mathbf{U} = \mathbf{0}$ and $\mathbf{v} \cdot \mathbf{u}_0 < 0$ ($\mathbf{v}$ is indifferently acceptable to all the individuals and strictly unacceptable to society), in violation of Pareto indifference.

Case 2: Semistrong Pareto. Let $\mathbf{x} = \pi$, let $\mathbf{A} = \mathbf{U}$, let $\mathbf{b} = \mathbf{u}_0$, let $\mathbf{c} = \mathbf{0}$, and let $\mathbf{y} = \mathbf{v}$. Then the LP's become:

[12]Many generalizations of this result have been explored. For example, Danan et al. (2015) consider the case in which the expected-utility preferences of the individuals and those of society may all be incomplete.

[13]Further discussion of variants of Harsanyi's theorem for different Pareto conditions can be found in Weymark (1993, 1995).

[14]The primal solution cannot be unbounded if $\mathbf{b} \neq \mathbf{0}$, which is the case here.

Primal LP 2: find $\boldsymbol{\pi}$ to satisfy $\mathbf{U}\boldsymbol{\pi} = \mathbf{u}_0, \boldsymbol{\pi} \geq \mathbf{0}$

Dual LP 2: find $\mathbf{v}$ to minimize $\mathbf{v} \cdot \mathbf{u}_0$ subject to $\mathbf{v} \cdot \mathbf{U} \geq \mathbf{0}$

The primal LP seeks a set of *non-negative weights* for which society's utility function is a weighted sum of the individual utility functions. This is not feasible if and only if the dual LP solution is unbounded, i.e., there exists a social p-bet $\mathbf{v}$ such that $\mathbf{v} \cdot \mathbf{U} \geq \mathbf{0}$ and $\mathbf{v} \cdot \mathbf{u}_0 < 0$ ($\mathbf{v}$ is acceptable to all the individuals and strictly unacceptable to society), in violation of semistrong Pareto.

Case 3: Strong Pareto. Use the same setup as in case 2 except with $\mathbf{c} = \beta \mathbf{e}_j$, where $\mathbf{e}_j$ is the jth unit vector and $\beta > 0$.[15] Then the LP's become:

Primal LP 3: find $\boldsymbol{\pi}$ to maximize $\beta \pi_j$ subject to $\mathbf{U}\boldsymbol{\pi} = \mathbf{u}_0, \boldsymbol{\pi} \geq \mathbf{0}$

Dual LP 3: find $\mathbf{v}$ to minimize $\mathbf{v} \cdot \mathbf{u}_0$ subject to $\mathbf{v} \cdot \mathbf{U} \geq \beta \mathbf{e}_j$

The primal LP seeks a set of non-negative weights for which society's utility function is a weighted sum of the individual utility functions *and individual j's weight is as large as possible.* From case 2 this LP is feasible if and only if semistrong Pareto is satisfied. Its optimal objective value is β times the largest weight that can be assigned to individual j in a weighted sum of utility functions that matches society's, and the solution to the dual LP is bounded with the same optimal objective value. Now suppose that for some j the primal optimal objective value is zero, in which case the dual optimal objective value is also zero. Letting $\mathbf{v}^*$ denote the optimal dual solution, it must satisfy the constraint that $\mathbf{v}^* \cdot \mathbf{U} \geq \beta \mathbf{e}_j$ ($\mathbf{v}^*$ is strictly acceptable to individual j and acceptable to everyone else), but meanwhile $\mathbf{v}^* \cdot \mathbf{u}_0 = 0$ (society is merely indifferent to $\mathbf{v}^*$), in violation of strong Pareto. Therefore strong Pareto requires that for every j there should be a feasible solution to the primal LP that assigns strictly positive weight to individual j, and a strictly convex combination of these solutions is a feasible solution assigning strictly positive weight to everyone. The primal LP is not feasible if and only if the dual solution is unbounded, in violation of both semistrong and strong Pareto.

---

[15] The scaling of $\mathbf{v}$ in the dual solution is determined by the value of β in the constraint $\mathbf{v} \cdot \mathbf{U} \geq \beta \mathbf{e}_j$, and β can be chosen arbitrarily so that the elements of $\mathbf{v}$ meet the lower and upper bound constraints of a p-bet.

In general the solutions to the primal LP's are not unique because of the possibility that one individual's utility function could be a linear or affine combination of others. An additional assumption of "affine independence"[16] of the utility functions is needed to rule this out. Under this condition the relative weights assigned to different individuals are uniquely determined for any particular normalization of the individual and societal utility functions. The social aggregation theorem then implies that if the utility function of an individual is rescaled, her weight in society's utility function is changed in a reciprocal fashion, so she has the same influence regardless of the scaling of her own utility function.

Of course, this simple and powerful result is predicated on some very strong assumptions, namely that objectively randomized policies can be contemplated and society's preferences with respect to them satisfy the same vNM independence condition as those of the individuals. The emergence of unique utility weights in Harsanyi's theorem follows from the fact that the individual utility functions and the societal utility function are conjured into existence simultaneously without regard for other considerations such as ex ante or ex post egalitarianism (Mongin and Pivato (2016), Fleurbaey (2018)). The weights are "profile-dependent": they have whatever values they need to have in order to fit an additive representation to the given utility profiles. A much deeper problem in social choice theory, highlighted by Condorcet's paradox and Arrow's impossibility theorem, is whether and under what conditions a societal utility function—even an ordinal one—will exist at all. As soon as a cardinal societal utility function is pulled out of the air, some of the most difficult problems in social choice are assumed away. Harsanyi argues, nevertheless, that interpersonal comparisons of cardinal utility differences are meaningful and useful. They provide a rationale for redistributive social policies such as progressive taxation, in which an extra dollar is assumed to be worth more to a poor person than a rich one, or public health and safety initiatives in which government funds are allocated so as to yield the greatest social benefit in terms of quality-adjusted-life-years saved. However, the same argument can be used to take from the poor and give to the rich. It depends on who is doing the social planning. The social aggregation theorem merely rationalizes societal preferences that are already given.[17]

---

[16]Affine independence of vectors means precisely that no one of them is an affine combination of the others (a linear combination plus a constant). Here the utility vectors have been additively adjusted so that the sums of their elements are zero, so affine independence for them is the same as linear independence.

[17]Harsanyi also uses a version of the common prior assumption to defend the additive utilitarian position in the context of his "impartial observer" theorem. He proposes that, when participating in social decisions, individuals should imagine themselves to be behind a "veil of ignorance" in which they do not yet know their own "types," i.e., who they are in society. From this impartial starting point, they should make the choices that would maximize the expected value of their utility based on the commonly known distribution of types in the population, which is equivalent to maximizing the sum of everyone's utilities.

Generalizations of this theorem that apply to subjective expected utility have been widely studied. A central finding is that in general there is no way to separately aggregate the subjective probability distributions and vNM utility functions of individuals so as to obtain an SEU representation of a group's preferences that satisfies a Pareto condition. It is only possible if the individuals agree on probabilities and/or if they agree on utilities (Seidenfeld et al. (1989), De Meyer and Mongin (1995), Mongin (1995)). This "impossibility" result can be circumvented if the group utility function is allowed to be state-dependent, an issue that has been explored by Schervish et al. (1991), Mongin (1998) and Blackorby et al. (1999), among others.[18] The 1991 paper of Schervish et al. and the 1998 paper by Mongin use the Anscombe-Aumann version of the SEU model and show that state-dependent utilities can be aggregated in a linear/additive fashion, but subjective probabilities cannot be uniquely separated from utilities. This is an issue for the SEU model all by itself. It was raised in Section 1.3 and will be discussed further in Chapter 5 in the context of both complete and incomplete preferences. If a group's preferences are permitted to be incomplete, represented by a convex set of state-dependent utilities, a Pareto condition naturally forces that set to be the convex hull of those of the individuals. These results all lend weight to the view that completeness of preferences and state-independence of utility ought not to be imposed on individuals or on groups.

A more "informationally parsimonious" version of Harsanyi's impartial-observer theorem is given by Karni and Weymark (1998). The common prior assumption is also part of the foundation of Aumann's (1974, 1987) correlated equilibrium concept for noncooperative games and Harsanyi's (1967, 1968ab) Bayesian equilibrium concept for games of incomplete information. These models will be discussed in terms of no-arbitrage, without an up-front common prior assumption, in Chapter 8. The as-if existence of a common prior distribution is itself an implication of no-arbitrage.

[18] Keeney and Nau (2011) consider a variation on the aggregation result in which the group has expected-utility preferences among gambles whose outcomes are vectors of expected utilities of individuals. The result is still that the preferences of the group have a state-dependent utility representation with respect to the original set of acts, which is a weighted sum of the state-independent utilities of the individuals.

Chapter 5

Subjective expected utility

5.1   Joint elicitation of beliefs and tastes

In preceding chapters, two models of rational individual behavior were introduced: one that applies to choices among alternatives with objectively evaluated outcomes and subjectively evaluated probabilities, and another that applies to choices among alternatives with objectively evaluated probabilities and subjectively evaluated outcomes. Most real decisions with uncertainty don't fit neatly into one of these boxes: they require subjective evaluations of both the probabilities and the outcomes. Insofar as the two models have exactly the same mathematical structure, merely with their parts interchanged, it might be expected that they could be easily integrated into a single model with the same structure that allows for both kinds of subjectivity, and indeed this is the case.

To set the stage, consider another example in which you are the last remaining contestant in a game show called "Flip or Guess," and you are scheduled to appear on live television tomorrow night and be given the opportunity to choose among two risky alternatives whose possible prizes are $\$1000K^+$, $\$100K^+$, or $\$0K^+$, where the $+$'s indicate some additional non-monetary attributes. (Assume these are the same prizes as in the example of the previous chapter.)   Suppose that one alternative is to flip a coin and get $\$100K^+$ if it comes up heads and $\$0K^+$ if it comes up tails. The other alternative is a test of knowledge in which you must guess the answers to two trivia questions. If you fail to answer the first one correctly, the game ends and you get $\$0K^+$; if you answer the first correctly but not the second you get

$100K^+$; and if you answer both correctly you get $1000K^+$, The questions will be drawn from each of two categories, "garden tools" and "black holes" in that order. Questions from the same two categories (among others) were asked in earlier rounds of the game, to yourself or other contestants, and you found that you knew the answers to only 2 out of 5 garden tools questions and 1 out of 5 black holes questions. You expect the remaining questions to be of a generally similar level of difficulty. The sets of questions in these and other categories were made up in advance by a team of experts, and those used in the game are drawn randomly from these sets, so you don't feel that the organizers will be able to manipulate the level of difficulty either for you or against you in the final round.

If your batting averages for the two categories of questions in the earlier rounds are representative, and if you get to answer the garden tools question first, your expected monetary payoff for guessing is approximately $0.6 \times \$0 + 0.4 \times (0.8 \times \$100K + 0.2 \times \$1000K) = \$130K$. That's a lot more than the expected monetary payoff of flipping, which is $50K$, but it doesn't take into account your appetite for risk and for consumption of the three outcomes. Furthermore, your empirical batting averages do not necessarily represent your beliefs concerning your ability to correctly answer tomorrow's questions, because they are based on very small samples, and privately you feel that a couple of your correct answers in the earlier rounds were just lucky guesses. That's what got you to the final round! However, you've done some web-searching this evening and learned a lot more about both subjects than you knew yesterday, so it's a bit hard to say whether you should expect to do better or worse on the final round of questions than you did before. If you go with the coin flip, you will at least be sure of the odds. Another potential issue is that the $\$0K^+$ and $\$100K^+$ outcomes may not be psychologically the same when they are received as a result of guessing the answers to trivia questions as when they are received as a result of flipping a coin. If you end up with $\$0K^+$ by failing to answer a question about garden tools, you will not only be unlucky, you will also look stupid on national television and YouTube. On the other hand, winning $\$100K+$ by demonstrating some self-confidence and factual knowledge might be more satisfying than winning it by a coin flip.

For simplicity, let's suppose that the values you attach to the prizes are independent of how they are obtained and that your appetite for risk is independent of the source of the risk. To evaluate and compare the two alternatives with any precision, it is necessary to embed them both in a much larger set of hypothetical acts and to assume that you have complete preferences on this larger set. The most convenient way to do this is that of Anscombe and Aumann (1963), in which the objects of choice are *lottery acts*, otherwise known as *horse lotteries*, which are mappings from events to objective lotteries over the prizes. Imagine a horse race in which each horse can be associated with objective lotteries over some set of prizes. The objects of choice consist of assignments of lotteries to horses, so that the outcome of a choice depends not only on which horse wins but also on the result of the lottery that it carries, which determines the final prize. These might appear to be rather strange–in fact, bizarre—objects of choice among which to have ready-made preferences, but there is no way to avoid a highly stretched imagination if the result is to be a representation of your beliefs and tastes in terms of crisp numbers.[1] The flipping and guessing alternatives are two degenerate kinds of lottery acts: flipping yields the same objective lottery over the prizes regardless of whether you know the answers to the questions, while guessing yields a prize that depends only on whether you know the answers, without any randomization. In a more general lottery act in this scenario, you might receive one of several different objective lotteries over the prizes depending on whether you know the answers to the questions. The advantage of this setup is that a probabilistic mixture of lottery acts is itself another lottery act. The mixing operation convexifies the set of alternatives, setting the stage for another application of the separating hyperplane theorem. It also conveniently allows the independence axiom of expected-utility theory (EU4^a) to be re-used without modification, which once again will greatly simplify your analysis.

A lottery act (henceforth called merely an act) can be represented by a matrix $\mathbf{F}$ whose rows correspond to states and whose columns correspond to consequences (prizes), in which the element in the mth row and n^{th} column, f_{mn}, is the objective probability of receiving consequence c_n when the state turns out to be s_m. The elements of such a matrix are non-negative and sum to 1 in each row, rather than having an overall sum of 1. By the fundamental theorem of expected utility, an individual's preferences over such acts satisfy the EU axioms if and only if there exists a matrix $\mathbf{V}$, unique up to positive affine transformations, such that $\mathbf{F} \succsim \mathbf{G}$ if and only $E_{\mathbf{V}}(\mathbf{F}) \geq E_{\mathbf{V}}(\mathbf{G})$, where $E_{\mathbf{V}}(\mathbf{F}) := \sum_{m=1}^{M} \sum_{n=1}^{N} f_{mn} v_{mn}$. This matrix is interpretable as a *state-dependent utility function* that assigns utility v_{mn} to consequence c_n when state s_m occurs, as if every consequence-state pair is counted as a distinct experience. As such it represents both the individual's beliefs about events and her tastes for consequences, and in general her tastes may depend on the events in which consequences are received. If we assume that beliefs and tastes are psychologically distinct and that consequences can be defined in such a way that the individual's

---

[1] In Savage's original axiomatization of the SEU model the acts are simply mappings from states to consequences. There, a mucher richer state space needs to be assumed in order to get the same fine-grained numerical representation of preferences that Anscombe and Aumann get via the introduction of objective randomization into the structure of the acts. The Anscombe-Aumman version is mathematically simpler and more transparent in its relation to the SP and EU models. Its use of objective randomization in the measurement of subjective probabilities and utilities is also in line with what is done in applied decision analysis.

tastes for them are the same in all states of the world, then we might hope to decompose $\mathbf{V}$ into a subjective probability distribution $\mathbf{p}$ and a utility function $\mathbf{u}$ such that $v_{mn} = p_m u_n$. We would then have $\mathbf{F} \succsim \mathbf{G}$ if and only if $E_{\mathbf{p},\mathbf{u}}(\mathbf{F}) \geq E_{\mathbf{p},\mathbf{u}}(\mathbf{G})$, where

$$E_{\mathbf{p},\mathbf{u}}(\mathbf{F}) := \mathbf{p} \cdot \mathbf{Fu} = \sum_{m=1}^{M} p_m \sum_{n=1}^{N} f_{mn} u_n \qquad (5.1)$$

is the expected utility of $\mathbf{F}$ determined by the subjective probability distribution $\mathbf{p}$ and the utility function $\mathbf{u}$. This is the *subjective expected utility* (SEU) model of preferences under uncertainty. It is obtained from the assumption that preferences among lottery acts satisfy the EU axioms *plus* one new axiom that allows utilities to be defined so as to be state-independent, which will be introduced shortly.

In the Flip-or-Guess game, an act can be represented by a matrix whose rows correspond to the states in which 0, 1, or 2 questions are correctly answered, whose columns correspond to the prizes ($1000K^+$, $100K^+$, $0K^+$), and whose generic element f_{mn} is the objective probability of getting prize c_n in state s_m:

TABLE 5.1: Structure of an act for the Flip-or-Guess game

	Prize:		
	$1000K^+$ (c_1)	$100K^+$ (c_2)	$0K^+$ (c_3)
No correct answers (s_1)	f_{11}	f_{12}	f_{13}
One correct answer (s_2)	f_{21}	f_{22}	f_{23}
Two correct answers (s_3)	f_{31}	f_{32}	f_{33}

In the game, "Flip" is the act whose parameters are:

$$\mathbf{F} := \begin{bmatrix} 0 & 1/2 & 1/2 \\ 0 & 1/2 & 1/2 \\ 0 & 1/2 & 1/2 \end{bmatrix},$$

and "Guess" is the act whose parameters are:

$$\mathbf{G} := \begin{bmatrix} 0 & 0 & 1 \\ 0 & 1 & 0 \\ 1 & 0 & 0 \end{bmatrix}.$$

In order to determine your preference between these two acts by separately reflecting on your beliefs and tastes in the fashion of SEU theory, the first step is to assess your utilities for the prizes. As in the previous chapter, let $\mathbf{L}_n$ denote a reference lottery that yields prize c_n with certainty, i.e.,

$$\mathbf{L}_1 := \begin{bmatrix} 1 & 0 & 0 \\ 1 & 0 & 0 \\ 1 & 0 & 0 \end{bmatrix}, \quad \mathbf{L}_2 := \begin{bmatrix} 0 & 1 & 0 \\ 0 & 1 & 0 \\ 0 & 1 & 0 \end{bmatrix}, \quad \mathbf{L}_3 := \begin{bmatrix} 0 & 0 & 1 \\ 0 & 0 & 1 \\ 0 & 0 & 1 \end{bmatrix},$$

and let $\mathbf{L}[u]$ denote a reference lottery that yields the best prize with objective probability u and the worst prize with probability $1 - u$, i.e.,

$$\mathbf{L}[u] := u\mathbf{L}_1 + (1-u)\mathbf{L}_3 = \begin{bmatrix} u & 0 & 1-u \\ u & 0 & 1-u \\ u & 0 & 1-u \end{bmatrix}$$

Its utility is equal to u by definition. Then the relative utility of the prize c_n on a 0-to-1 scale can be defined as the unique number u_n for which you are indifferent between $\mathbf{L}[u_n]$ and $\mathbf{L}_n$, i.e., the numbers (u_1, u_2, u_3) satisfy:

$$\mathbf{L}[u_1] \sim \mathbf{L}_1$$
$$\mathbf{L}[u_2] \sim \mathbf{L}_2$$
$$\mathbf{L}[u_3] \sim \mathbf{L}_3$$

By assumption $u_1 = 1$ and $u_3 = 0$, so the only number that actually needs to be assessed in a 3-prize situation is u_2, which must lie between 0 and 1.

The next step is to assess your subjective probabilities for events. Let $\mathbf{I}_m$ denote an "indicator act" in which you get the best prize in state s_m and the worst prize otherwise,[2] i.e.,

$$\mathbf{I}_1 := \begin{bmatrix} 1 & 0 & 0 \\ 0 & 0 & 1 \\ 0 & 0 & 1 \end{bmatrix}, \quad \mathbf{I}_2 := \begin{bmatrix} 0 & 0 & 1 \\ 1 & 0 & 0 \\ 0 & 0 & 1 \end{bmatrix}, \quad \mathbf{I}_3 := \begin{bmatrix} 0 & 0 & 1 \\ 0 & 0 & 1 \\ 1 & 0 & 0 \end{bmatrix}$$

Then your subjective probability for state s_m can be defined as the unique number p_m for which you are indifferent between $\mathbf{L}[p_m]$ and $\mathbf{I}_m$, i.e., the numbers (p_1, p_2, p_3) satisfy:

$$\mathbf{L}[p_1] \sim \mathbf{I}_1$$
$$\mathbf{L}[p_2] \sim \mathbf{I}_2$$
$$\mathbf{L}[p_3] \sim \mathbf{I}_3$$

So far, so good. You have assessed a vector of utilities $\mathbf{u} = (u_1, u_2, u_3)$ and a vector of subjective probabilities $\mathbf{p} = (p_1, p_2, p_3)$. It remains to show that you are justified in using these to calculate subjective expected utilities in order to construct your preferences among all other acts.

---

[2]Note that $\mathbf{I}_1$ is not only counterfactual, but also it would create a moral hazard situation if you possessed it, because you would then want to give a wrong answer to the first question. Because this is only a thought experiment that you are using to help quantify your beliefs, let's suppose you imagine you will answer the questions to the best of your ability regardless of the stakes that may be attached. Alternatively, it could be supposed that you are betting on someone else's performance, unbeknownst to her.

The next step is to observe that the utility vector can be used to evaluate arbitrary constant acts. For example, suppose $\mathbf{X}$ is a constant act, i.e., an act that yields the same lottery $\mathbf{x} = (x_1, x_2, x_3)$ in each state:

$$\mathbf{X} := \begin{bmatrix} x_1 & x_2 & x_3 \\ x_1 & x_2 & x_3 \\ x_1 & x_2 & x_3 \end{bmatrix}.$$

Then $\mathbf{X}$ can be decomposed as a convex combination of the lotteries $\mathbf{L}_1$, $\mathbf{L}_2$, and $\mathbf{L}_3$:

$$\mathbf{X} := x_1 \mathbf{L}_1 + x_2 \mathbf{L}_2 + x_3 \mathbf{L}_3. \tag{5.2}$$

The independence axiom of EU theory implies that an equally-preferred act can be obtained by substituting the equally-preferred lotteries $\mathbf{L}[u_1]$, $\mathbf{L}[u_2]$, and $\mathbf{L}[u_3]$ for $\mathbf{L}_1$, $\mathbf{L}_2$, and $\mathbf{L}_3$, respectively:

$$\mathbf{X} \sim x_1 \mathbf{L}[u_1] + x_2 \mathbf{L}[u_2] + x_3 \mathbf{L}[u_3] = \begin{bmatrix} E_\mathbf{u}(\mathbf{x}) & 0 & 1 - E_\mathbf{u}(\mathbf{x}) \\ E_\mathbf{u}(\mathbf{x}) & 0 & 1 - E_\mathbf{u}(\mathbf{x}) \\ E_\mathbf{u}(\mathbf{x}) & 0 & 1 - E_\mathbf{u}(\mathbf{x}) \end{bmatrix} = \mathbf{L}[E_\mathbf{u}(\mathbf{x})].$$
$$\tag{5.3}$$

It follows that if $\mathbf{Y}$ is some other constant act yielding the lottery $\mathbf{y} = (y_1, y_2, y_3)$ in every state, then $\mathbf{X} \succsim \mathbf{Y}$ if and only $E_\mathbf{u}(\mathbf{x}) \geq E_\mathbf{u}(\mathbf{y})$ because $\mathbf{L}[E_\mathbf{u}(\mathbf{x})] \succsim \mathbf{L}[E_\mathbf{u}(\mathbf{y})]$ if and only if $E_\mathbf{u}(\mathbf{x}) \geq E_\mathbf{u}(\mathbf{y})$ (a larger total probability of getting the best prize rather than the worst prize is preferred). So, the utility vector $\mathbf{u}$ can at least be used to evaluate *constant* acts on the basis of their expected utilities.

To obtain the SEU representation of preferences for more general acts, in which the lotteries may vary by state and the subjective probabilities of states must be taken into account, an additional assumption is needed. Let $\mathbf{F}$ denote an arbitrary act (not necessarily the "Flip" act in the game). The additional assumption that is needed is that the same substitution that was performed in each row of $\mathbf{X}$ in the preceding example, using the utility vector $\mathbf{u}$, can also be performed in each row of $\mathbf{F}$, despite the fact that its rows are not necessarily identical. This means that *for any act* $\mathbf{F}$, *an equally-preferred act can be obtained by replacing the lottery that it yields in a given state with a lottery that would be equally preferred in a comparison of constant acts*. For example, consider the lottery (f_{11}, f_{12}, f_{13}) that is the first row of $\mathbf{F}$. If a constant act is constructed in which this lottery is received in every state, then by the independence axiom, as shown above, it follows that a preferentially equivalent constant act can be constructed by

$$\begin{bmatrix} f_{11} & f_{12} & f_{13} \\ f_{11} & f_{12} & f_{13} \\ f_{11} & f_{12} & f_{13} \end{bmatrix} \sim \begin{bmatrix} E_\mathbf{u}(\mathbf{f}_1) & 0 & 1 - E_\mathbf{u}(\mathbf{f}_1) \\ E_\mathbf{u}(\mathbf{f}_1) & 0 & 1 - E_\mathbf{u}(\mathbf{f}_1) \\ E_\mathbf{u}(\mathbf{f}_1) & 0 & 1 - E_\mathbf{u}(\mathbf{f}_1) \end{bmatrix} = \mathbf{L}[E_\mathbf{u}(\mathbf{f}_1)]. \tag{5.4}$$

The new assumption, which will be called "separability," is that because the lottery whose generic row is $\mathbf{f}_1 := (f_{11}, f_{12}, f_{13})$ is equally preferred to the lottery $(E_\mathbf{u}(\mathbf{f}_1), 0, 1 - E_\mathbf{u}(\mathbf{f}_1))$ in a comparison of constant acts, the same substitution can be performed within the context of the original act, where it is only one of several rows, i.e.,

$$
\begin{bmatrix} f_{11} & f_{12} & f_{13} \\ f_{21} & f_{22} & f_{23} \\ f_{31} & f_{32} & f_{33} \end{bmatrix} \sim \begin{bmatrix} E_\mathbf{u}(\mathbf{f}_1) & 0 & 1 - E_\mathbf{u}(\mathbf{f}_1) \\ f_{21} & f_{22} & f_{23} \\ f_{31} & f_{32} & f_{33} \end{bmatrix}. \tag{5.5}
$$

Repeating this step for states s_2 and s_3 yields:

$$
\begin{bmatrix} f_{11} & f_{12} & f_{13} \\ f_{21} & f_{22} & f_{23} \\ f_{31} & f_{32} & f_{33} \end{bmatrix} \sim \begin{bmatrix} E_\mathbf{u}(\mathbf{f}_1) & 0 & 1 - E_\mathbf{u}(\mathbf{f}_1) \\ E_\mathbf{u}(\mathbf{f}_2) & 0 & 1 - E_\mathbf{u}(\mathbf{f}_2) \\ E_\mathbf{u}(\mathbf{f}_3) & 0 & 1 - E_\mathbf{u}(\mathbf{f}_3) \end{bmatrix}. \tag{5.6}
$$

Thus we obtain a preferentially equivalent act which in each state yields a reference lottery whose expected utility is determined by the previously-assessed utility vector $\mathbf{u}$. The final step, which factors in the assessed subjective probabilities, is to rewrite the latter act as a convex combination of $\mathbf{I}_1$, $\mathbf{I}_2$, $\mathbf{I}_3$, and $\mathbf{L}[0]$ and invoke the independence axiom again to replace the $\mathbf{I}_m$'s with their own preferentially equivalent reference lotteries:

$$
\begin{bmatrix} E_\mathbf{u}(\mathbf{f}_1) & 0 & 1 - E_\mathbf{u}(\mathbf{f}_1) \\ E_\mathbf{u}(\mathbf{f}_2) & 0 & 1 - E_\mathbf{u}(\mathbf{f}_2) \\ E_\mathbf{u}(\mathbf{f}_3) & 0 & 1 - E_\mathbf{u}(\mathbf{f}_3) \end{bmatrix} \tag{5.7}
$$

$$
= \quad E_\mathbf{u}(\mathbf{f}_1)\mathbf{I}_1 + E_\mathbf{u}(\mathbf{f}_2)\mathbf{I}_2 + E_\mathbf{u}(\mathbf{f}_3)\mathbf{I}_3 + (1 - E_\mathbf{u}(\mathbf{f}_1 + \mathbf{f}_2 + \mathbf{f}_3))\mathbf{L}[0]
$$

$$
\sim \quad E_\mathbf{u}(\mathbf{f}_1)\mathbf{L}[p_1] + E_\mathbf{u}(\mathbf{f}_2)\mathbf{L}[p_2] + E_\mathbf{u}(\mathbf{f}_3)\mathbf{L}[p_3] + (1 - E_\mathbf{u}(\mathbf{f}_1 + \mathbf{f}_2 + \mathbf{f}_3))\mathbf{L}[0]
$$

$$
= \quad \mathbf{L}[p_1 E_\mathbf{u}(\mathbf{f}_1) + p_2 E_\mathbf{u}(\mathbf{f}_2) + p_3 E_\mathbf{u}(\mathbf{f}_3)]
$$

$$
= \quad \mathbf{L}[E_{\mathbf{p},\mathbf{u}}(\mathbf{F})],
$$

whose utility is $E_{\mathbf{p},\mathbf{u}}(\mathbf{F})$ by definition, which is the subjective expected utility of $\mathbf{F}$ that is determined by $\mathbf{p}$ and $\mathbf{u}$.[3] It follows that for any two acts $\mathbf{F}$ and $\mathbf{G}$, $\mathbf{F} \succsim \mathbf{G}$ if and only if $E_{\mathbf{p},\mathbf{u}}(\mathbf{F}) \geq E_{\mathbf{p},\mathbf{u}}(\mathbf{G})$ as advertised.

---

[3]The coefficient of $\mathbf{L}[0]$ in the decomposition, which is $(1 - E_\mathbf{u}(\mathbf{f}_1 + \mathbf{f}_2 + \mathbf{f}_3))$, is negative if $E_\mathbf{u}(\mathbf{f}_1 + \mathbf{f}_2 + \mathbf{f}_3) > 1$, in which case the act is a linear but not convex combination of $\mathbf{I}_1$, $\mathbf{I}_2$, $\mathbf{I}_3$, and $\mathbf{L}[0]$. This is unimportant: the utility vector $\mathbf{u}$ could just as well have been scaled so that its maximal element is $1/3$ rather than 1, which would have ensured non-negativity of all the coefficients.

Now let's use this result to evaluate the acts $\mathbf{F}$ and $\mathbf{G}$ in the Flip-or-Guess game. Let $\mathbf{u} = (1, u_2, 0)$ where u_2 is your assessed value for the utility of the intermediate prize (i.e., the number for which you judge that $\mathbf{L}[u_2] \sim \mathbf{L}_2$), and let p_1, p_2, and p_3 denote your assessed subjective probabilities of the 3 states (i.e., the numbers for which you judge that $\mathbf{L}[p_1] \sim \mathbf{I}_1$, $\mathbf{L}[p_2] \sim \mathbf{I}_2$, and $\mathbf{L}[p_3] \sim \mathbf{I}_3$). Then $E_{\mathbf{p},\mathbf{u}}(\mathbf{F}) = \frac{1}{2}u_2$ and $E_{\mathbf{p},\mathbf{u}}(\mathbf{G}) = p_1 \times 0 + p_2 \times u_2 + p_3 \times 1 = p_2 u_2 + p_3$. For example, suppose you judge that $u_2 = 0.6$ (a highly risk-averse value) while $p_1 = 0.6$ and $p_2 = 0.3$ (indicating that you think you have a 40% chance of correctly answering the first question and a 25% chance of correctly answering the second question if you get to it). Then the expected utility of $\mathbf{F}$ is 0.3 and your subjective expected utility of $\mathbf{G}$ is 0.28, so you would rather flip. If instead $u_2 = 0.5$, which is a bit less risk-averse, both alternatives are equally preferred, with expected utilities of 0.25, and if $u_2 < 0.5$ you would strictly prefer to guess.

This constructive approach to the analysis of choices under uncertainty, in which the additional assumption of separability is used to reduce the problem to one of assessing subjective probabilities for states and utilities for prizes, is formalized in the next section.

5.2   The fundamental theorem of subjective expected utility

The modeling framework to be used here is that of Anscombe and Aumann (1963), in which an act is a mapping from a finite set of states to objective probability distributions over a finite set of consequences, as illustrated in the example above. This is merely a state-dependent extension of the EU model. Let $\mathcal{S} := \{s_1, ..., s_M\}$ denote the set of states of the world and let $\mathcal{C} := \{c_1, ..., c_N\}$ denote the set of consequences. Let $\mathbf{A}$ and $\mathbf{B}$ denote events (subsets of $\mathcal{S}$) as well as their respective $M \times N$ indicator matrices with generic elements a_{mn} and b_{mn} ($a_{mn} = 1[0]$ for all n if event $\mathbf{A}$ is true[false] in state m), and let $\mathbf{E}_m$ denote the indicator matrix for state m. Let $\mathbf{F}$, $\mathbf{G}$, and $\mathbf{H}$, denote "SEU-acts" that are mappings from states of the world to objective lotteries over consequences, which are represented by $M \times N$ matrices. (For the remainder of this chapter these will just be referred to as "acts.") f_{mn} is the probability with which $\mathbf{F}$ yields consequence c_n in state s_m. The elements of an act are non-negative and sum to 1 in every row, i.e., in every state. The set of all acts is denoted by $\mathcal{F}$. Acts can be *objectively mixed* in a statewise manner by independent randomization: for any acts $\mathbf{F}$ and $\mathbf{H}$ and probability α, $\alpha\mathbf{F} + (1 - \alpha)\mathbf{H}$ is an act that yields consequence c_n with probability $\alpha f_{mn} + (1 - \alpha)h_{mn}$ in state s_m. Acts can be *subjectively mixed* in a statewise manner by conditioning them on events. Let $\mathbf{F_A}$ denote the matrix $\mathbf{F}$ with its elements zeroed-out in rows corresponding to states where $\mathbf{A}$ is not true, i.e., $[\mathbf{F_A}]_{mn} = f_{mn}a_{mn}$. Then for any acts $\mathbf{F}$ and $\mathbf{H}$ and event $\mathbf{A}$, $\mathbf{F_A} + \mathbf{H}_{1-\mathbf{A}}$ is an act that yields consequence c_n with probability f_{mn} if $a_{mn} = 1$ and with probability h_{mn} if $a_{mn} = 0$. (This can also be written as $\mathbf{AF} + (1 - \mathbf{A})\mathbf{H}$.) Let $\mathbf{V}$ denote an $M \times N$ matrix having the interpretation of a state-dependent utility function, whose mn^{th} element, v_{mn}, is the utility associated with consequence c_n in state s_m and let $E_\mathbf{V}[\mathbf{F}] := \sum_{m=1}^{M} \sum_{n=1}^{N} f_{mn}v_{mn}$ denote the utility it assigns to act $\mathbf{F}$.

Observe that in this setup the counterfactualism of the EU model has been raised to the Mth power, so to speak. Not only is it imagined that consequences of a personal nature can be received with any objective probabilities, but also that this can be done differently and arbitrarily in every state of the world. Whereas, in any concrete application, or even a non-crazy thought experiment, a consequence is literally the "con-sequence" of a feasible action by the decision-maker and a possible state of nature. You can't change the state of nature, holding the decision-maker's action constant, and get the same physical or psychological experience. Even more strangely, in the standard SEU model the utility of a consequence is forced to be independent of the state to which it is counterfactually assigned. A step back in the direction of realism can be obtained by allowing the utilities of consequences to be state-dependent. There is a large literature on the topic of state-dependent preferences and state-dependent utility, in which many modeling approaches have been explored in different analytic frameworks, e.g., by Arrow (1951c, 1974), Baccelli (2017, 2019, 2021), Drèze (1987), Drèze and Rustichini (2004), Karni (1985, 1993, 2003), Karni and Mongin (2000), Karni et al. (1983), Karni and Schmeidler (1993, 2016), Nau (1995a, 2001, 2003, 2011b), Rubin (1987), Schervish et al. (1990, 1991), Seidenfeld et al. (1990a), Wakker (1987) and Wakker and Zank (1999).

In what follows, both the state-dependent and state-independent versions of the Anscombe-Aumann model will be considered, where the latter version includes a 6th axiom to separate beliefs about events from tastes for consequences. Let the consequences be labeled so that c_1 is weakly most-preferred and c_N is weakly least-preferred and strictly less preferred than c_1 when they are received with certainty. These will be used to establish endpoints of the utility scale which can be assigned values of 1 and 0.[4] Then define dominance of acts as follows:

---

[4]This assumption follows Luce and Raiffa (1957) and Anscombe and Aumann (1963), and it is technically without loss of generality in the sense that the preference order can always be extended to a larger domain that includes two additional consequences which by construction are better and worse, respectively, than all the original consequences.

Definitions and symbols (strict and weak dominance): $\mathbf{F}$ *strictly dominates* $\mathbf{G}$, written as $\mathbf{F} \gg \mathbf{G}$, if, for every state m, $f_{mn} > g_{mn}$ for $n = 1, ..., N-1$, which means that $\mathbf{G}$ is obtained from $\mathbf{F}$ by shifting some positive probability mass to c_N from each of the other consequences *in every state*. $\mathbf{F}$ *weakly dominates* $\mathbf{G}$, written as $\mathbf{F} \geqq \mathbf{G}$, if, for every m, $f_{mn} \geq g_{mn}$ for $n = 1, ..., N-1$.

Rational beliefs and tastes can be jointly characterized in terms of consistency conditions that should be satisfied by an individual's preferences among acts. Suppose they satisfy (at least) the following axioms for any acts $\mathbf{F}$, $\mathbf{G}$, and $\mathbf{H}$:

Axiom SEU1^a (completeness): $\mathbf{F} \succsim \mathbf{G}$ or $\mathbf{G} \succsim \mathbf{F}$ or both

Axiom SEU2^a (continuity): the sets $\{\mathbf{G} : \mathbf{F} \succsim \mathbf{G}\}$ and $\{\mathbf{G} : \mathbf{G} \succsim \mathbf{F}\}$ are closed.

Axiom SEU3^a (transitivity): $\mathbf{F} \succsim \mathbf{G}$ and $\mathbf{G} \succsim \mathbf{H} \Longrightarrow \mathbf{F} \succsim \mathbf{H}$.

Axiom SEU4^a (independence): $\mathbf{F} \succsim \mathbf{G} \Longleftrightarrow$
$\alpha\mathbf{F} + (1-\alpha)\mathbf{H} \succsim \alpha\mathbf{G} + (1-\alpha)\mathbf{H}$ for all $\mathbf{H}$ and all $\alpha \in [0, 1]$.

Axiom SEU5^a (strict monotonicity): $\mathbf{F} \gg \mathbf{G} \Longrightarrow \mathbf{F} \succ \mathbf{G}$.

These are point-by-point identical to the "a" axioms of subjective probability and those of expected utility, and they deliver a corresponding representation theorem, although (as will be seen) it does not separate the decision-maker's probabilities from her utilities. Rather, it yields a state-dependent-utility representation of preferences. To achieve a state-independent representation, an additional axiom (accompanied by a few more definitions) is needed to separate the two hypothetical cognitive dimensions of preference.

Definition (constant acts): $\mathbf{F}$ is a *constant act* if it yields the same lottery over consequences in every state of the world.

Definition (conditional preference): $\mathbf{F}$ is weakly preferred to $\mathbf{G}$ *conditional on event* $\mathbf{A}$, denoted by $\mathbf{F} \succsim_{\mathbf{A}} \mathbf{G}$, if $\mathbf{F_A} + \mathbf{H_{1-A}} \succsim \mathbf{G_A} + \mathbf{H_{1-A}}$ for all $\mathbf{H}$.

Definition (null events): Event $\mathbf{A}$ is *null* if $\mathbf{F} \succsim_{\mathbf{A}} \mathbf{G}$ for all $\mathbf{F}$ and $\mathbf{G}$, otherwise it is *non-null*.

Axiom SEU6^a (separability of beliefs and tastes): if $\mathbf{F}$ and $\mathbf{G}$ are constant acts and $\mathbf{A}$ is non-null, then $\mathbf{F} \succsim \mathbf{G} \Longleftrightarrow \mathbf{F} \succsim_{\mathbf{A}} \mathbf{G}$.

Let a v-bet be defined as a difference between two acts, represented by an $M \times N$ matrix $\mathbf{X}$ whose elements sum to zero in each row. These are interpretable as amounts of probability added to or subtracted from the consequences experienced in each state, relative to whatever is the status quo lottery, when the bet is enforced. In other words, a v-bet is a state-dependent p-bet.

> **Definition (feasible v-bets):** $\mathbf{X}$ is a *feasible* v-bet if and only if $\mathbf{X} = \mathbf{F} - \mathbf{G}$ for some $\mathbf{F} \in \mathcal{F}$ and $\mathbf{G} \in \mathcal{F}$. That is, a feasible v-bet is an exchange of one SEU act for another. The set of all feasible v-bets is denoted by $\mathcal{X}^{\mathcal{F}}$.

> **Definition (preferable and acceptable v-bets):** $\mathbf{X}$ is a *preferable* v-bet if and only if $\mathbf{X} \in \mathcal{X}^{\mathcal{F}}$ and $\mathbf{F} + \alpha\mathbf{X} \succsim \mathbf{F}$ for every $\mathbf{F}$ and $\alpha > 0$ such that both $\mathbf{F} \in \mathcal{F}$ and $\mathbf{F} + \alpha\mathbf{X} \in \mathcal{F}$ and the set of such $\mathbf{F}$ is non-empty. That is, a preferable v-bet is a feasible change in state-dependent utilities assigned to consequences that is weakly desirable, regardless of the status quo. It is *acceptable* if it is preferable *and* the individual is publicly willing to accept the change at the discretion of an observer who chooses α. The set of acceptable v-bets is denoted by $\mathcal{X}$.

> **Definition (negative v-bets):** $\mathbf{X}$ is a *negative* v-bet if $\mathbf{X} \ll \mathbf{0}$, which means that it transfers positive amounts of probability to c_N from all the other consequences in every state.

The counterparts to axioms $\text{SEU1}^a - \text{SEU5}^a$ for elements of $\mathcal{X}$ are given by:

Theorem 5.1: Equivalence of SEU axioms for preferences and acceptable bets

The preference relation $\succsim$ on SEU-acts satisfies SEU1^a-SEU5^a if and only if the set $\mathcal{X}$ of acceptable v-bets satisfies:

Axiom SEU1^b (completeness): either $\mathbf{X} \in \mathcal{X}$ or $-\mathbf{X} \in \mathcal{X}$ or both.

Axiom SEU2^b (continuity): $\mathcal{X}$ is closed.

Axiom SEU3^b (convexity): $\mathbf{X} \in \mathcal{X}$ and $\mathbf{Y} \in \mathcal{X} \Longrightarrow$
$\alpha\mathbf{X} + (1 - \alpha)\mathbf{Y} \in \mathcal{X}$ for any $\alpha \in (0, 1)$.

Axiom SEU4^b (linear extrapolation): $\mathbf{X} \in \mathcal{X} \Longrightarrow \alpha\mathbf{X} \in \mathcal{X}$
for any $\alpha > 1$ such that $\alpha\mathbf{X} \in \mathcal{X}^{\mathcal{F}}$.

Axiom SEU5^b (no arbitrage): $\mathbf{X} \ll \mathbf{0} \Longrightarrow \mathbf{X} \notin \mathcal{X}$.

These are point-by-point identical to the "*b*" axioms of subjective probability and those of expected utility. In this setup, as in the EU setup, no-arbitrage means to not accept bets that transfer positive amounts of probability to the worst consequence from all of the other consequences. Here that requirement is applied within every state.

Lastly, the betting version of the separability axiom and its attendant notation are as follows:

> **Definition (conditionally acceptable bets):** $\mathbf{X}$ is an acceptable v-bet *conditional on event* $\mathbf{A}$ if $\mathbf{X_A} \in \mathcal{X}$, i.e., $\mathbf{AX} \in \mathcal{X}$. (This is $\mathbf{X}$ with its values zeroed-out in states where $\mathbf{A}$ is not true.)
>
> **(Re-)Definition (null events):** Event $\mathbf{A}$ is *null* if $\mathbf{X_A} \in \mathcal{X}$ for all $\mathbf{X} \in \mathcal{X}^{\mathcal{F}}$, otherwise it is *non-null*.
>
> **Definition (constant v-bets):** $\mathbf{X}$ is a *constant v-bet* if it is proportional to the difference between two constant acts, in which case it has constant columns.
>
> **Axiom SEU6^b (separability of beliefs and tastes):** if $\mathbf{X}$ is a constant v-bet and $\mathbf{A}$ is a non-null event, then $\mathbf{X} \in \mathcal{X} \Longleftrightarrow \mathbf{X_A} \in \mathcal{X}$.

From these we obtain both state-dependent[5] and state-independent versions of Anscombe and Aumann's (1963) theorem, which merges the fundamental theorems of subjective probability and expected utility (Theorems 3.2 and 4.2) by making lotteries and p-bets state-contingent:

Theorem 5.2: Fundamental theorem of subjective expected utility

(a) **State-dependent case:**   the preference relation $\succsim$ on SEU-acts satisfies SEU1^a–SEU5^a, and equivalently the set $\mathcal{X}$ of acceptable v-bets satisfies SEU1^b–SEU5^b, if and only if there exists a unique, non-negative state-dependent utility function $\mathbf{V}$ satisfying $1 = \sum_{m=1}^{M} v_{m1} \geq \sum_{m=1}^{M} v_{mn} \geq 0$ for consequences $n = 2, ..., N-1$, and $v_{mN} = 0$ for all states m, with the following properties:

(i) $v_{mn} = 0$ for all n if and only if state m is null,

(ii) $\mathbf{F} \succsim \mathbf{G}$ if and only if $E_{\mathbf{V}}(\mathbf{F}) \geq E_{\mathbf{V}}(\mathbf{G})$, and equivalently

$\mathbf{X} \in \mathcal{X}$ if and only if $E_{\mathbf{V}}(\mathbf{X}) \geq 0$.

---

[5] Arrow (1951c) describes a model for preferences among hypothesis-dependent mixtures of income functions, citing work by Chernoff and Rubin, which has the same structure as the state-dependent form of the Anscombe-Aumann model.

(b) State-independent case: $\succsim$ satisfies SEU1^a–SEU6^a, and equivalently $\mathcal{X}$ satisfies SEU1^b–SEU6^b, if and only if there exists a unique subjective probability distribution $\mathbf{p}$ and a unique utility function $\mathbf{u}$ satisfying $1 = u_1 \geq u_n \geq u_N = 0$ for $n = 2, ..., N-1$, with the following properties:

   (i) $p_m = 0$ if and only if state m is null,

    (ii) $\mathbf{F} \succsim \mathbf{G}$ if and only if $E_{\mathbf{p},\mathbf{u}}(\mathbf{F}) \geq E_{\mathbf{p},\mathbf{u}}(\mathbf{G})$, and equivalently

      $\mathbf{X} \in \mathcal{X}$ if and only if $E_{\mathbf{p},\mathbf{u}}(\mathbf{X}) \geq 0$.

Proof: Because the first 5 axioms are the same as those of expected utility, merely applied to matrices instead of vectors, it follows from Theorem 4.2 that they are satisfied if and only if there exists a matrix $\mathbf{V}$, having the interpretation of a state-dependent utility function in this case, such that $\mathbf{F} \succsim \mathbf{G}$ if and only $E_{\mathbf{V}}(\mathbf{F}) \geq E_{\mathbf{V}}(\mathbf{G})$, and equivalently $\mathbf{X} \in \mathcal{X}$ if and only if $E_{\mathbf{V}}(\mathbf{X}) \geq 0$. $\mathbf{V}$ satisfies $v_{mn} \geq v_{mN}$ for all m and all $n < N$ by virtue of the no-arbitrage requirement. $\mathbf{V}$ is unique up to positive scaling and also the row-wise addition of constants, because the elements of each row of an element of $\mathcal{F}$ must sum to 1 and correspondingly each row of an element of $\mathcal{X}$ must sum to zero. It follows from the definition of null states that rows of $\mathbf{V}$ corresponding to null states must have constant elements, which can be taken to be zeros without loss of generality. Rows of $\mathbf{V}$ corresponding to non-null states (of which there must be at least one, otherwise all v-bets are acceptable) must not have constant elements and can be normalized by the addition of constants so that $v_{mN} = 0$ for all m, and $\mathbf{V}$ can then be made unique by scaling it so that $\sum_{m=1}^{M} v_{m1} = 1$. To additionally prove part (b), consider the subset of v-bets that are constant. If $\mathbf{X}$ is constant, then $E_{\mathbf{V}}(\mathbf{X}) = \sum_{n=1}^{N} x_n (\sum_{m=1}^{M} v_{mn}) = \mathbf{x} \cdot \mathbf{u}$, where $\mathbf{x}$ is the p-bet vector that is a generic row of $\mathbf{X}$ (i.e., $x_n = x_{mn}$ for all m) and $\mathbf{u}$ is the vector that is the sum of the rows of $\mathbf{V}$. Therefore, the sum of the rows of $\mathbf{V}$ is a utility function that represents the set of acceptable constant v-bets and satisfies $1 = u_1 \geq u_n \geq u_N = 0$ for $n = 2, ..., N-1$. Next consider a constant v-bet that is conditioned on a non-null state s_m, which is represented by a matrix $\mathbf{X}$ whose elements are zero everywhere except in row m. By the separability axiom, the acceptability of constant v-bets is preserved when they are conditioned on non-null events, and in particular when they are conditioned on non-null states. It follows that the set of acceptable v-bets conditioned on

s_m must be represented by the same utility function $\mathbf{u}$, and it also must be represented by the mth row of $\mathbf{V}$ because that is all that comes into play when the sum $\sum_{m=1}^{M} \sum_{n=1}^{N} x_{mn} v_{mn}$ is evaluated. This means that $\mathbf{u}$ and the mth row of $\mathbf{V}$ are equivalent utility functions, and it follows that there exists a non-negative vector $\mathbf{p}$ whose m^{th} element is strictly positive if state s_m is non-null (and zero otherwise), such that $v_{mn} = p_m u_n$ and $\sum_{m=1}^{M} p_m = 1$. This decomposition of $\mathbf{V}$ yields $E_{\mathbf{V}}(\mathbf{F}) = \sum_{m=1}^{M} \sum_{n=1}^{N} p_m f_{mn} u_n = E_{\mathbf{p},\mathbf{u}}(\mathbf{F})$, whence $\mathbf{F} \succsim \mathbf{G}$, and equivalently $\mathbf{X} \in \mathcal{X}$, if and only if $E_{\mathbf{p},\mathbf{u}}(\mathbf{F}) \geq E_{\mathbf{p},\mathbf{u}}(\mathbf{G})$.

5.3 (In)separability of beliefs and tastes (state-dependent utility)

As noted in Sections 1.3 and 5.2, the unique separation of beliefs and tastes in terms of subjective probabilities and utilities comes at a high cost in counterfactualism and mental gymnastics. The decision-maker supposedly can imagine preferences among acts that are by definition impossible to imagine. In usage of the SEU model in applied decision analysis there is a waving of hands in the general direction of the axioms, and subjective probabilities and utilities are forcibly separated through the contemplation of hypothetical bets for money and visual aids such as "probability wheels" and analytic tools such as exponential utility functions and decision trees and statistical analysis and Monte Carlo simulation. These techniques require training and are often used in collaboration with colleagues and experts when the stakes are large, because they do not come naturally. The model is a tool of choice, not a representation of something that is latent in the mind.

The separability axioms require preferences among *objective lotteries* to be the same in all states of the world, which implies that preferences can be given a representation in which consequences have utilities that are state-independent, but those axioms do not imply that the utilities really *are* state-independent. The use of state-independent utility functions in subjective expected utility theory is a convention that is made possible by axioms SEU6[a] and SEU6[b] on an as-if basis, but it is not required. If you multiply the probabilities of states by some arbitrary set of scale factors and renormalize so they again sum to 1, then divide the utilities by the same factors on a state-by-state basis, you get the same representation of preferences but with state-dependent utilities and different probabilities for the states. Therefore it is impossible in general to determine whether a decision-maker really has state-independent utilities for consequences nor to infer unique values for her subjective probabilities. You can't *truly* separate beliefs from possibly-state-dependent tastes, even with the use of counterfactual acts, within the context of the state-independent or state-dependent SEU models (Nau (2001)).

A separation of beliefs from tastes can be achieved by adding more dimensions to the objects of choice, but this requires a much larger and stranger space of counterfactual acts among which the decision-maker is assumed to have preferences. One avenue is to add a "moral hazard" dimension to the model (Drèze (1987), Karni (2008), Baccelli (2017, 2019, 2021)), in which the decision-maker can take actions that influence the probabilities of events as well as the assignment of consequences to events. Another avenue is to introduce "state-prize lotteries" as objects of preference, which are lotteries over pairwise combinations of prizes and states, as opposed to state-dependent lotteries over prizes (Karni et al. (1983), Karni and Schmeidler (2016)). Thus, it is as if the decision-maker is able to directly compare the experience of getting consequence c_n in state s_m to the experience of getting consequence c_j in state s_k.

The very elegance of these solutions to the problem of uniquely identifying the decision-maker's beliefs is a warning flag. If it is really this hard to do it in principle, how do we expect real decision-makers to do it in practice except by deliberate use of analytical tools such as those of applied decision analysis? It will be shown in the chapter on state-preference theory that the inseparability of probabilities and utilities is not a problem for rational thinking about uncertainty in situations where money is present. All that matters there is betting behavior that can be observed or which at least is possible to imagine.

5.4 Incomplete preferences with state-dependent utilities

In subjective expected utility theory, as in the other branches of rational choice theory, the most problematic assumption is that of completeness of preferences, which is needed for uniqueness of numerical representations of states of mind. This issue was raised earlier in regard to subjective probability theory and its dual image, expected utility theory. Completeness is vastly *more* problematic in subjective expected utility theory, where acts are mostly counterfactual assignments of arbitrary consequences (states of the person) to states of nature. (Seidenfield et al. (1990a, 1995), Nau (2006b), Ok et al. (2012), Galaabaatar and Karni (2012, 2013)).

As was seen in the subjective probability and expected utility chapters, dropping of the completeness axiom creates the need for axioms of weak monotonicity and acceptability[6]:

Axiom SEU1^{a*} (weak monotonicity): $\mathbf{F} \geqq \mathbf{G} \Longrightarrow \mathbf{F} \succsim \mathbf{G}$

Axiom SEU1^{b*} (weak acceptability): $\mathbf{X} \geqq \mathbf{0} \Longrightarrow \mathbf{X} \in \mathcal{X}$,

The following special cases of acts that were introduced in Section 5.1 will be used again here:

- For $u \in [0, 1]$, $\mathbf{L}[u]$ denotes the constant act that yields the best consequence c_1 with probability u and the worst consequence c_N with probability $1 - u$ in every state. It has a utility of u by definition.

- For every $n \in \{1, ..., N\}$, $\mathbf{L}_n$ denotes the constant act that yields consequence c_n with certainty in every state. $\mathbf{L}_n \succsim \mathbf{L}[u]$ means that c_n has a utility greater than or equal to u.

- For every $m \in \{1, ..., M\}$, $\mathbf{I}[s_m]$ denotes an indicator act that yields the best consequence c_1 with probability 1 in state s_m and yields the worst consequence c_N with probability 1 in all other states. $\mathbf{I}[s_m] \succsim \mathbf{L}[p]$ means that state m has a probability greater than or equal to p.

- For every event $\mathbf{A}$, $\mathbf{I}[\mathbf{A}]$ denotes an indicator act that yields consequence c_1 with probability 1 in states where $\mathbf{A}$ is true and yields consequence c_N in all other states. $\mathbf{I}[\mathbf{A}] \succsim \mathbf{L}[p]$ means that event $\mathbf{A}$ has a probability greater than or equal to p.

The original definition of a null event cannot be applied when preferences are incomplete, so the following notion will be substituted:

Definition (not potentially null events): Event $\mathbf{A}$ is *not potentially null* if $\mathbf{I}[\mathbf{A}] \succsim \mathbf{L}[p]$ for some $p > 0$, i.e., $\mathbf{A}$ has a strictly positive lower probability.

A bet in which you get c_1 if a not-potentially-null event occurs and c_N otherwise has strictly positive expected utility.

---

[6]$\mathbf{F}$ weakly dominates $\mathbf{G}$, written as $\mathbf{F} \geqq \mathbf{G}$, if, for every i, $f_{ij} \geq g_{ij}$ for $j = 1, ..., n - 1$. The symbol "$\geqq$" rather than "$\geq$" is used here, as in the preceding chapter, because the inequality applies only to the first $n - 1$ consequences (the non-worst ones).

Definition (basis for a preference relation): A finite collection of asserted preferences $\{\mathbf{F}_k \succsim \mathbf{G}_k\}$ is a *basis* for a preference relation $\succsim$ under a given axiom system if all other preferences can be derived from it by application of those axioms.

In these terms we have:

Theorem 5.3: Fundamental theorem of subjective expected utility with incomplete preferences[7]

(a) State-dependent case: the preference relation $\succsim$ on SEU-acts satisfies SEU1^{a*} and SEU2^a–SEU5^a, and equivalently the set $\mathcal{X}$ of acceptable v-bets satisfies SEU1^{b*} and SEU2^b–SEU5^b, if and only if there exists a *nonempty closed convex set* $\mathcal{V}$ of state-dependent utility functions $\{\mathbf{V}\}$ which satisfy $1 = \sum_{m=1}^{M} v_{m1} \geq \sum_{m=1}^{M} v_{mn} \geq 0$ for $n = 2, ..., N-1$, and $v_{mN} = 0$ for all m, such that $\mathbf{F} \succsim \mathbf{G}$ if and only if $E_{\mathbf{V}}(\mathbf{F}) \geq E_{\mathbf{V}}(\mathbf{G})$, and equivalently $\mathbf{X} \in \mathcal{X}$ if and only if $E_{\mathbf{V}}(\mathbf{X}) \geq 0$.

(b) Basis for an incomplete state-dependent preference relation: If $\{\mathbf{F}_k \succsim \mathbf{G}_k\}$ is a basis for $\succsim$ under the assumptions of part (a), then $\mathcal{X}$ is the closed convex hull of the rays whose directions are $\{\mathbf{F}_k - \mathbf{G}_k\}$ for all k, together with $\{\mathbf{L}_1 - \mathbf{L}_n\}$ and $\{\mathbf{L}_n - \mathbf{L}_N\}$ for all $n = 2, ..., N-1$.

(c) State-independent case: in addition to the assumptions of part (a), $\succsim$ satisfies SEU6^a, and equivalently $\mathcal{X}$ satisfies SEU6^b, if and only if there exists a subjective probability distribution $\mathbf{p}$ and a state-independent utility function $\mathbf{u}$ satisfying $1 = u_1 \geq u_n \geq u_N = 0$ for $n = 2, ..., N-1$, such that $\mathbf{F} \succsim \mathbf{G}$ if and only if $E_{\mathbf{p},\mathbf{u}}(\mathbf{F}) \geq E_{\mathbf{p},\mathbf{u}}(\mathbf{G})$, and equivalently $\mathbf{X} \in \mathcal{X}$ if and only if $E_{\mathbf{p},\mathbf{u}}(\mathbf{X}) \geq 0$

(d) Lower and upper state-independent expected utilities for constant acts: The assumptions of part (c) imply that the lower and bounds on the expected utility of every *constant* act are achieved at elements of $\mathcal{V}$ in which utilities are state-independent.

Proof: Part (a) follows from the proof of Theorem 5.2(a): when completeness is relaxed. the unique state-dependent utility function merely gets replaced by a convex set thereof. Part (b) is true by construction. In part (c), "if" follows from the fact that the probability/utility pair $\{\mathbf{p}, \mathbf{q}\}$

---

[7]Part (a) of this theorem is a combination of Theorems 1 and 2 in Nau (2006b), and part (c) corresponds to Theorem 3 there.

determines a state-independent utility function that satisfies the same conditions as in part (a). For "only if" it suffices to show that a preference relation satisfying the given assumptions can be extended to assign a state-independent utility to any consequence between 1 and consequence N (those endpoints having utilities of 1 and 0 by definition). This extension can be performed for $n = 2, ..., N - 1$ (in any order) to obtain a state-independent utility function. The result is not necessarily unique because the order of consequences could matter. (See the appendix for details of how to carry out this extension.) For part (d), the same method of extension, namely, adding a preference which asserts that the greatest lower utility is also an upper utility, can be applied to an arbitrary *constant* lottery $\mathbf{F}$ prior to doing so for $\{\mathbf{L}_n\}$. This shows that the maximum or minimum expected utility of a constant lottery must be achieved by an agreeing probability/utility pair. Axiom SEU6^b applies to any constant lottery, not merely those that yield a single consequence with probability 1.

5.5 Representation by sets of probability/utility pairs

A preference relation can satisfy the axioms above and yet not be represented only by utilities that are state-independent. Some of the extreme points of the representing set $\mathcal{V}$ could be state-dependent, which would place looser bounds on subjective expected utilities of some acts than would be obtained by further restriction to a set of agreeing probability/utility pairs. When preferences are incomplete, axiom SEU6^a is insufficient to guarantee that they are represented by a set of probability/utility pairs or their convex hull.[8] A broader state-independence condition is needed, such as:

Axiom SEU6^{a*} (strong state-independence for preferences): If $\mathbf{F}$ and $\mathbf{G}$ are constant acts, and $\mathbf{F}'$ and $\mathbf{G}'$ are arbitrary acts satisfying $\mathbf{F}' \succsim \mathbf{G}'$, and event $\mathbf{A}$ is not potentially null, and

$$\alpha(\mathbf{F_A} + \mathbf{H_{1-A}}) + (1 - \alpha)\mathbf{F}' \succsim \alpha(\mathbf{G_A} + \mathbf{H_{1-A}}) + (1 - \alpha)\mathbf{G}' \quad (5.8)$$

---

[8] A numerical example is given in Nau (2006b).

for some $\mathbf{H}$ and some $\alpha \in (0, 1]$, then for any other not-potentially-null event $\mathbf{B}$ there exists $\beta \in (0, 1]$ such that

$$\beta(\mathbf{F_B} + \mathbf{H_{1-B}}) + (1 - \beta)\mathbf{F'} \succsim \beta(\mathbf{G_B} + \mathbf{H_{1-B}}) + (1 - \beta)\mathbf{G'}.$$

for all $\mathbf{H}$. In particular, this holds for $\beta = 1$ if $\alpha = 1$ (implying SEU6^a) and otherwise for all β such that $\beta/(1 - \beta) \leq (\alpha/(1 - \alpha))(p/q)$ where p is a positive lower probability for $\mathbf{A}$ and q is an upper probability for $\mathbf{B}$, and it holds for any positive β if $\mathbf{F'} \sim \mathbf{G'}$.

The corresponding axiom in betting terms is:

Axiom SEU6^{b*}(strong state-independence for acceptable bets): If $\mathbf{X'} \in \mathcal{X}$ and $\mathbf{X'} + \mathbf{X_A} \in \mathcal{X}$ where $\mathbf{X}$ is constant and $\mathbf{A}$ is not potentially null, then $\mathbf{X'} + \beta\mathbf{X_B} \in \mathcal{X}$ for any other not-potentially-null event $\mathbf{B}$ for some $\beta > 0$. In particular, this holds for $\beta = p/q$ where p is a positive lower probability for $\mathbf{A}$ and q is an upper probability for $\mathbf{B}$, and it holds for any positive β if both $\mathbf{X'}$ and $-\mathbf{X'}$ are acceptable.

In words, axiom SEU6^{b*} says that if $\mathbf{X'}$ is an acceptable bet and it remains acceptable when added to a constant bet $\mathbf{X}$ conditioned on some not-potentially-null event $\mathbf{A}$, then it also remains acceptable when added to a positive multiple of the same constant bet $\mathbf{X}$ conditioned on some other not-potentially-null event $\mathbf{B}$, where the multiplier is adjusted appropriately for bounds on the probabilities of the conditioning events. Another way to express this idea is the following. The usual state-independence condition requires that $\beta\mathbf{X_B}$ should be an acceptable bet if $\mathbf{X_A}$ is an acceptable bet. (In fact this holds for any positive β.) Axiom SEU6^{b*} requires that this condition must still hold when $\mathbf{X_A}$ and $\beta\mathbf{X_B}$ are both contaminated by the same "background noise" $\mathbf{X'}$, where the value of β is generally determined from bounds on the probabilities of $\mathbf{A}$ and $\mathbf{B}$ but is arbitrary if both $\mathbf{X'}$ and $-\mathbf{X'}$ are acceptable. The reason why this strengthening is needed is that in the absence of completeness the set of acceptable bets may not include a comprehensive set of bets that are constant, in which case the weaker separability condition SEU6^b may have limited applicability.

The main result (Nau 2006b), which generalizes this example, can now be stated:

Theorem 5.4: Representation of incomplete preferences with state-independent utilities

The preference relation $\succsim$ on SEU-acts satisfies SEU1*^{a} and SEU2^a— SEU5^a and SEU6*^{a} and equivalently the set $\mathcal{X}$ of acceptable v-bets satisfies SEU1$*^b$ and SEU2^b–SEU5^b and SEU6$*^b$ if and only if they are represented by a nonempty set $\mathcal{V}$ of s.d.e.u. functions that is the convex hull of a set of probability/utility pairs

Here is a sketch of the proof. (Details are in the appendix.) For an arbitrary bet **X**, compute the *lower* [upper] bound on its expected utility over all **V** in $\mathcal{V}$, and call it d. The strengthened axioms can be shown to imply that d can also be asserted to be an *upper* [lower] bound without producing incoherence. By Theorem 5.3(d), which holds under the weaker axioms SEU6^a and SEU6^b, the supporting set of state-dependent utilities must still contain at least one probability/utility pair, and its expected utility achieves the same extremal value d. Thus, a tight lower or upper bound on the expected utility of **X** is achieved at an element of $\mathcal{V}$ that is a probability/utility pair, QED.

If the decision-maker's preferences are revealed in the form of a finite set of direct assertions $\{\mathbf{F}_k \succsim \mathbf{G}_k\}$, the construction of $\mathcal{V}$ in Theorem 5.4 can be carried out as follows. First, form the convex polyhedron defined by the constraints $\{E_v(\mathbf{F}_k) \geq E_v(\mathbf{G}_k)\}$. Next, take the intersection of this polyhedron with the nonconvex surface consisting of all probability/utility pairs. (If the latter intersection is empty, the preferences do not satisfy all the axioms: they are incoherent.) Finally, take the convex hull of what remains: this is the set $\mathcal{V}$.

The model of Theorem 5.4 can be viewed as a representation of consensus among a group of SEU-maximizing agents, each of whom has complete preferences according to the standard model, which are represented by his or her own personal probabilities and utilities. It determines lower and upper bounds for subjective expected utilities upon which they can agree, i.e., with respect to which they would appear as one individual with incomplete preferences as seen by an observer. The beliefs and tastes of each agent are separable, but those of the group are not: the set $\mathcal{V}$ generally cannot be described by a set of probability distributions and a set of utility functions that can be paired up arbitrarily. A model of the latter kind would describe the viewpoint of a single agent with imprecise beliefs and, separately and independently, imprecise tastes. Such an agent might carry out decision analysis in the usual fashion by using different kinds of elicitation devices for probabilities and utilities, thus allowing the assessed values of each to have their own lower and upper bounds which would then be multiplied out to obtain bounds on subjective expected utilities.

Seidenfeld et al. (1995) give an alternative (and more elegant) axiomization for sets of probability/utility pairs. Galaabaatar and Karni (2013) axiomatize all four ways in which incomplete beliefs and tastes can be separated from each other to varying extents: (i) multiple subjective probability distributions paired with a unique utility function, (ii) multiple utility functions paired with a unique subjective probability distribution, (iii) multiple utility functions each paired with multiple subjective probability distributions, and (iv) multiple subjective probability distributions and multiple utility functions that are paired in all possible combinations. Model (iii) is equivalent to the model of Theorem 4 above (each can be viewed as a special case of the other), and Galaabaatar and Karni derive it elegantly from an application of a dominance axiom that is a variation on the sure-thing-principle. Model (iv) achieves a complete separation of probabilities and utilities when both are imprecise. It describes a situation in which decision analysis is performed in the usual fashion via separate kinds of mental exercises for eliciting probabilities and utilities, with a relaxation of the requirement that both exercises should converge to unique parameter estimates.

Chapter 6

State-preference theory, uncertainty aversion, and risk-neutral probabilities

6.1   The state-preference framework for choice under uncertainty

State-preference theory (Arrow (1953), Debreu (1959), Hirshleifer (1965, 1966), Radner (1968)) provides an alternative model of choice under uncertainty whose primitives are more concrete than those of subjective expected utility, and it lends itself easily to relaxations of conventional restrictions on preferences that have been challenged in recent decades. It is merely an application of neoclassical consumer theory (indifference-curve analysis) to situations in which consumption is uncertain and perhaps also distributed over time. Objects of choice are mappings from states of the world (which could be indexed by time) to amounts of money (which could be bundled with other goods). The payments or goods received by the decision-maker in different events at different times are not "consequences" in the sense that the term is used in SEU theory. They need not be complete descriptions of the state of the person, nor even the entire contents of her portfolio of investments: they are merely the publicly visible *accompaniments* of whatever else the decision-maker may experience under the given conditions. There is no intrinsic counterfactualism in the objects of choice. It might be the case that some scenarios within the mathematical scope of a model are not actually available—for example, choices that violate a

DOI: 10.1201/9781003527145-6

budget constraint—but they are not illogical to contemplate, such as enjoying life in a state where you are dead.

Insofar as state-preference theory gives a distinguished role to money, it addresses the issues raised in Chapter 1 about the importance of having a yardstick for quantifying beliefs and tastes on a precise numerical scale and doing it in a way that is interpersonally observable. As such it is the natural framework for modeling what goes on in markets, and this is the setting in which it has been most often applied. However, the theory can also be adapted to the modeling of personal decisions and multi-player games in which money is not necessarily the bottom line but rather a medium for making side bets through which the subjective parameters of the situation can be publicly revealed in a credible way (Nau and McCardle (1990, 1991), Nau (1995b, 2015)). This extension will be discussed in the context of game theory in Chapter 8. In all of these settings, rationality and common knowledge thereof can be defined in terms of no-arbitrage requirements.

Another benefit of working within the state-preference framework is that it does not attach intrinsic importance to the independence axiom, so it easily accommodates non-expected-utility preferences such as those that are evoked in Ellsberg's paradox. The independence axiom forces a utility function to be additive across its arguments, which is not normatively or descriptively compelling in the context of consumer theory insofar as different goods may serve as complements for each other, and some are more complementary than others. Why shouldn't the same sorts of effects arise in some choices under uncertainty? In situations where probabilities are not well-defined or the decision-maker justifiably worries that the game may be rigged by others with better information, it is perfectly reasonable to suppose that amounts of money received in complementary events could be treated as complementary goods ex ante, and more for some events than others (never mind the traditional argument that ex post you get only one or the other). The even-more-restrictive assumption of state-independent utility is not compelling in this framework either (the decision-maker need not exhibit the same degree of diminishing marginal utility toward everything), and little modeling power is lost by dropping it.

A seeming limitation of this approach is that it cannot be applied to situations where monetary exchange and contracting are impossible to imagine, but this just underscores the fact that without some reference to money—the real stuff—it is hard to argue that what is going on in the decision-maker's head can be represented by real numbers, let alone real numbers that others can know, and let alone with many digits of precision. The state-preference framework is the best case scenario for attaching numbers to properties of mind, so its limitations carry lessons for theories that try to quantify the psychology of the decision-maker in more abstract settings where numbers are not part of ordinary discourse. Moreover, it generalizes to settings in which the events of interest have non-monetary consequences, as noted above. The important thing is that money should be available for small side bets on events that follow decisions.

This chapter describes the basic tools and principles of state-preference theory that apply to monetary decisions, especially with regard to the modeling of aversion to risk and uncertainty.[1] It will emphasize the foundational importance of *risk-neutral probabilities* (state prices or betting rates) as parameters of the most general models. Derivatives of risk-neutral probabilities (rather than derivatives of utility functions) will be shown to be the basic carrier of information about attitudes toward uncertainty. The application of these tools to non-expected-utility preferences and to personal decisions and noncooperative games with extra-monetary consequences will be considered in later chapters.

6.2   Examples of utility functions for uncertainty-averse agents

To illustrate some of the issues that will be discussed, consider the situation diagrammed in Figure 6.1, which shows indifference curves of four individuals with respect to distributions of wealth over two states of the world. The curves are aligned so that the origin of coordinates is the status quo for everyone, so a point (x_1, x_2) in the plane can be interpreted as a change in state-dependent wealth for a given individual. The true (but possibly unobservable) status quo wealth distribution of that individual will be denoted by (f_1, f_2). Coincidentally, everyone has the same marginal rate of substitution for wealth between states 1 and 2, namely that \$3 in state 1 is equivalent to \$2 in state 2, perhaps because they have already engaged in betting or trade with each other, or perhaps because they are in contact with an external market in which the price of a state-1 Arrow security is 0.4.

ALICE claims to be a state-independent expected utility maximizer whose utility function in (x_1, x_2) coordinates is

$$U_A(x_1, x_2) = -p \exp(-r(x_1 + f_1)) - (1 - p) \exp(-r(x_2 + f_2)), \qquad (6.1)$$

---

[1]It has become conventional to use the terms "risk" and "risk aversion" and "risk premium" only in settings where probabilities are objective and to use "uncertainty" and "uncertainty aversion" and "uncertainty premium" in more general settings where probabilities could be either objective or subjective or imprecise or inseparable from marginal utilities for money or otherwise undefined. This chapter and those that follow deal only with settings of the latter kind, but in some places the "risk" variants of the terms will be used in connecting with classic literature

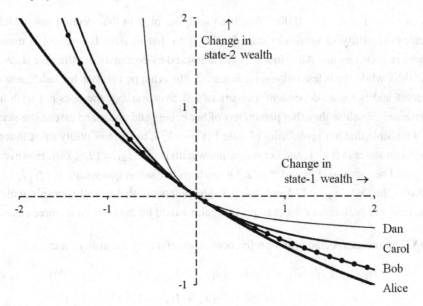

FIGURE 6.1: Indifference curves of 4 individuals for wealth in two states

with $p = 1/3$, $r = 0.2$, and $(f_1, f_2) = (2.8, 4.3)$. Thus, Alice exhibits constant absolute aversion to risk with an Arrow-Pratt risk aversion measure of 0.2, and it is as if she assigns probability 1/3 to state 1 and her status quo wealth is 2.8 in state 1 and 4.3 in state 2. Those probabilities and wealth amounts are not uniquely revealed by her preferences, however. We must rely on her self-report. Perhaps she has constructed her preferences through a formal process of decision analysis.

BOB has a "Cobb-Douglas" utility function

$$U_B(x_1, x_2) = (x_1 + f_1)^p (x_2 + f_2)^{1-p} \qquad (6.2)$$

with $p = 0.355$ and $(f_1, f_2) = (2.5, 3.0)$. This is equivalent to having expected-utility preferences with a logarithmic utility function and a probability of 0.355 for state 1 and a status quo wealth of 2.5 in state 1 and 3.0 in state 2. However, Bob does not subscribe to expected utility theory and has no opinion concerning the probabilities of the states and does not know what all his other assets are worth at the moment; he merely feels that this function adequately describes his preferences for deviations from the status quo.

CAROL has state-dependent expected utility preferences represented by the utility function

$$U_C(x_1, x_2) = -p\exp(-r_1(x_1 + f_1)) - (1 - p)\beta\exp(-r_2(x_2 + f_2)) \qquad (6.3)$$

with $r_1 = 1$ and $r_2 = 0.05$. (Note the multiplier of β in the second term, which scales the utility of wealth in state 2 relative to that in state 1.)   Carol is much more risk-averse than Alice in state 1 (as measured by concavity of utility for state-1 wealth), while she is less risk-averse in state 2. She is happy to reveal her indifference curves and her state-dependent measure of risk aversion, but she does not wish to separately disclose the other parameters of her beliefs and tastes and assets: she says it's possible that her probability of state 1 is $p = 1/3$, her rate of utility substitution between states is $\beta = 1$, and her status quo wealth is $(f_1, f_2) = (2.8, 2.0)$. However, it could be that $p = 1/2$ and $\beta = 2$. Or, perhaps her status quo wealth is $(f_1, f_2) = (3.023, 2.0)$ while $p = 1/3$ and $\beta = 0.8$. She points out that it really doesn't matter because her preferences for changes in wealth would be the same in all three cases.

DAN has non-expected utility preferences represented by the utility function

$$
\begin{aligned}
U_D(x_1, x_2) = \ & -\exp(-r(p(x_1 + f_1) + (1 - p)(x_2 + f_2))) \qquad (6.4) \\
& -\exp(-r(q(x_1 + f_1) + (1 - q)(x_2 + f_2)))
\end{aligned}
$$

with $p = 0.05$, $q = 0.95$, $r = 2$, and $(f_1, f_2) = (1.25, 1.0)$, which is not additively separable across x_1 and x_2, in violation of the independence axiom.[2]

Suppose that we wish to measure and compare the degrees of local risk aversion of these four individuals in order to determine their relative propensity to gamble, purchase risky assets, or take other decisions under uncertainty, starting from their status quo wealth positions rather than hypothetical riskless positions. The usual definitions of risk aversion and risk premia do not apply to anyone other than Alice, because their probabilities and current wealth positions are ill-defined or unobservable and their risk aversion measures are state-dependent or non-additive. Yet all four are plainly uncertainty-averse in the sense of having preferences that are payoff-convex[3], and they can even be strictly ordered in terms of their local aversion to uncertainty: Alice<Bob<Carol<Dan. The indifference curves passing through their status quo wealth positions are the frontiers of their respective sets of acceptable gambles—which are observable—and following Yaari (1969), an individual

---

[2]This is an example of a "smooth" model of non-expected utility preferences (Nau 2001, 2006a; Klibanoff et al. 2005), and it can be interpreted as follows. It is as if Dan has linear utility for money but he is uncertain about the probabilities of the states and is averse to this second-order uncertainty. In particular, he thinks it is equally likely that the probability of state 1 is 0.05 or that it is 0.95, and his aversion to the corresponding uncertainty about the expected value of his wealth is described by an exponential utility function whose risk aversion parameter is equal to 2. In settings with more than 2 states, models with a similar structure can be used to rationalize Ellsberg's paradox as will be shown in the next chapter.

[3]Payoff-convexity means that the set of payoff distributions that are preferred to any given status quo distribution is a convex set.

with a strictly smaller set of acceptable gambles is "more risk-averse." To compare local risk preferences in these terms, it is not necessary to know anyone's probabilities, utility functions for money, or prior wealth: it suffices to consider the slope and curvature of their respective indifference curves. Yaari explored the two-dimensional case, showing that the degree of local risk aversion could be measured by the second derivative of the parameterized indifference curve. However, the latter measure does not generalize directly to higher dimensions. One that does will be introduced below.

6.3 The fundamental theorem of state-preference theory

Let M be the number of states and let the set $\mathcal{F}$ of acts be a set of M-vectors denoted $\mathbf{f}$, $\mathbf{g}$, etc., whose elements are state-dependent amounts of wealth that lie in some interval $[w_{\min}, w_{\max}]$. The axioms of state-preference theory are merely the axioms of consumer theory applied to such objects. They can also be viewed as a weakening of the axioms of subjective probability in which the independence axiom is replaced by a convexity axiom that allows indifference curves to be nonlinear.

Axiom SPT1 (completeness): $\mathbf{f} \succsim \mathbf{g}$ or $\mathbf{g} \succsim \mathbf{f}$ or both.

Axiom SPT2 (continuity): the sets $\{\mathbf{g} : \mathbf{f} \succsim \mathbf{g}\}$ and $\{\mathbf{g} : \mathbf{g} \succsim \mathbf{f}\}$ are closed.

Axiom SPT3 (transitivity): $\mathbf{f} \succsim \mathbf{g}$ and $\mathbf{g} \succsim \mathbf{h} \Longrightarrow \mathbf{f} \succsim \mathbf{h}$.

Axiom SPT4 (strong monotonicity): $\mathbf{f} \geq \mathbf{g}$ and $\mathbf{f} \neq \mathbf{g} \Longrightarrow \mathbf{f} \succ \mathbf{g}$.

Axiom SPT5 (strict payoff-convexity): $\mathbf{f} \sim \mathbf{g}$ and $\mathbf{f} \neq \mathbf{g} \Longrightarrow$
$\alpha \mathbf{f} + (1 - \alpha)\mathbf{g} \succ \mathbf{f}$ for all $\alpha \in (0, 1)$.

The following result is standard (e.g., Debreu (1954, 1964)), and in the present setting it is appropriate to consider it as the fundamental theorem of state-preference theory:

Theorem 6.1: Fundamental theorem of utility for monetary acts

$\succsim$ satisfies SPT1^a-SPT5^a if and only if there exists a continuous utility function U that is strictly monotone and strictly quasiconcave such that $\mathbf{f} \succsim \mathbf{g}$ if and only if $U(\mathbf{f}) \geq U(\mathbf{g})$

In particular, for any $\mathbf{f}$ there exists an amount of money $C(\mathbf{f}) \in [w_{\min}, w_{\max}]$ such that $\mathbf{f} \sim (C(\mathbf{f}), ..., C(\mathbf{f}))$, which is called its *certainty equivalent*, and the utility function can be constructed as $U(\mathbf{f}) = \varphi(C(\mathbf{f}))$ where φ is any increasing, continuous function. The strict payoff-convexity assumption implies that the decision-maker is strictly risk-averse in the sense that she prefers to diversify among heterogeneous assets that are equally desirable when considered separately. U is (only) an *ordinal* utility function because any strictly increasing transformation of it represents the same preferences.

In light of its quasiconcavity, the utility function is differentiable almost everywhere, and without too much more loss of generality it may be assumed to be twice differentiable, which will be done for the remainder of this chapter. That is, the decision-maker's preferences will be assumed to be "smooth," as is common in consumer theory. Several arguments justify this assumption. First, if the decision is one that takes place outside the laboratory or casino, and it involves events that are of intrinsic importance to the decision-maker, then she may have significant and possibly unobservable prior stakes in events, i.e., background uncertainty. If this is the case, then there is a very low probability that her status quo position will be astride or even close to a kink in an indifference curve. Second, if indifference curves do have kinks in some locations, they could be rounded off without much of an effect on the modeling of decisions involving gains and losses that are substantial in magnitude. Third, and most importantly, to the extent that the decision-maker's preferences are not adequately representable by a smooth utility function, a more likely explanation is that her preferences are incomplete.

Stronger assumptions can be (and usually are) imposed on preferences to restrict the form of the utility function. Let $\mathbf{f}_{-m}a$ denote the act that yields the amount a in state m and agrees with $\mathbf{f}$ in all other states, and consider:

Axiom SPT6 (coordinate independence): $\mathbf{f}_{-m}a \succsim \mathbf{g}_{-m}a \iff$
$\mathbf{f}_{-m}b \succsim \mathbf{g}_{-m}b$ for all $\mathbf{f}$, $\mathbf{g}$, a, and b.

In other words, if the payoffs of two acts agree in some state, it doesn't matter *how* they agree there. This axiom together with the others yields a *cardinal* utility function:

Corollary 6.1 (additive utility):

If $M \geq 3$,[4] then $\succsim$ satisfies SPT6 in addition to SPT1-SPT5 if and only if the utility function has the form:

$$U(\mathbf{f}) = \sum_{m=1}^{M} u_m(f_m), \qquad (6.5)$$

---

[4]If $n = 2$, the hexagon condition illustrated by Figure 2.2 must also be satisfied to obtain this result, or else a stronger form of the independence axiom known as generalized triple cancellation must be used.

where the evaluation functions $\{u_m\}$ are strictly convex and unique up to positive affine transformations with a common scale factor (Wakker (1989)).

This can be viewed as a model of *state-dependent expected utility without separation of probabilities and utilities*, because it is possible to decompose an evaluation function as $u_m(f_m) = p_m v_m(f_m)$, where p_m is a probability assigned to state m and v_m is a utility function for money received in state m, but insofar as v_m can have an arbitrary scale factor, the probabilities are not unique. The following additional assumption (Köbberling and Wakker (2003)) makes a unique decomposition possible:

Axiom SPT7 (tradeoff consistency): $\mathbf{f}_{-m}a \sim \mathbf{g}_{-m}b$ and
$\mathbf{f}_{-m}c \sim \mathbf{g}_{-m}d$ and $\mathbf{f}^*_{-k}a \sim \mathbf{g}^*_{-k}b \Longrightarrow \mathbf{f}^*_{-k}c \sim \mathbf{g}^*_{-k}d$.

To see how this works, consider two acts $\mathbf{f}$ and $\mathbf{g}$, and suppose that substituting a for f_m and substituting b for g_m yields an indifference, and the same is true when c and d are substituted for f_m and g_m respectively. This means that the tradeoff between a and b is equivalent to the tradeoff between c and d when they are to be received in state m in the context of a comparison of $\mathbf{f}$ and $\mathbf{g}$. The requirement is that this equivalence of tradeoffs should be the same in any state in the context of any comparison of acts. It implies that the evaluation functions in all states are the same up to positive linear scaling, and the (normalized) scale factors can be given the interpretation of subjective probabilities:

Corollary 6.2 (state-independent additive utility):

$\succsim$ satisfies SPT7 in addition to SPT1-SPT6 if and only if the utility function has the form:

$$U(\mathbf{f}) = \sum_{m=1}^{M} p_m v(f_m),\qquad\qquad (6.6)$$

where $\mathbf{p}$ is a unique strictly positive probability distribution and is v is strictly convex.

$\mathbf{p}$ behaves just like a measure of pure belief under these assumptions, so it is conventionally interpreted as one, and a similar convention is followed in models of choice under uncertainty that use the Savage or Anscombe-Aumann frameworks rather than the state-preference framework.

Alas, even under these strong conditions it is generally impossible to be sure that $\mathbf{p}$ correctly measures the decision-maker's beliefs, because it still could be distorted by state-dependent utility scale factors arising from intrinsic state-dependence of value for money and/or unobservable background uncertainty. For example, suppose that the Bernoulli utility function has the exponential form $v(x) = -\exp(-\frac{1}{2}rx)$, in which r is the Arrow-Pratt measure of risk aversion, and the decision-maker's status quo wealth is the act $\mathbf{f}$, and let $\alpha = \sum_{m=1}^{M} \exp(-\frac{1}{2}rf_m))$. Then the utility function according to which she evaluates a bet with observable payoff vector $\mathbf{x}$ is:

$$U(\mathbf{f} + \mathbf{x}) \quad = \quad -\sum_{m=1}^{M} p_m v(f_m + x_m) \tag{6.7}$$

$$= \quad -\sum_{m=1}^{M} p_m \exp(-\frac{1}{2}rx_m) v(f_m) \tag{6.8}$$

$$= \quad -\sum_{m=1}^{M} p_m^* v^*(f_m) \tag{6.9}$$

where $p_m^* := p_m \exp(-\frac{1}{2}rf_m))/\alpha$ is a distorted probability distribution and $v^*(x) := \alpha v(x)$ is equivalent to the original utility function. Thus, her revealed probability distribution is $\mathbf{p}^*$ when evaluating bets whose payoffs are received in the context of background uncertainty $\mathbf{f}$.

6.4   Risk-neutral probabilities and their matrix of derivatives

The utility function U determined by the first five axioms above is unique only up to strictly increasing transformations, which means that its derivatives are not unique either. However, the gradient of U at $\mathbf{f}$, which will be denoted here by $DU(\mathbf{f})$,[5] is uniquely determined upon normalization so that its elements sum to 1. Those elements are strictly positive by virtue of the strict monotonicity assumption, hence the normalized gradient is a probability distribution on states, in particular:

> **Definition of risk-neutral probabilities:** The normalized gradient of U is the decision-maker's local *risk-neutral probability distribution*. The risk-neutral probability of state j at act $\mathbf{f}$ is:

$$\pi_j(\mathbf{f}) := \frac{[DU(\mathbf{f})]_j}{\sum_{m=1}^{M}[DU(\mathbf{f})]_m}, \tag{6.10}$$

---

[5]In other words, $DU(\mathbf{f})$ is the vector of first partial derivatives of the scalar function U, whose i^{th} element is $[DU(\mathbf{f})]_i := \partial U(\mathbf{f})/\partial f_i$. Similarly $D^2U(\mathbf{f})$ is the matrix of second partial derivatives of U, known as the "Hessian matrix," whose ij^{th} element is $[D^2U(\mathbf{f})]_{ij} := \partial^2 U(\mathbf{f})/\partial f_i \partial f_j$,

In vector notation, $\pi(\mathbf{f}) = DU(\mathbf{f})/(\mathbf{1} \cdot DU(\mathbf{f}))$ where $\mathbf{1}$ denotes a vector of 1's.

The local risk-neutral distribution represents the first-order properties of the decision-maker's local preferences in the sense that she behaves as if she is risk-neutral with $\pi(\mathbf{f})$ as her subjective probability distribution when evaluating small bets. More precisely, if $\mathbf{x}$ is the payoff vector of a bet whose risk-neutral expected value, $\pi(\mathbf{f}) \cdot \mathbf{x}$, is strictly positive, then there exists some $\epsilon > 0$ such that $U(\mathbf{f} + \delta\mathbf{x}) > U(\mathbf{f})$ for all δ between 0 and ϵ. Hence, the decision-maker is willing to accept[6] any sufficiently small bet whose risk-neutral expected value is positive, and on the margin she will accept a bet whose risk-neutral expected value is zero.

In the special case where the decision-maker has *additive utility* preferences, which requires the additional axiom SPT6 above, her risk-neutral probability for state j is:

$$\text{additive utility version:} \quad \pi_j(\mathbf{f}) = \frac{u_j'(f_j)}{\sum_{m=1}^{M} u_m'(f_m)}. \tag{6.11}$$

If the additive utility function is interpreted to be product of a subjective probability distribution $\mathbf{p}$ and a *state-dependent expected utility* function $v_j(f_j)$, then $\pi_j(\mathbf{f})$ is expressible as:

$$\text{state-dependent utility version:} \quad \pi_j(\mathbf{f}) = \frac{p_j v_j'(f_j)}{\sum_{m=1}^{M} p_m v_m'(f_j)}, \tag{6.12}$$

i.e., it is the renormalized product of the decision-maker's subjective probability of state j and her marginal utility for money in state j. However, this representation is not unique insofar as the probabilities are confounded with state-dependent utility scale factors. If the decision-maker has *state-independent expected utility* preferences ($v_j(f) \equiv v(f)$), which requires axiom SPT7 above, this simplifies to:

$$\text{state-independent utility version:} \quad \pi_j(\mathbf{f}) = \frac{p_j v'(f_j)}{\sum_{m=1}^{M} p_m v'(f_m)}. \tag{6.13}$$

A problem for the uniqueness of the representation in the latter two versions is that the probabilities and utilities are confounded with potentially-unobservable background uncertainty.

Regardless of whether the representation is additive or state-independent, second-order properties of the decision-maker's local preferences are represented

---

[6]Here, as elsewhere, it will be assumed that every preferred bet is publicly acceptable at the discretion of an observer.

by the *matrix of derivatives of risk-neutral probabilities*, which will be denoted as $D\pi(\mathbf{f})$:

$$[D\pi(\mathbf{f})]_{jk} := \partial\pi_j(\mathbf{f})/\partial f_k. \qquad (6.14)$$

Its elements are measured in units of 1/money, e.g., 1/\$, and they are uniquely determined by preferences for a given choice of currency.[7] This matrix is an object of fundamental importance in state-preference theory, yet it is hardly mentioned in the literature. By definition it determines the marginal change in the decision-maker's risk-neutral probability distribution that results from a change $d\mathbf{x}$ in state-dependent wealth:

$$D\pi(\mathbf{f})d\mathbf{x} = d\pi. \qquad (6.15)$$

The matrix of derivatives of risk-neutral probabilities measures the decision-maker's local aversion to uncertainty, and its structure determines whether she is ambiguity averse and more generally whether her aversion to uncertainty is source-dependent, which will be shown in the next chapter.

$D\pi(\mathbf{f})$ can be regarded as a behavioral primitive insofar as it is directly observable—its elements can be estimated from small-but-finite bets the decision-maker is willing to accept in the vicinity of the status quo—but it also can be expressed in terms of the first and second derivatives of $U(\mathbf{f})$ as follows:

$$[D\pi(\mathbf{f})]_{jk} = \frac{[D^2U(\mathbf{f})]_{jk} - \pi_j(\mathbf{f})\sum_{m=1}^{M}[D^2U(\mathbf{f})]_{mk}}{\sum_{m=1}^{M}[DU(\mathbf{f})]_m}. \qquad (6.16)$$

The matrix of second derivatives of the utility function (the so-called "Hessian matrix"), denoted by $D^2U(\mathbf{f})$, has rank M for a decision-maker who is strictly risk-averse, and $D\pi(\mathbf{f})$ has rank $M - 1$ because its columns sum to zero. The latter is uniquely determined by preferences while the former is not, because U is unique only up to an increasing transformation.

---

[7] A subtlety arises here when the states of the world refer to different possible exchange rates between two currencies, as noted earlier. In this case, changing the currency in terms of which risk-neutral probabilities are elicited does not merely rescale the elements of $D\pi(\mathbf{x})$ by the same conversion factor, while leaving $\pi(\mathbf{x})$ unaffected. Rather, it changes them both in a state-dependent way. This is not a problem. It does not change the results of any analysis of the decision-maker's preferences among assets, whatever their denomination. In fact, this is one more reason why beliefs ought to be measured in terms of risk neutral probabilities, without any pretense of divorcing them from a financial context. As noted by Seidenfeld et al. (1990), it is impossible to apply de Finetti's definition of subjective probability in this situation, even under an assumption that the decision-maker is risk-neutral.

In the example of Figure 6.1, the $D\pi$ matrices of the four actors at the status quo are:

TABLE 6.1: $D\pi$ matrices for 2 states

Alice		Bob	
−0.048	0.048	−0.096	0.080
0.048	−0.048	0.096	−0.080

Carol		Dan	
−0.240	0.012	−0.385	0.385
0.240	− 0.012	0.385	−0.385

The absolute values of the largest elements in these matrices increase from left to right,[8] reflecting the ordering of the agents in terms of their degree of local risk aversion. The elements of $D\pi$ are always negative on the main diagonal in any number of dimensions. $D\pi$ happens to be symmetric for Alice and Dan in this example, and its second row is −1 times its first row for Bob and Carol.

The following table shows the formulas from which these values were calculated. The utility functions of Alice, Bob, and Carol are all of the form $U(\mathbf{f} + \mathbf{x}) = u_1(x_1 + f_1) + u_2(x_2 + f_2)$, which is additively separable between states, consistent with expected utility maximization, and they are determined by the univariate functions $u_1(.)$ and $u_2(.)$ in the first rows of their respective tables. The variable x_j denotes an observable receipt of money in state j and f_j denotes possibly-unobserved background uncertainty. x_1^* stands for $x_1 + f_1$ and x_2^* stands for $x_2 + f_2$. When u_1 and u_2 are written without arguments for these agents, they stand for $u_1(x_1^*)$ and $u_2(x_2^*)$. Dan's utility function is not additively separable in this way, indicative of non-expected-utility preferences. In his formulas u_1 and u_2 stand for the two bivariate functions $u_1(x_1^*, x_2^*)$ and $u_2(x_1^*, x_2^*)$. With these conventions, $U(\mathbf{f} + \mathbf{x}) = u_1 + u_2$ is total utility in all cases. Only the π_1 formula is shown because $\pi_2 = 1 - \pi_1$.

The fact that Dan has non-expected-utility preferences is not observable from the structure of $D\pi(\mathbf{x})$ because there are only 2 states and hence only one source of uncertainty. In higher dimensions, non-expected-utility preferences (e.g., ambiguity aversion) would be revealed by indifference curves that do not bend in the same way with respect to different sources of risk, as will be discussed in the next chapter.

---

[8] The values in the right column of Carol's matrix are not larger than Bob's, but this is compensated by the much larger values in the left column of her matrix, which result in calculations of higher risk premia for neutral assets as will be shown in the next section.

TABLE 6.2: Parameters of utility functions and risk-neutral probabilities and their derivatives for the four individuals

Alice	State-independent SEU with exponential utility
$u_1 =$	$-p\exp(-rx_1^*)$
$u_2 =$	$-(1-p)\exp(-rx_2^*)$
$\pi_1 =$	$u_1/(u_1+u_2)$
$[D\pi]_{jj} =$	$-ru_1u_2/(u_1+u_2)^2 \quad j=1,2$
$[D\pi]_{jk} =$	$-[D\pi]_{jj} \quad k\neq j$

Bob	State-independent SEU with logarithmic utility
$u_1 =$	$p\ln(x_1^*)$
$u_2 =$	$(1-p)\ln(x_2^*)$
$\pi_1 =$	$px_2^*/(px_2^*+(1-p)x_1^*)$
$[D\pi]_{11} =$	$-p(1-p)x_2^*/(px_2^*+(1-p)x_1^*)^2$
$[D\pi]_{22} =$	$-p(1-p)x_1^*/(px_2^*+(1-p)x_1^*)^2$
$[D\pi]_{21} =$	$-[D\pi]_{11}$
$[D\pi]_{12} =$	$-[D\pi]_{22}$

Carol	State-*dependent* SEU with exponential utility
$u_1 =$	$-p\exp(-r_1x_1^*)$
$u_2 =$	$-(1-p)\beta\exp(-r_2x_2^*)$
$\pi_1 =$	$r_1u_1/(r_1u_1+r_2u_2)$
$[D\pi]_{11} =$	$r_1^2r_2u_1u_2/(r_1u_1+r_2u_2)^2$
$[D\pi]_{22} =$	$r_1r_2^2u_1u_2/(r_1u_1+r_2u_2)^2$
$[D\pi]_{21} =$	$-[D\pi]_{11}$
$[D\pi]_{12} =$	$-[D\pi]_{22}$

Dan	Smooth non-expected utility (source-dependent)
$u_1 =$	$-\exp(-rpx_1^*+(1-p)x_2^*)$
$u_1 =$	$-\exp(-rqx_1^*+(1-q)x_2^*)$
$\pi_1 =$	$(pu_1+qu_2)/(u_1+u_2)$
$[D\pi]_{jj} =$	$-ru_1u_2(p-q)^2/(u_1+u_2)^2 \quad j=1,2$
$[D\pi]_{jk} =$	$-[D\pi]_{jj} \quad k\neq j$

Total utility $= u_1+u_2$
x_j^* denotes x_j+f_j where x_j is observable money
received and f_j is background uncertainty

6.5   The risk aversion matrix

The matrix of derivatives of the risk-neutral probabilities, $D\pi(\mathbf{f})$, can also be expressed in terms of a *matrix-valued Arrow-Pratt measure of local risk aversion*, denoted here by $\mathbf{R}(\mathbf{f})$, whose jk^{th} element is

$$r_{jk}(\mathbf{f}) := \frac{[D^2U(\mathbf{f})]_{jk}}{[DU(\mathbf{f})]_j}. \tag{6.17}$$

Then Equation (6.16) can be rewritten as:

$$[D\pi(\mathbf{f})]_{jk} = -\pi_j(\mathbf{f}) \left(r_{jk}(\mathbf{f}) - \sum_{m=1}^{M} \pi_m(\mathbf{f})r_{mk}(\mathbf{f}) \right) \tag{6.18}$$

$\mathbf{R}(\mathbf{f})$ is not uniquely determined if the independence axiom is not satisfied, because $U(\mathbf{f})$ is unique only up to increasing transformations in that case, and such transformations also affect $D^2U(\mathbf{f})$ and $\mathbf{R}(\mathbf{f})$, although their effects ultimately cancel out in formulas for measuring local risk aversion. In particular, if $U^*(\mathbf{f}) = \varphi(U(\mathbf{f}))$, where φ is strictly increasing and twice differentiable, then U^* represents the same preferences as U but the corresponding risk aversion matrix $\mathbf{R}^*(\mathbf{f})$ differs from $\mathbf{R}(\mathbf{f})$ by an additive constant in each column:

$$\mathbf{R}^*(\mathbf{f}) = \mathbf{R}(\mathbf{f}) + \alpha\overline{\boldsymbol{\Pi}}(\mathbf{f}), \tag{6.19}$$

where $\overline{\boldsymbol{\Pi}}(\mathbf{f})$ is the matrix whose rows are all equal to $\pi(\mathbf{f})$, i.e., the matrix whose elements in the kth column are all equal to $\pi_k(\mathbf{f})$, and $\alpha = (1 \cdot \nabla U(\mathbf{f}))(\varphi''(U(\mathbf{f})/\varphi'(U(\mathbf{f}))$.

To eliminate the arbitrary constants, define the *normalized risk aversion matrix* $\overline{\mathbf{R}}(\mathbf{f})$ to be the matrix whose jk^{th} element is:

$$\overline{r}_{jk}(\mathbf{f}) := r_{jk}(\mathbf{f}) - \sum_{m=1}^{M} \pi_m(\mathbf{f})r_{mk}(\mathbf{f}), \tag{6.20}$$

which is the jk^{th} element of $\mathbf{R}(\mathbf{f})$ minus the risk-neutral expected value of its kth column, thus getting rid of the arbitrary additive constants. Plugging this into (6.18) yields

$$[D\pi(\mathbf{f})]_{jk} = -\pi_j(\mathbf{f})\overline{r}_{jk}(\mathbf{f}). \tag{6.21}$$

In matrix notation these last two equalities are:

$$\overline{\mathbf{R}}(\mathbf{f}) := \mathbf{R}(\mathbf{f}) - \overline{\boldsymbol{\Pi}}(\mathbf{f})\mathbf{R}(\mathbf{f}). \tag{6.22}$$

and

$$D\pi(\mathbf{f}) = -\Pi(\mathbf{f})\overline{\mathbf{R}}(\mathbf{f}) \tag{6.23}$$

where $\Pi(\mathbf{f}) = \mathrm{diag}(\pi(\mathbf{f}))$. Equivalently

$$\overline{\mathbf{R}}(\mathbf{f}) = -\Pi(\mathbf{f})^{-1}D\pi(\mathbf{f}) \tag{6.24}$$

from which

$$\overline{r}_{jk}(\mathbf{f}) = -(\partial\pi_j(\mathbf{f})/\partial f_k)/\pi_j(\mathbf{f}). \tag{6.25}$$

That is, *the jk^{th} element of the normalized risk aversion matrix is minus the relative rate of change of the risk-neutral probability of state j as wealth increases in state k.* The quantities on the right are observable, hence so is $\overline{\mathbf{R}}(\mathbf{f})$, which means it is invariant to monotonic transformations of U. It satisfies $\overline{\Pi}(\mathbf{f})\overline{\mathbf{R}}(\mathbf{f}) = 0$, hence it has rank $M - 1$, as does $D\pi(\mathbf{f})$.

In terms of $\overline{\mathbf{R}}(\mathbf{f})$ the structure of $D\pi$ in the 4-state case is

$$D\pi = \begin{bmatrix} -\pi_1\overline{r}_{11} & -\pi_1\overline{r}_{12} & -\pi_1\overline{r}_{13} & -\pi_1\overline{r}_{14} \\ -\pi_2\overline{r}_{21} & -\pi_2\overline{r}_{22} & -\pi_2\overline{r}_{23} & -\pi_2\overline{r}_{24} \\ -\pi_3\overline{r}_{31} & -\pi_3\overline{r}_{32} & -\pi_3\overline{r}_{33} & -\pi_3\overline{r}_{34} \\ -\pi_4\overline{r}_{41} & -\pi_4\overline{r}_{42} & -\pi_4\overline{r}_{43} & -\pi_4\overline{r}_{44} \end{bmatrix}, \tag{6.26}$$

suppressing the dependence of all terms on $\mathbf{f}$.

If the decision-maker has additive utility, i.e., if she satisifies the independence axiom, then $\mathbf{R}(\mathbf{f})$ is a unique diagonal matrix whose j^{th} diagonal element is a *state-dependent Arrow-Pratt*[9] *measure of local risk aversion:*

$$r_j(\mathbf{f}) := -u_j''(\mathbf{f})/u_j'(\mathbf{f}), \tag{6.27}$$

which is measured in units of 1/money and uniquely determined by preferences up to the choice of currency. In this special case $D\pi(\mathbf{f})$ satisfies

$$[D\pi(\mathbf{f})]_{jk} = \pi_j(\pi_k - 1_{j=k})r_k, \tag{6.28}$$

where $1_{j=k} := 1$ if $j = k$ and $1_{j=k} := 0$ otherwise. For example, it looks like this for 4 states:

$$D\pi = \begin{bmatrix} \pi_1(\pi_1 - 1)r_1 & \pi_1\pi_2r_2 & \pi_1\pi_3r_3 & \pi_1\pi_4r_4 \\ \pi_2\pi_1r_1 & \pi_2(\pi_2 - 1)r_2 & \pi_2\pi_3r_3 & \pi_2\pi_4r_4 \\ \pi_3\pi_1r_1 & \pi_3\pi_2r_2 & \pi_3(\pi_3 - 1)r_3 & \pi_3\pi_4r_4 \\ \pi_4\pi_1r_1 & \pi_4\pi_2r_2 & \pi_4\pi_3r_3 & \pi_4(\pi_4 - 1)r_4 \end{bmatrix} \tag{6.29}$$

---

[9]The relevance of the quantity $-u''(x)/u'(x)$ as a measure of local risk aversion was first noted by de Finetti (1952) and then independently by Pratt (1964) and Arrow (1965).

If $D\pi$ does not have this structure for some set of numbers $\{r_j\}$, then the decision-maker's local indifference curves at $\mathbf{f}$ bend in a way that is inconsistent with additive utility.[10] An example is given by the source-dependent utility model for Ellsberg's two-urn problem that will be discussed in the next chapter.

In the special cases of state-dependent expected utility ($u_j(f_j) = p_j v_j(f_j)$) or state-independent expected utility ($u_j(f_j) = p_j v(f_j)$), the probabilities in the equation $r_j(f_j) = -u_j''(f_j)/u_j'(f_j)$ can be canceled between numerator and denominator, yielding $r_j(f_j) = -v_j''(f_j)/v_j'(f_j)$ or $r(f_j) = -v''(f_j)/v'(f_j)$, respectively. Notice that even if the decision-maker has state-*independent* expected utility preferences, the local Arrow-Pratt measure in state j is generally state-*dependent* due to its dependence on f_j, which is the local background uncertainty.

In the example we have $\pi_1 = 0.4$ for everyone, $r_1 = r_2 = 0.2$ for Alice, $r_1 = 0.4$ and $r_2 = 1/3$ for Bob, and $r_1 = 1$ and $r_2 = 0.05$ for Carol. For Dan, we can compute $r_1 = r_2 = 0.385/(0.4 \times 0.6) = 1.6042$ as his apparent state-dependent measures of local risk aversion, but these numbers do not behave in the same way as those of the other agents as his wealth increases in one state or the other. For someone with an additive utility function, an increase in wealth in state 1, ceteris paribus, will affect r_1 but not r_2, and vice versa. Dan's utility function does not satisfy this property

To sum up things so far:

> In the most general case the properties of local preferences in the vicinity of the status quo payoff vector $\mathbf{f}$ are determined up to second-order effects by the vector of risk-neutral probabilities, $\pi(\mathbf{f})$, and the matrix of its derivatives, $D\pi(\mathbf{f})$, or alternatively a matrix-valued generalization of the Arrow-Pratt measure, $\mathbf{R}(\mathbf{f})$. $\pi(\mathbf{f})$ and $D\pi(\mathbf{f})$ are uniquely determined by preferences and are directly measurable in terms of acceptable small-but-finite bets. $\mathbf{R}(\mathbf{f})$ is not unique if utility is not additive, but it can be normalized to be so. In the various special cases of additive utility the risk aversion measure is a unique vector $\mathbf{r}(\mathbf{f})$.

---

[10]This test can only be applied if $n > 2$. If $n = 2$, as in the example of Figure 6.1, then $\pi_2 = (1-\pi_1)$, and because the columns of $D\pi$ always sum to zero, it is possible to choose r_1 and r_2 so that $D\pi$ has this special structure regardless of the form of the utility function.

6.6   A generalized uncertainty premium measure

Pratt (1964) defined a risk premium as the amount of expected value that a decision-maker with state-independent expected-utility preferences *would be willing to pay* to dispose of all the risk she faces. In the general state-preference setting, there may no such thing as an objective or subjective probability distribution from which to calculate expected values, and there may be no such thing as a riskless position that is feasible to achieve (or even locate) by selling off risky assets. Nevertheless, it is straightforward to generalize the risk premium concept to deal with these issues by defining it in terms of the amount the decision-maker *would need to be paid* to take on a new risk, on top of whatever risk she already faces. When this is done, the correct distribution with which to calculate expected values and variances turns out to be the decision-maker's local risk-neutral probability distribution rather than a probability distribution that is a hypothetical measure of pure belief.

Suppose that a decision-maker in possession of an act $\mathbf{f}$ contemplates taking on some additional risk in the form of a small bet whose payoff vector is $\mathbf{x}$. The *marginal price* of the bet $\mathbf{x}$ in the vicinity of act $\mathbf{f}$ is its *risk-neutral expected value* $\pi(\mathbf{f}) \cdot \mathbf{x}$, because on the margin a decision-maker who already possesses $\mathbf{f}$ would be willing to pay $\pi(\mathbf{f}) \cdot \mathbf{x}$ per unit of $\mathbf{x}$. The bet is *neutral* in the vicinity of act $\mathbf{f}$ if its risk-neutral expected value is zero, i.e., $\pi(\mathbf{f}) \cdot \mathbf{x} = 0$. The *buying price* for $\mathbf{x}$ in its entirety in the vicinity of act $\mathbf{f}$, denoted by $B(\mathbf{x}; \mathbf{f})$, is the maximum amount the decision-maker would be willing to pay for $\mathbf{x}$, i.e., it satisfies

$$\mathbf{f} + \mathbf{x} - B(\mathbf{x}; \mathbf{f}) \sim \mathbf{f}, \tag{6.30}$$

or equivalently

$$U(\mathbf{f} + \mathbf{x} - B(\mathbf{x}; \mathbf{f})) = U(\mathbf{f}). \tag{6.31}$$

> **Definition (uncertainty premium):** The *uncertainty premium*[11] for $\mathbf{x}$ at $\mathbf{f}$, denoted by $b(x; \mathbf{f})$, is the difference between the risk-neutral expected value of x and its buying price:
>
> $$b(\mathbf{x}; \mathbf{f}) := \pi(\mathbf{f}) \cdot \mathbf{x} - B(\mathbf{x}; \mathbf{f}) \tag{6.32}$$

If the decision-maker is strictly risk-averse, this quantity is strictly positive for all $\mathbf{x} \neq \mathbf{0}$. If $\mathbf{x}$ is a neutral bet, then $\mathbf{f} + \mathbf{x} + b(\mathbf{x}; \mathbf{f}) \sim \mathbf{f}$ as illustrated below.

---

[11]This was called the "overall uncertainty premium" in Nau (2011b) and "total risk premium" in Nau (2003) and Nau (2006a).

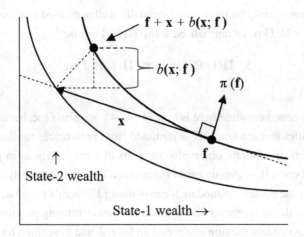

FIGURE 6.2: $b(\mathbf{x}; \mathbf{f})$ is the uncertainty premium for the neutral bet $\mathbf{x}$ at act $\mathbf{f}$.

The uncertainty premium is a generalization of Pratt's definition of a risk premium that extends it from the expected-utility framework to the non-expected utility framework of state-preference theory. It can be locally approximated by a formula that is a multivariate generalization of Pratt's formula[12]:

Theorem 6.2: Calculation of the uncertainty premium[13]

The uncertainty premium of a neutral bet $\mathbf{x}$ in the vicinity of act $\mathbf{f}$ has a quadratic approximation which is given by any of the following formulas:

$$b(\mathbf{x};\mathbf{f}) \quad \approx \quad \widehat{b}(\mathbf{x};\mathbf{f}) := -\frac{1}{2}\,\mathbf{x}\cdot D\pi(\mathbf{f})\,\mathbf{x} \qquad (6.33)$$

$$= \quad -\frac{1}{2}\frac{\mathbf{x}\cdot D^2 U(\mathbf{f})\,\mathbf{x}}{\mathbf{1}\cdot DU(\mathbf{f})} \qquad (6.34)$$

$$= \quad \frac{1}{2}\,\mathbf{x}\cdot \boldsymbol{\Pi}(\mathbf{f})\,\mathbf{R}(\mathbf{f})\,\mathbf{x} \qquad (6.35)$$

$$= \quad \frac{1}{2}\,\mathbf{x}\cdot \boldsymbol{\Pi}(\mathbf{f})\,\overline{\mathbf{R}}(\mathbf{f})\mathbf{x} \qquad (6.36)$$

---

[12] A multivariate generalization of the risk premium formula and a corresponding matrix-valued measure of local risk aversion also arise naturally in the setting where the decision-maker's utility function is a von Neumann-Morgenstern utility function over vectors of commodities, which has been studied by Duncan (1977) and Karni (1979). There the objects of choice are objective probability distributions over commodity vectors rather than vectors of monetary payoffs across states whose probabilities are subjective or undefined, and the risk premium is vector-valued rather than scalar-valued. In both settings the risk aversion measure takes the form of a matrix of second partial derivatives divided by first partial derivatives of the appropriate utility function, although the applications are otherwise quite different.

[13] Proofs of this theorem and its corollary are given in Nau (2003).

Note that the uncertainty premium can be equally well expressed in terms of the risk aversion matrix $\mathbf{R}(\mathbf{f})$ or its normalized form $\overline{\mathbf{R}}(\mathbf{f})$, because[14]:

$$\mathbf{x} \cdot \mathbf{\Pi}(\mathbf{f})\,\mathbf{R}(\mathbf{f})\mathbf{x} = \mathbf{x} \cdot \mathbf{\Pi}(\mathbf{f})\,\overline{\mathbf{R}}(\mathbf{f})\mathbf{x}. \qquad (6.37)$$

for any neutral bet $\mathbf{x}$.

The fundamental equation here is $\widehat{b}(\mathbf{x};\mathbf{f}) := -\frac{1}{2}\,\mathbf{x} \cdot D\pi(\mathbf{f})\,\mathbf{x}$, because it refers only to quantities that are uniquely determined from preferences (unlike U and R) and its elements are directly observable in terms of local variations in preferences among small bets. ($\overline{\mathbf{R}}$ is also uniquely determined but observationally it is derived from $D\pi$.) In the context of models for explaining Ellsberg's paradox, which will be discussed in the next chapter, the formula for the uncertainty premium turns out to be decomposable into the sum of a premium for risk and a premium for ambiguity.

> **Corollary 6.3** In the special case where $U(\mathbf{f})$ is additive, $\mathbf{R}(\mathbf{f})$ is a diagonal matrix that is uniquely determined by preferences, and its jth diagonal element is $r_j(f) = -u_j''(f)/u_j'(f)$, which is a *state-dependent local Arrow-Pratt risk aversion measure* at $\mathbf{f}$ in state j. The approximation formula for the uncertainty premium therefore simplifies to[15]
>
> $$\widehat{b}(\mathbf{x};\mathbf{f}) = \frac{1}{2}\sum_{j=1}^{M}\pi_j(\mathbf{f})r_j(f_j)x_j^2 = \frac{1}{2}\mathbf{x}\cdot\mathbf{\Pi}(\mathbf{f})\,\mathbf{R}(\mathbf{f})\mathbf{x}. \qquad (6.38)$$
>
> In the *very* special case where the decision-maker has state-*independent* expected utility preferences and also zero background uncertainty ($f_j = f$ for all j), this can be further simplified to Pratt's formula
>
> $$\widehat{b}(\mathbf{x};\mathbf{f}) = \frac{1}{2}r(f)\sum_{j=1}^{M}p_jx_j^2 = \frac{1}{2}r(f)\mathrm{Var}(\mathbf{x}) = \frac{1}{2}r(f)\mathbf{x}\cdot\mathbf{P}\,\mathbf{x} \qquad (6.39)$$
>
> where p_j is interpretable as a subjective probability of state j (which is uniquely determined under these conditions), $\mathbf{P}$ is the diagonal matrix whose jj^{th} element is p_j, and $r(f)$ is the Arrow-Pratt risk aversion measure at riskless wealth f.

So, in the passage from state-independent expected-utility preferences and zero background uncertainty to more general settings in which the decision-maker merely has locally smooth preferences, *two* things happen. First, the state-independent Arrow-Pratt measure $r(f)$ gets replaced by a state-dependent measure $r(f_j)$ or

---

[14]This is Corollary 2.1 in Nau (2003).
[15]This is equation (2.8) in Karni (1985).

$r_j(f_j)$ or $r_{jk}(\mathbf{f})$. Second, and less obviously, the decision-maker's true subjective probability p_j gets replaced by the local risk-neutral probability $\pi_j(\mathbf{f})$.

To sum up these results:

> In the calculation of uncertainty premia, the correct probabilities to use are the decision-maker's risk-neutral probabilities, not her true probabilities (if those are even defined). True probabilities appear only in the special cases where the two coincide. In the most general case, the uncertainty premium of a neutral bet is a quadratic form in which the vector of payoffs pre- and post-multiplies the matrix of derivatives of the risk-neutral probabilities.

6.7   Risk-neutral probabilities and the Slutsky matrix

$D\pi$ is closely related to the Slutsky matrix of consumer theory. Suppose that there are M commodities (which could be financial assets or other goods), let $\mathbf{p} := (p_1, p_2)$ denote the vector of market prices for them, and let $\pi(\mathbf{f})$ denote the consumer's relative marginal rates of substitution at $\mathbf{f}$, which are risk-neutral probabilities if the commodities are Arrow securities. Then $\mathbf{f}$ is an optimal act for the consumer when $\pi(\mathbf{f})$ coincides with the vector of relative prices (prices normalized to sum to 1), because the first-order condition for an optimum is that relative marginal rates of substitution should equal relative prices. In the case of Arrow securities, this is known as the *fundamental theorem of risk bearing* as illustrated here:

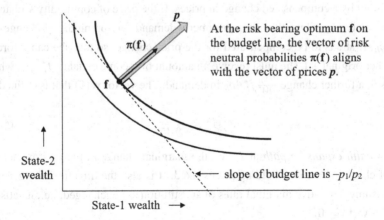

At the risk bearing optimum $\mathbf{f}$ on the budget line, the vector of risk neutral probabilities $\pi(\mathbf{f})$ aligns with the vector of prices $\mathbf{p}$.

slope of budget line is $-p_1/p_2$

FIGURE 6.3: The fundamental theorem of risk bearing

The Slutsky matrix at $\mathbf{f}$, which will be denoted here as $\mathbf{S}(\mathbf{f})$,[16] determines the marginal change in the consumption vector that is induced by a compensated change in the relative-marginal-price vector. That is, $\mathbf{S}(\mathbf{f})$ has the defining property that the optimal act changes from $\mathbf{f}$ to $\mathbf{f} + \mathbf{S}(\mathbf{f})d\pi$ when relative prices change from $\pi(\mathbf{f})$ to $\pi(\mathbf{f}) + d\pi$ and the consumer is compensated for the change in prices, or equivalently, if she already owns $\mathbf{f}$ and can reallocate it at the new relative prices. In a financial market, the Slutsky matrix and the matrix of derivatives of the risk-neutral probabilities are therefore linked by the following equivalence:

$$\mathbf{S}(\mathbf{f})d\pi = d\mathbf{f} \quad \Longleftrightarrow \quad D\pi(\mathbf{f})d\mathbf{f} = d\pi. \tag{6.40}$$

There are several analytic formulas that can be used to compute the Slutsky matrix. The one that is commonly seen nowadays in microeconomics textbooks is stated in terms of the consumer's *Marshallian demand function*, denoted here as $\mathbf{f}^*(\mathbf{p}, W)$, which is the utility-maximizing act that a consumer with disposable wealth W will choose to purchase when the price vector is $\mathbf{p}$. In other words, $\mathbf{f}^*(\mathbf{p}, W)$ is the solution to $\max_{\mathbf{f} \in \mathcal{F}} U(\mathbf{f})$ subject to $\mathbf{p} \cdot \mathbf{f} \leq W$. The first-order condition for an optimum is that $\pi(\mathbf{f}) \propto \mathbf{p}$, i.e., a consumer with strictly convex preferences in possession of $\mathbf{f}$ is at an optimum if and only her relative marginal rates of substitution at $\mathbf{f}$ are proportional to market prices.

If $\mathbf{f}$ is an optimum at prices $\mathbf{p}$ and wealth W, the Slutsky matrix at $\mathbf{f}$ can be determined from the Marshallian demand function as follows:

$$[\mathbf{S}(\mathbf{f})]_{jk} := \frac{\partial f_j^*}{\partial p_k} + \frac{\partial f_j^*}{\partial W} f_k^*. \tag{6.41}$$

The terms on the RHS of this equation are called the *price effect* and the *income effect* caused by a compensated change in prices. If the price of commodity k changes by dp_k, ceteris paribus, then the consumer's demand for commodity j changes by $\frac{\partial f_j^*}{\partial p_k} dp_k$. If she is to be compensated for the price change, so that she can afford to retain her original act, she must receive an amount of income equal to $f_k^* dp_k$, which results in a further change $\frac{\partial f_j^*}{\partial W} f_k^* dp_k$ in demand. The vector $\mathbf{w}(\mathbf{f})$ that is defined by

$$w_j(\mathbf{f}) := \frac{\partial f_j^*}{\partial W} \tag{6.42}$$

is the *wealth expansion path* at $\mathbf{f}$, i.e., the marginal change in the optimal act per unit of change in wealth, holding prices fixed. It is also the direction along which the consumer's relative marginal rates of substitution are unchanged, i.e., it satisfies $D\pi(\mathbf{f}) \cdot \mathbf{w}(\mathbf{f}) = \mathbf{0}$.

---

[16]Here "at f" means "at f under conditions where it is an optimum."

The Slutsky matrix and the wealth expansion vector can also be obtained by inverting the Hessian matrix after bordering it with the price vector, which is the derivation used by Slutsky and Hicks[17]:

$$
\begin{bmatrix} D^2U(\mathbf{f}) & \mathbf{p} \\ \mathbf{p}^T & 0 \end{bmatrix}^{-1} = \begin{bmatrix} \alpha\mathbf{S}(\mathbf{f}) & \mathbf{w}(\mathbf{f}) \\ \mathbf{w}(\mathbf{f})^T & \beta_f \end{bmatrix}. \tag{6.43}
$$

Here, α^{-1} is the marginal utility of income (the ratio of marginal utilities to marginal prices), and β_f is the income derivative of the marginal utility of income (the derivative of the marginal utility of income along the wealth expansion path). Equation (6.43) can be rewritten as

$$
\begin{bmatrix} \alpha D^2U(\mathbf{f}) & \mathbf{p} \\ \mathbf{p}^T & 0 \end{bmatrix}^{-1} = \begin{bmatrix} \mathbf{S}(\mathbf{f}) & \mathbf{w}(\mathbf{f}) \\ \mathbf{w}(\mathbf{f})^T & \alpha\beta_f \end{bmatrix}, \tag{6.44}
$$

which cancels the units of utility on both sides. When prices are normalized to sum to 1, $\alpha^{-1} = \mathbf{1} \cdot DU(\mathbf{f})$, in which case $\alpha D^2U(\mathbf{f})$ is the same normalized Hessian matrix that is the first term in the formula (6.16) for $D\pi(\mathbf{f})$.

The Slutsky matrix and the $D\pi$ matrix have inverse properties insofar as $\mathbf{S}(\mathbf{f})d\pi = d\mathbf{f}$ if and only if $D\pi(\mathbf{f})d\mathbf{f} = d\pi$, as noted above. However, they are not literally inverses of each other, because both are singular. $\mathbf{S}(\mathbf{f})$ is singular because it satisfies

$$
\mathbf{S}(\mathbf{f})\pi(\mathbf{f}) = \mathbf{0},
$$

$$
\pi(\mathbf{f}) \cdot \mathbf{S}(\mathbf{f}) = \mathbf{0}, \tag{6.45}
$$

and $D\pi(\mathbf{f})$ is singular because it satisfies

$$
D\pi(\mathbf{f})\mathbf{w}(\mathbf{f}) = \mathbf{0},
$$

$$
\mathbf{1} \cdot D\pi(\mathbf{f}) = \mathbf{0}. \tag{6.46}
$$

Also, $\mathbf{S}(\mathbf{f})$ is symmetric, while $D\pi(\mathbf{f})$ is not.

The main result of this section is that *the Slutsky matrix and the matrix of derivatives of risk-neutral probabilities are submatrices of the inverses of bordered versions of each other*. In particular, $\mathbf{S}(\mathbf{f})$ and $\mathbf{w}(\mathbf{f})$ can be obtained by inverting $D\pi$ after bordering it with the vector of risk-neutral probabilities.

---

[17]This equation is discussed by Stern (1986). The bordered Hessian matrix was used by Pareto as well as Slutsky and Hicks in modeling the utility function of the consumer (Afriat (1980), Dooley (1983)).

Theorem 6.3: Relationship of the Slutsky matrix to the matrix of derivatives of risk-neutral probabilities

If the decision-maker is strictly risk-averse, then

$$\begin{bmatrix} D\boldsymbol{\pi}(\mathbf{f}) & \boldsymbol{\pi}(\mathbf{f}) \\ \boldsymbol{\pi}(\mathbf{f})^T & 0 \end{bmatrix}^{-1} = \begin{bmatrix} \mathbf{S}(\mathbf{f}) & \mathbf{w}(\mathbf{f}) \\ \mathbf{1}^T & 0 \end{bmatrix} \qquad (6.47)$$

If the prices are not normalized to sum to 1,[18] the equation is generalized to:

$$\begin{bmatrix} D\boldsymbol{\pi}(\mathbf{f}) & \mathbf{p} \\ \mathbf{p}^T & 0 \end{bmatrix}^{-1} = \begin{bmatrix} \mathbf{S}(\mathbf{f}) & \mathbf{w}(\mathbf{f}) \\ ((\mathbf{1}\cdot\mathbf{p})^{-1})\mathbf{1}^T & 0 \end{bmatrix}. \qquad (6.48)$$

The vector $\mathbf{w}(\mathbf{f})$ in the latter equation is the wealth expansion path with respect to real prices rather than normalized prices. This follows from the fact that if $\mathbf{p}$ is rescaled by some positive constant, the length of $\mathbf{w}$ and the constant in the bottom row of the matrix on the right are rescaled in a reciprocal fashion. The Slutsky matrix is unaffected.

Theorem 6.3 can be used to rewrite the relation $\mathbf{S}(\mathbf{f})d\boldsymbol{\pi} = d\mathbf{f}$ in terms of $D\boldsymbol{\pi}(\mathbf{f})$, yielding the result that the change $d\mathbf{f}$ in the optimal act that follows from an infinitesimal change $d\boldsymbol{\pi}$ in relative marginal prices is determined by:

$$\begin{bmatrix} d\mathbf{f} \\ 0 \end{bmatrix} = \begin{bmatrix} D\boldsymbol{\pi}(\mathbf{f}) & \boldsymbol{\pi}(\mathbf{f}) \\ \boldsymbol{\pi}(\mathbf{f})^T & 0 \end{bmatrix}^{-1} \begin{bmatrix} d\boldsymbol{\pi} \\ 0 \end{bmatrix} \qquad (6.49)$$

However, the same formula does not provide the best first-order approximation to the effect of a *finite* change $\Delta\boldsymbol{\pi}$. This is given instead by:

Corollary 6.4:

A first-order approximation for the change $\Delta\mathbf{f}$ in the optimal act that results from a small finite change $\Delta\boldsymbol{\pi}$ in relative marginal prices is given by[19]:

$$\begin{bmatrix} \Delta\mathbf{f} \\ 0 \end{bmatrix} \approx \begin{bmatrix} D\boldsymbol{\pi}(\mathbf{x}) & \boldsymbol{\pi}(\mathbf{f}) + \Delta\boldsymbol{\pi} \\ (\boldsymbol{\pi}(\mathbf{f}) + \Delta\boldsymbol{\pi})^T & 0 \end{bmatrix}^{-1} \begin{bmatrix} \Delta\boldsymbol{\pi} \\ 0 \end{bmatrix} \qquad (6.50)$$

---

[18]This could be the case if payoffs are to be received at a future date, requiring some amount of discounting of Arrow securities.

[19]This is a simplification of Proposition 4(b) in Nau (2003). See the appendix for the proof of this result and Theorem 6.3.

Chapter 7

Ambiguity and source-dependent uncertainty aversion

7.1   Introduction

For the first few decades after it was introduced into the literature of Bayesian decision theory, the independence axiom (discussed in Section 2.2) was generally considered to be an unassailable requirement of rational preferences under uncertainty. In a comparison between two acts that assign consequences to states of the world or states of a random device, if there are some states in which the acts agree, it shouldn't matter how they agree there. Those states ought not to come into play in the determination of which act is preferred. Paradoxical thought-experiments devised by Allais (1953) and Ellsberg (1961) showed that many persons willfully and predictably—even cheerfully—violate the independence axiom in some situations. These paradoxes were at first treated mainly as curiosities (exceptions that prove the rule), but many more such violations of the EU and SEU axioms began to gain visibility in the literature of experimental economics and behavioral decision theory, inspiring a reappraisal and a search for alternatives. In the 1970s and early 1980s an explosion of new work on models of non-expected-utility preferences began to emerge, and much of it centered on relaxations of the independence axiom. Some of this work sought to generalize the EU model (which was challenged by Allais) while other work took aim at the SEU model. The phenomenon of aversion to ambiguity,

DOI: 10.1201/9781003527145-7

which is highlighted by Ellsberg's paradox, has inspired a huge literature, of which surveys can be found in Etner et al. (2012), Trautmann and van de Kuilen (2015), and Gilboa and Marinacci (2016). This chapter will show how ambiguity aversion, considered as a form of source-dependent uncertainty aversion, can be modeled within the state-preference framework using the tools developed in the previous chapter. This is not to say that this framework provides a comprehensive way to deal with the phenomenon, but merely that it has its own way to accomodate it within its sphere of application. A further review of the broader literature of non-expected-utility will be deferred to the end of this chapter.

7.2 Ellsberg's paradox and smooth non-expected-utility preferences

In the context of state-preference theory, the independence axiom yields a cardinal utility function that is additive across states, and it could be state-dependent. The functional forms of the statewise components of the utility function could differ from each other in arbitrary ways. (Another axiom can be added to force state-independence of utility.) An important and not-quite-obvious implication of an additive utility function is that it requires the decision-maker to be (locally) equally averse toward all sources of uncertainty, even if utility is state-dependent. The reason for this is that, under additive utility, the risk aversion matrix $\mathbf{R}(\mathbf{f})$ is a diagonal matrix, and the quadratic approximation formula for the uncertainty premium of a neutral bet in Corollary 6.3 therefore depends only on the squared values of payoffs in individual states. If two neutral bets yield equal-magnitude payoffs in each state, differing only in the signs of those payoffs, then they must have the same uncertainty premium.

A violation of this requirement is revealed in the two "urn" experiments of Daniel Ellsberg (1961), famously performed upon subjects who were leading decision theorists rather than naive college students. Most of them displayed a systematic aversion toward ambiguity in the probabilities of events on which payoffs depend. In the first experiment there are two urns, each of which contains balls that may be red or black. Urn A contains 100 balls that are red and black in unknown (ambiguous) proportions, while urn B contains exactly 50 red and 50 black balls. One ball will be drawn at random from each urn. Let $A_1[A_2]$ denote the event that the ball drawn from urn A is red [black], and let $B_1[B_2]$ denote the event that the ball drawn from urn B is red [black], so that the relevant state space is $\{A_1B_1, A_1B_2, A_2B_1, A_2B_2\}$.

The subject is asked to state her preferences within two pairs of bets[1] such as the following, where amounts in the cells are payoffs in dollars received according to the colors drawn:

TABLE 7.1: Ellsberg's two-urn paradox

	$A_1 B_1$ *red, red*	$A_1 B_2$ *red, black*	$A_2 B_1$ *black, red*	$A_2 B_2$ *black, black*
Bet R_A	100	100		
Bet R_B	100		100	
Bet B_A			100	100
Bet B_B		100		100

Here R_A is a bet that pays \$100 if the ball drawn from urn A is red, and zero otherwise, and R_B is a bet that pays \$100 if the ball drawn from urn B is red, and zero otherwise, and similarly for B_A and B_B. The typical response pattern is $R_B \succ R_A$ and $B_B \succ B_A$, i.e., the subject strictly prefers to win \$100 based on the color of the ball drawn from the less ambiguous urn, regardless of whether the winning color is red or black, which is a direct violation of independence. To see this, change the payoffs of both B_A and B_B from 0 to 100 in state $A_1 B_1$, and change them from 100 to 0 in state $A_2 B_2$. The independence axiom requires that the direction of preference between B_A and B_B should be unaffected. However, the effect of these substitutions is to change B_A to R_B and to change B_B to R_A. The preference $R_B \succ R_A$ therefore requires $B_A \succ B_B$, contrary to typical behavior.

It is straightforward to construct a smooth utility function within the state-preference framework that exhibits this pattern of ambiguity aversion (Nau (2001)). Let $\mathbf{x}$ denote a doubly subscripted vector of monetary payoffs of a bet, with x_{jk} representing the payoff in state $A_j B_k$.[2] Then define the utility function as follows:

$$U(\mathbf{x}) = \frac{1}{2}u\left(\frac{1}{2}x_{11} + \frac{1}{2}x_{12}\right) + \frac{1}{2}u\left(\frac{1}{2}x_{21} + \frac{1}{2}x_{22}\right), \tag{7.1}$$

---

[1] In the examples of this section, the bets are not neutral (following Ellsberg's original design), but they could be modified to be so by assuming symmetric risk-neutral probabilities and subtracting 50 from the payoffs in Table 7.1, subtracting 100 from the payoffs in the first set of bets in Table 7.3, and subtracting 200 from the payoffs in the second set of bets there. These amounts are the risk-neutral expected values of the original bets. In each case this causes the payoffs in the first and second set of bets to have the same magnitudes but opposite signs, so strict preferences in the first set ought to be reversed in the second.

[2] Throughout this chapter, as in the previous one, the symbol $\mathbf{f}$ will denote an act, i.e., a distribution of total wealth across states of the world, which is an object of the decision-maker's preferences. The possibly-unobservable status quo act will often be referred to as "background risk." $\mathbf{x}$ will denote a vector of payoffs of a bet, i.e., deviations from the status quo. Thus, a decision-maker who accepts the bet $\mathbf{x}$ moves from a wealth distribution of $\mathbf{f}$ to $\mathbf{f} + \mathbf{x}$. Her local risk neutral probabilities are determined by $\mathbf{f}$. In equations 7.1 and 7.2, $\mathbf{f}$ is taken to be $\mathbf{0}$ for simplicity, so $U(\mathbf{f} + \mathbf{x})$ reduces to $U(\mathbf{x})$.

where u is a strictly concave function. This yields $U(R_A) = U(R_B) = \frac{1}{2}u(100) + \frac{1}{2}u(0)$ and $U(R_B) = U(B_B) = u(50)$, and the latter value is greater than the former by virtue of strict concavity. *It is as if the decision-maker assigns probability $\frac{1}{4}$ to all 4 outcomes and she bets on draws from urn A as if she is uncertainty-averse with Bernoulli utility function u and she bets on draws from urn B as if she is uncertainty-neutral.* Notice that this model does not make any gratuitous assumptions about how to parameterize ambiguity itself, potentially saving some head-scratching for someone analyzing her own decision. The probabilities of the four outcomes are all taken to be $\frac{1}{4}$ by symmetry, and the individual's deviation from expected utility preferences is modeled entirely by u, which could be (say) an exponential or power function with a single parameter.

In Ellsberg's second experiment, there is a single urn containing 30 red balls and 60 balls that are yellow and black in unknown proportions. A single ball will be randomly drawn, and the subject is asked to state her preference orders within two sets of bets, $\{R, Y, B\}$ and $\{YB, RB, RY\}$, whose payoff tables have this structure:

TABLE 7.2: Ellsberg's single-urn paradox

	red	*yellow*	*black*
Bet R	300		
Bet Y		300	
Bet B			300

	red	*yellow*	*black*
Bet YB		300	300
Bet RB	300		300
Bet RY	300	300	

The typical response pattern is $R \succ Y \sim B$ and $YB \succ RB \sim RY$. That is, a subject prefers to stake a \$300 prize on the draw of a red ball rather than either yellow or black, but if the prize is to be won when either of *two* specified colors is drawn, she prefers yellow-or-black over either red-or-yellow or red-or-black, which again is a direct violation of the independence axiom. If the payoffs of bets R and Y are converted from 0 to 300 in the *Black* outcome, they are changed to RB and YB respectively, hence $R \succ Y$ ought to imply $RB \succ YB$ if independence holds, and similarly $R \succ B$ ought to imply $RY \succ YB$.

The single-urn experiment does not immediately fit into the model that was given above for the 2-urn experiment because there are only three payoff-relevant states. However, without loss of generality it can be translated to a 4-element state space by drawing a second ball from the urn if a red ball is drawn first. Let $B_1[B_2]$ denote the event that the *first* ball is red [not red], and let $A_1[A_2]$ denote the event that the *last* ball is yellow [black]. Thus we have the mapping $A_1B_1 \to Red$, $A_2B_1 \to Red$, $A_1B_2 \to Yellow$, and $A_2B_2 \to Black$, and the payoff tables of the bets are:

TABLE 7.3: Ellsberg's single-urn paradox in 4-state form

	A_1B_1 *red, yellow*	A_1B_2 *yellow*	A_2B_1 *red, black*	A_2B_2 *black*
Bet R	300		300	
Bet Y		300		
Bet B				300
Bet YB		300		300
Bet RB	300		300	300
Bet RY	300	300	300	

A strict preference for R over either Y or B, together with a strict preference for YB over either RY or RB can be explained by a utility function of same general form as (7.1) but with different probabilities:

$$U(\mathbf{x}) = \frac{1}{2}u(\frac{1}{3}x_{11} + \frac{2}{3}x_{12}) + \frac{1}{2}u(\frac{1}{3}x_{21} + \frac{2}{3}x_{22}) \qquad (7.2)$$

A decision-maker with this utility function will evaluate the bets as follows: $U(R) = u(100)$, $U(Y) = U(B) = \frac{1}{2}u(200) + \frac{1}{2}u(0)$, and $U(YB) = u(200)$, $U(RB) = U(RY) = \frac{1}{2}u(300) + \frac{1}{2}u(100)$. Concavity of u then implies $U(R) > U(Y)$, $U(R) > U(B)$, $U(YB) > U(RB)$, and $U(YB) > U(RY)$, consistent with the usual responses.

The models presented above are merely the simplest examples of smooth utility functions that explain Ellsberg's paradox when it is set in the state-preference modeling framework. The numeric parameters in them have the appearance of probabilities, but the model is actually agnostic about whether decision-maker thinks in terms of probabilities or whether those probabilities are revealed by preferences. The examples could be fancied-up to include complications such as state-dependent utility and unobservable background uncertainty, thus rendering it impossible to separate beliefs from tastes for uncertainty and ambiguity, while still retaining the ability to detect source-dependent attitudes toward uncertainty, as will be shown later.

As it happens, both of the above 4-outcome utility functions are examples of the Klibanoff et al. ("KMM," 2005) version of the smooth model, in which the decision-maker's perception of ambiguity is modeled by a second-order probability distribution over first-order probabilities, and her attitude toward such ambiguity is modeled with a second-order utility function. In particular, in the model for the 2-urn problem, it is as if there are two first-order probability distributions, $(\frac{1}{2}, \frac{1}{2}, 0, 0)$

and $(0, 0, \frac{1}{2}, \frac{1}{2},)$, and they have equal second-order probabilities as if the ambiguous urn contains either all red balls or all black balls with equal probability. Further, the decision-maker's first-order utility function for money is linear and her second-order utility function for ambiguity is u. The model for the single-urn problem has the same structure with first-order probability distributions $(\frac{1}{3}, \frac{2}{3}, 0, 0)$ and $(0, 0, \frac{1}{3}, \frac{2}{3})$, as if the urn contains 60 yellow balls or 60 black balls with equal probability, along with 30 red balls.

7.3 Source-dependent utility revealed by risk-neutral probabilities

The behavior typically observed in experiments such as Ellsberg's is most commonly explained by models in which the decision-maker's first-order beliefs are to some extent indeterminate, represented by a set of subjective probability distributions (as in models of incomplete preferences or multiple priors) or by a second-order probability distribution as in the KMM model, which will be discussed in detail later. A more parsimonious explanation is that whether or not the decision-maker has determinate subjective probabilities, she may be more averse to some sources of uncertainty than others, perhaps due to differences in the quality of information on which her beliefs are based or due to differences in her tastes for different kinds of uncertainties or due to social norms that condemn some forms of gambling. Such preferences are incompatible with the independence axiom but don't necessarily expose the decision-maker to exploitation. In fact, they may protect her from exploitation by others with better information or more cleverness (like Ellsberg himself). This section presents axiomatic models of such behavior within the state-preference modeling framework, yielding a general class of utility functions of which (7.1) and (7.2) are special cases

In the simplest form of the model there are two distinct sources of uncertainty represented by two logically independent partitions of events, $\mathcal{A} := \{A_1, ..., A_J\}$ and $\mathcal{B} := \{B_1, ..., B_K\}$, so that the state space is $\mathcal{A} \times \mathcal{B}$.[3] Events that are measurable with respect to $\mathcal{A}$ and $\mathcal{B}$ will be called $\mathcal{A}$-events and $\mathcal{B}$-events, respectively. Suppose that the decision-maker's utility function has the compound cardinal form:

$$U(\mathbf{f}) = \sum_{j=1}^{J} u_j \left(\sum_{k=1}^{K} v_{jk}(f_{jk}) \right),$$ (7.3)

---

[3] In general there could be more than two qualitatively different sources of risk, in which case a more deeply nested utility function of this form could be used.

where j and k refer to A_j and B_k, and the evaluation functions $\{u_j\}$ and $\{v_{jk}\}$ are strictly concave. The inner summation is a state-dependent additive utility function for monetary payoffs ("first order utility") and the outer summation is an $\mathcal{A}$-event-dependent additive utility function ("second-order utility") for first-order utilities. Concavity of the second-order functions $\{u_j\}$ amplifies the decision-maker's aversion to uncertainties that depend on the $\mathcal{A}$-events. If the coordinate independence axiom (SPT6) is applied globally, the latter functions must be linear, and the decision-maker merely has state-dependent utility for money determined by $\{v_{jk}\}$. To obtain the nested functional form with nonlinear $\{u_j\}$ which allows total utility to be source-dependent, it is necessary to weaken this axiom so that it treats $\mathcal{A}$-events differently from $\mathcal{B}$-events. This will be done in the axiomatization below. In the formulas that follow, the arguments $\mathbf{f}$ and f_{jk} and v_j will be suppressed in most places where they would otherwise appear. Thus, for example, the terms $\mathbf{R}$, π_{jk}, v_{jk}, v_j, u_j, $r_{jk,mn}$, etc., stand for $\mathbf{R}(\mathbf{f})$, $\pi_{jk}(\mathbf{f})$, $v_{jk}(f_{jk})$, $\sum_{k=1}^{K} v_{jk}$, $u_j(v_j)$, and $r_{jk,mn}(\mathbf{f})$.

The model's local *risk-neutral probabilities at* $\mathbf{f}$ satisfy

$$\pi_{jk}(\mathbf{f}) := \frac{[DU(\mathbf{f})]_{jk}}{\sum\limits_{m=1}^{J} \sum\limits_{n=1}^{K} [DU(\mathbf{f})]_{mn}} \propto u_j' v_{jk}', \text{ where } u_j' := u_j'(v_j) \text{ and } v_j' := \sum_{k=1}^{K} v_{jk}'.$$

(7.4)

The corresponding *marginal* risk-neutral probabilities of $\mathcal{A}$-events are $\pi_j := \sum_{k=1}^{K} \pi_{jk}$ and the *conditional* risk-neutral probabilities of $\mathcal{B}$-events given $\mathcal{A}$-events are $\pi_{k|j} := \pi_{jk}/\pi_j$. In principle these are all observable in terms of acceptable small bets. Using the risk-neutral probabilities, a simple test that is a state-dependent generalization of Ellsberg's 2-urn experiment can be designed to reveal whether the decision-maker is locally more uncertainty-averse toward bets on $\mathcal{A}$-events than toward bets on $\mathcal{B}$-events Let A denote some $\mathcal{A}$-event and let B denote some $\mathcal{B}$-event, and let $\overline{A}$ and $\overline{B}$ denote their complements, so the four possible joint realizations of the two events are expressible as AB, $A\overline{B}$, $\overline{A}B$, and $\overline{A}\,\overline{B}$. Let their risk-neutral probabilities be denoted π_{AB}, $\pi_{A\overline{B}}$, etc., and assume they are all non-zero.

> **Definition of $A{:}B$-bets and $B{:}A$-bets:** An $A{:}B$-*bet of size* Δ, denoted $\mathbf{x}_{A:B}(\Delta)$, is a bet in which the payoff in each joint event is equal to Δ divided by the corresponding risk-neutral probability, with a positive sign attached if A occurs and a negative sign if $\overline{A}$ occurs. (Δ itself may be positive or negative.) A $B{:}A$-*bet of size* Δ, denoted $\mathbf{x}_{B:A}(\Delta)$, has payoffs of the same magnitudes as $\mathbf{x}_{A:B}(\Delta)$ in every state but with signs that depend on whether B occurs rather than whether A occurs

The payoffs of the two kinds of bets look like this in contingency-table form:

TABLE 7.4: Payoffs of $A{:}B$-bets and $B{:}A$-bets

	Bet $\mathbf{x}_{A:B}(\Delta)$		Bet $\mathbf{x}_{B:A}(\Delta)$	
	B	$\overline{B}$	B	$\overline{B}$
A	$+\Delta/\pi_{AB}$	$+\Delta/\pi_{A\overline{B}}$	$+\Delta/\pi_{AB}$	$-\Delta/\pi_{A\overline{B}}$
$\overline{A}$	$-\Delta/\pi_{\overline{A}B}$	$-\Delta/\pi_{\overline{A}\,\overline{B}}$	$+\Delta/\pi_{\overline{A}B}$	$-\Delta/\pi_{\overline{A}\,\overline{B}}$

If the risk-neutral probabilities are all equal by virtue of symmetries of the decision-maker's beliefs and prior investments, then the payoffs are all equal in magnitude, in which case the $A{:}B$-bet and the corresponding $B{:}A$-bet are precisely the choices in Ellsberg's 2-urn problem, merely recentered so that they are all neutral bets in which losses as well as gains are possible. In general the risk-neutral probabilities of the four events will vary, but because the payoff magnitudes are scaled in inverse proportion to them, the bets remain neutral and symmetry is preserved among the four cells insofar as they each contribute $\pm\Delta$ to the risk-neutral expected value.

If the second-order utility functions are linear, the decision-maker merely has state-dependent expected utility preferences and will not exhibit a systematic preference for $B{:}A$-bets over $A{:}B$-bets: they will have equal uncertainty premia computed by the quadratic approximation formula in the preceding chapter. However, if the second-order utility functions are strictly concave, then for sufficiently small Δ the uncertainty premium of every $A{:}B$-bet will be strictly larger than the uncertainty premium of the corresponding $B{:}A$-bet, consistent with the pattern of source-dependent uncertainty aversion that arises in Ellsberg's paradox as shown in the theorem below. This is true regardless of background uncertainty.

We now proceed to axiomatize the representation given above and to derive analytic formulas for uncertainty premia that quantify the phenomenon of source-dependent uncertainty aversion. For any event E and acts $\mathbf{f}$ and $\mathbf{g}$, let $\mathbf{f}_A\mathbf{g}_{-A}$ denote the act that agrees with $\mathbf{f}$ on A and agrees with $\mathbf{g}$ on $\overline{A}$. Now suppose that preferences among acts satisfy the following partition-specific independence conditions for a source-dependent utility (SDU) representation:

Axiom SDU1 ($\mathcal{A}$-independence): $\mathbf{f}_A\mathbf{g}_{-A} \succsim \mathbf{f}_A^*\mathbf{g}_{-A} \Leftrightarrow \mathbf{f}_A\mathbf{g}_{-A}^* \succsim \mathbf{f}_A^*\mathbf{g}_{-A}^*$ for all acts $\mathbf{f}, \mathbf{f}^*, \mathbf{g}, \mathbf{g}^*$ and every $\mathcal{A}$-event A. If $J = 2$, then $\succsim$ also satisfies the corresponding hexagon condition: $\mathbf{f}_A\mathbf{g}_{-A} \sim \mathbf{f}_A^*\mathbf{g}_{-A}^*$ and $\mathbf{f}_A\mathbf{g}_{-A}^{**} \sim \mathbf{f}_A^*\mathbf{g}_{-A}$ and $\mathbf{f}_A^*\mathbf{g}_{-A} \sim \mathbf{f}_A^{**}\mathbf{g}_{-A}^* \Rightarrow \mathbf{f}_A^*\mathbf{g}_{-A}^{**} \succsim \mathbf{f}_A^{**}\mathbf{g}_{-A}$.

Definition of $\mathcal{A}$-conditional preference: For any $\mathcal{A}$-event A, $\mathbf{f} \succsim_A \mathbf{f}^*$ means that $\mathbf{f}_A\mathbf{g}_{-A} \succsim \mathbf{f}_A^*\mathbf{g}_{-A}$ for some $\mathbf{g}$, and $\succsim_j$ denotes the special case of $\succsim_A$ when $A = A_j$.

Axiom SDU2 (conditional $\mathcal{B}$-independence): $f_B g_{-B} \succsim_j f_B^* g_{-B} \Leftrightarrow$ $f_B g_{-B}^* \succsim_j f_B^* g_{-B}^*$ for all acts f, f^*, g, g^*, every j, and every B-event B. If $K = 2$, then $\succsim_j$ also satisfies the corresponding hexagon condition: $f_B g_{-B} \sim_j f_B^* g_{-B}^*$ and $f_B g_{-B}^{**} \sim_j f_B^* g_{-B}$ and $f_B^* g_{-B} \sim_j$ $f_B^{**} g_{-B}^* \Rightarrow f_B^* g_{-B}^{**} \sim_j f_B^{**} g_{-B}$.

In other words, the decision-maker's preferences satisfy coordinate independence unconditionally with respect to $\mathcal{A}$-events and conditionally with respect to $\mathcal{B}$-events given any element of $\mathcal{A}$. $\mathcal{A}$-independence and conditional $\mathcal{B}$-independence are similar to the time-0 and time-1 substitution axioms of Kreps and Porteus (1979), adapted to a setting of uncertainty rather than uncertainty and stripped of their temporal interpretation. (See also Segal (1987, 1990) and Grant et al. (1998) for related ideas.)

The intuition for $\mathcal{A}$-independence is similar to the intuition for restrictions of independence in other axiomatic models that rationalize ambiguity aversion (e.g., comonotonic independence, certainty independence, weak certainty independence), namely that it requires the direction of preference between two acts to be preserved only under common payoff substitutions or mixtures that do not hedge the ambiguity of one act more than the other. Coordinate independence requires that when two acts agree in some states, it doesn't matter how they agree there: if the agreeing payoffs are replaced by some other agreeing payoffs, the direction of preference must remain the same. This is violated in Ellsberg's experiments because the agreeing payoffs are changed precisely so as to hedge the ambiguity in one of the acts while injecting ambiguity into the alternative act. For example, in the two-urn problem of Table 7.1, if the agreeing payoffs of R_A and R_B in the first and last rows are replaced by the opposite payoffs, then R_A becomes B_B and R_B becomes B_A, so coordinate independence requires $R_B \succ R_A$ if and only if $B_A \succ R_B$, which is violated by most subjects because this substitution eliminates the ambiguity in the expected payoff of R_A while adding ambiguity to the expected payoff of B_A. $\mathcal{A}$-independence fails to bite on this problem because R_A and R_B do not agree on either $\mathcal{A}$-event, so the typical preference pattern is not disallowed.

Similarly, in the single-urn problem of Table 7.2, if the agreeing payoffs of 0 for R and Y in the last row are replaced by agreeing payoffs of 300, then R becomes RB and Y becomes BY, so coordinate independence requires $R \succ Y$ if and only if $RY \succ YB$, which is violated by most subjects because this substitution eliminates the ambiguity in the expected payoff of Y while adding ambiguity to the expected payoff of R. Again, $\mathcal{A}$-independence fails to bite because R and Y (or B) do not agree on either $\mathcal{A}$-event.

The intuition for conditional $\mathcal{B}$-independence is that conditioning on a singleton in $\mathcal{A}$ resolves all of the ambiguity in the situation, leaving only the uncertainty of the $\mathcal{B}$-events to be considered, for which the decision-maker is assumed to have (state-dependent) SEU preferences.

The various other characteristics of the utility function (7.3) that will appear in the main theorem have the following functional forms derived from $\{u_j\}$ and $\{v_{jk}\}$ and $\{\pi_{jk}\}$. The local *risk aversion matrix* $\mathbf{R}$ is unique in this case and is the sum of a diagonal matrix and a block-diagonal matrix, with $r_{jk,mn} = 0$ if $j \neq m$ and otherwise

$$r_{jk,jn} = -\frac{u_j''}{u_j'}v_j' - \frac{v_{jk}''}{v_{jk}'}1_{kn} \tag{7.5}$$

where 1_{kn} is the indicator for $k = n$. The *normalized risk aversion matrix* is defined as before by $\overline{\mathbf{R}} = \mathbf{R} - \overline{\mathbf{\Pi}}\mathbf{R}$ where where $\overline{\mathbf{\Pi}}$ is the matrix whose rows are all equal to π.

The elements of the *matrix of derivatives of risk-neutral probabilities* $D\pi$ are:

$$[D\pi]_{jk,mn} = [-\mathbf{\Pi}\overline{\mathbf{R}}]_{jk,mn} = \pi_{jk}(\lambda_{mn} - r_{jk,mn}) \tag{7.6}$$

where $\lambda_{mn} := \sum_{j=1}^{J} \sum_{k=1}^{K} \pi_{jk}r_{jk,mn}$ is the *risk-neutral expectation of the mn^{th} col-*

umn of $\mathbf{R}$.

Let $\mathbf{s}$ and $\mathbf{t}$ denote JK-vectors uniquely defined by:

$$s_{jk} \quad : \ = -\frac{u_j''}{u_j'}v_{jk}' \tag{7.7}$$

$$t_{jk} \quad : \ = -\frac{v_{jk}''}{v_{jk}'},$$

which are positive numbers that measure local *second-order and first-order uncertainty aversion* in state jk. In particular t_{jk} is a first-order state-dependent Arrow-Pratt measure for money received in state jk. Then let s_j denote the marginal sum of s_{jk}:

$$s_j := \sum_{k=1}^{K} s_{jk}, \tag{7.8}$$

which is a second-order state-dependent Arrow-Pratt measure for first-order utility received in event j. These parameters are all measured in units of 1/money. (In the formula for s_{jk} the units of second-order utility cancel out in the quotient u_j''/u_j' and the units of first-order utility in the denominator are canceled by multiplication by v_{jk}'.)

For example, in the 2×2 case the risk aversion matrix is constructed as follows, with zeros not shown:

$$
\mathbf{R}(\mathbf{f}) =
\begin{bmatrix}
s_{11} & s_{12} & & \\
s_{11} & s_{12} & & \\
& & s_{21} & s_{22} \\
& & s_{21} & s_{22}
\end{bmatrix}
+
\begin{bmatrix}
t_{11} & & & \\
& t_{12} & & \\
& & t_{21} & \\
& & & t_{22}
\end{bmatrix}
\tag{7.9}
$$

Note the subscripts. In the s matrix the first two rows are the same and the last two rows are the same, and the diagonal elements have the same indices in both matrices.

In terms of these parameters of local utility, this chapter's main result is the following:

Theorem 7.1: Uncertainty premia in the two-source model

(a) Axioms SDU1 and SDU2 hold (in addition to SPT1–SPT5) if and only if $\succsim$ is represented by an observable, cardinal utility function U of the nested-additive form

$$
\text{Model I:} \quad U(\mathbf{f}) = \sum_{j=1}^{J} u_j \left(\sum_{k=1}^{K} v_{jk}(f_{jk}) \right), \tag{7.10}
$$

where j and k are indices for $\mathcal{A}$-events and $\mathcal{B}$-events, respectively. v_{jk} is a first-order evaluation function for money in state $A_j B_k$, u_j is a second-order evaluation function for total first-order value in event A_j, and all are strictly increasing, strictly convex, and twice differentiable. U is unique up to positive affine transformations, $\{v_{jk}\}$ are unique up to transformations of the form $\alpha_j v_{jk} + \beta_{jk}$ and $\{u_j\}$ are unique up to transformations of the form $\gamma u_j + \eta_j$ given $\{v_{jk}\}$, where $\{\beta_{jk}\}$ and $\{\eta_j\}$ are arbitrary constants and $\{\alpha_j\}$ and γ are arbitrary positive constants.

(b) Let $\mathbf{x}$ be the payoff vector of a small neutral bet, i.e., one that satisfies $\sum_{j=1}^{J} \sum_{k=1}^{K} \pi_{jk} x_{jk} = 0$, and let $\bar{\mathbf{x}}$ be the J-vector whose j^{th} element is

$$
\bar{x}_j := \sum_{k=1}^{K} \pi_{k|j} x_j \tag{7.11}
$$

i.e., the conditional risk-neutral expectation of $\mathbf{x}$ given event A_j. Then its uncertainty premium $b(\mathbf{x}; \mathbf{f})$ has the following quadratic approximations[4]:

$$b(\mathbf{x}; \mathbf{f}) \quad \approx \quad \widehat{b}(\mathbf{x}; \mathbf{f}) := -\frac{1}{2} \, \mathbf{x} \cdot D\boldsymbol{\pi}(\mathbf{f}) \, \mathbf{x} \qquad (7.12)$$

$$= \quad \frac{1}{2} \, \mathbf{x} \cdot \boldsymbol{\Pi}(\mathbf{f}) \, \mathbf{R}(\mathbf{f}) \, \mathbf{x} \qquad (7.13)$$

$$= \quad \frac{1}{2} \, \mathbf{x} \cdot \boldsymbol{\Pi}(\mathbf{f}) \, \overline{\mathbf{R}}(\mathbf{f}) \mathbf{x} \qquad (7.14)$$

$$= \quad \mathcal{AB}\text{-premium} \; + \; \mathcal{A}\text{-premium} \qquad (7.15)$$

where

$$\mathcal{AB}\text{-premium} \quad : \quad = \frac{1}{2} \sum_{j=1}^{J} \sum_{k=1}^{K} \pi_{jk} t_{jk} x_{jk}^2, \qquad (7.16)$$

$$\mathcal{A}\text{-premium} \quad : \quad = \frac{1}{2} \sum_{j=1}^{J} \pi_j s_j \overline{x}_j^2 \qquad (7.17)$$

(c) In the quadratic approximation formula of part (b), the $\mathcal{A}$-premium of a $B{:}A$-bet is *zero*.

(d) In the quadratic approximation formula of part (b), if the second-order evaluation functions $\{u_j\}$ are all strictly concave, then the $\mathcal{A}$-premium of an $A{:}B$-bet is *positive*, while if $\{u_j\}$ are all linear, then the $\mathcal{A}$-premia are zero for all bets.

Thus, under Model I the approximate uncertainty premium $\widehat{b}(\mathbf{x}; \mathbf{f})$ is decomposed into the sum of an $\mathcal{AB}$-premium and an $\mathcal{A}$-premium, where the latter may represent the effect of ambiguity. The formulas for both of them have an Arrow-Pratt functional form in which expectations of weighted squared payoffs are computed with respect to risk-neutral probabilities. In general the $\mathcal{A}$-premium could represent issues other than ill-defined probabilities. Strictly speaking it represents source-dependence of the decision-maker's attitude toward uncertainty.

By construction, an $A{:}B$-bet and its corresponding $B{:}A$-bet always have the same (approximate) $\mathcal{AB}$-premium, which is determined from the decision-maker's joint risk-neutral probability distribution exactly as in the case of state-dependent expected utility preferences. However, the $\mathcal{A}$-premium of the $B{:}A$-bet is zero while the $\mathcal{A}$-premium of the $A{:}B$-bet is strictly positive for an ambiguity averse decision-maker, i.e., one for whom u is a strictly concave function.

---

[4]The second of these expressions, which involves the (un-normalized) risk aversion matrix $\mathbf{R}$, is not strictly the same as the others but it coincides if $\mathbf{x}$ has zero risk-neutral expectation as is assumed here.

In this model the evaluation functions $\{u_j\}$ and $\{v_{jk}\}$ are state-dependent and there is no separation of probability from utility. A state-independent representation that separates probability from utility—at least in an "as if" sense—can be obtained by imposing stronger conditions of tradeoff consistency, which are a cross-state form of independence (Wakker (1989), Wakker and Tversky (1993)). Henceforth, let $a\mathbf{f}_{-jk}$ denote the act that yields a in state A_jB_k and which agrees with $\mathbf{f}$ in every other state. Let $a\mathbf{f}_{-j}$ denote the act that yields a in event A_j and which agrees with $\mathbf{f}$ in every other event. Define an act $\mathbf{f}$ to be "$\mathcal{A}$-constant" if its payoffs are constant within each event A_j (i.e., if its payoffs depend only on which element of $\mathcal{A}$ occurs).

Axiom SDU3 (conditional $\mathcal{B}$-tradeoff consistency): For any acts $\mathbf{f}$, $\mathbf{g}$, $\mathbf{y}$, $\mathbf{z}$, and any $j, j^* \in \{1, ..., J\}$ and $k, k^* \in \{1, ..., K\}$: If $a\mathbf{f}_{-jk} \preccurlyeq_j b\mathbf{g}_{-jk}$ and $c\mathbf{f}_{-jk} \succcurlyeq_j d\mathbf{g}_{-jk}$ and $a\mathbf{y}_{-j*k*} \succcurlyeq_{j*} b\mathbf{z}_{-j*k*}$ then $c\mathbf{y}_{-j*k*} \succcurlyeq_{j*} d\mathbf{z}_{-j*k*}$.

Axiom SDU4 (tradeoff consistency for $\mathcal{A}$-constant acts): For any $\mathcal{A}$-constant acts $\mathbf{f}$, $\mathbf{g}$, $\mathbf{y}$, $\mathbf{z}$, and any $j, j^* \in \{1, ..., J\}$: if $a\mathbf{f}_{-j} \preccurlyeq b\mathbf{g}_{-j}$ and $c\mathbf{f}_{-j} \succcurlyeq d\mathbf{g}_{-j}$ and $a\mathbf{y}_{-j*} \succcurlyeq b\mathbf{z}_{-j*}$ then $c\mathbf{y}_{-j*} \succcurlyeq d\mathbf{z}_{-j*}$.

Theorem 7.2: Separation of probabilities and utilities in the two-source model

Axioms SDU3-SDU4 hold (in addition to SPT1-SPT5, SDU1-SDU2) if and only if $\succsim$ is represented by a utility function of the form:

$$\text{Model II:} \quad U(\mathbf{f}) = \sum_{j=1}^{J} p_j u \left(\sum_{k=1}^{K} q_{k|j} v(f_{jk}) \right), \qquad (7.18)$$

where $\mathbf{p}$ is a unique marginal probability distribution on $\mathcal{A}$, $\mathbf{q}$ is a unique conditional probability distribution on $\mathcal{B}$ given A_j, v is a strictly increasing state-independent first-order Bernoulli utility function unique up to positive affine transformations, and u is a strictly increasing state-independent second-order Bernoulli utility function unique up to positive affine transformations.

Under Model II the decision-maker behaves as though she assigns probability $p_j q_{j|k}$ to state A_jB_k, and she bets on $\mathcal{A}$-events as though her utility function for money is $u(v(w))$. Meanwhile, conditional on any element of $\mathcal{A}$ (or unconditionally if $\mathcal{A}$ and $\mathcal{B}$ are probabilistically independent so that q_{jk} does not depend on j), she bets on $\mathcal{B}$-events as though her utility function for money is $v(x)$. (This result is very similar to Ergin and Gul's Theorem 3, except that $\mathcal{A}$ and $\mathcal{B}$ are not required to be probabilistically independent.) The usual caveats apply here, namely that the uniqueness of the probabilities depends on the conventional but unverifiable assumption that u and v do not have state-dependent scale factors, and the set of acts that are $\mathcal{A}$-constant may not be apparent to an observer if the decision-maker has unknown prior stakes in the $\mathcal{B}$-events.

If u is strictly concave in Model II, the decision-maker is uniformly more uncertainty-averse to bets on $\mathcal{A}$-events than to bets on $\mathcal{B}$-events, despite appearing to have equally precise probabilities for both: concavity of v models her aversion to uncertainty per se while concavity of u models additional aversion to betting on $\mathcal{A}$-events. In the illustrative utility functions (7.1) and (7.2) that rationalize Ellsberg's two-urn and single-urn paradoxes, v is linear and u is concave, hence the decision-maker is uncertainty-neutral but ambiguity averse.

These models could in principle be extended to 3-fold partitions, 4-fold partitions, etc., all having different degrees of uncertainty, via deeper nestings of the independence axiom, although a 2-fold partition suffices to demonstrate a source-dependent ambiguity effect.

7.4   A 3×3 example of a two-source model

For a more complex example, consider a decision problem with state space $\{A_1, A_2, A_3\} \times \{B_1, B_2, B_3\}$, and suppose that the decision-maker's beliefs and state-dependent marginal utilities combine to yield the following local risk-neutral probabilities in the 9 states:

TABLE 7.5: Risk-neutral probabilities of all states

	B_1	B_2	B_3
A_1	.05	.20	.05
A_2	.20	.10	.05
A_3	.10	.05	.20

Now consider a pair of $A{:}B$ and $B{:}A$ bets in which $A = A_1 \cup A_2$ and $B = B_1$. Collecting terms, the risk-neutral probabilities for these events are
and the payoff tables for the bets (in which the payoffs are inversely proportional to the risk-neutral probabilities) are

TABLE 7.6: Risk-neutral probabilities for a 2×2 partition

	B_1	$B_2 \cup B_3$
$A_1 \cup A_2$	.25	.40
A_3	.10	.25

TABLE 7.7: Payoffs of corresponding test bets

Bet $\mathbf{x}_{A:B}(\Delta)$			Bet $\mathbf{x}_{B:A}(\Delta)$		
	B_1	$B_2 \cup B_3$		B_1	$B_2 \cup B_3$
$A_1 \cup A_2$	$+4\Delta$	$+2.5\Delta$	$A_1 \cup A_2$	$+4\Delta$	-2.5Δ
A_3	-10Δ	-4Δ	A_3	$+10\Delta$	-4Δ

A decision-maker who is uniformly more uncertainty-averse toward bets on $\mathcal{A}$-events than toward bets on $\mathcal{B}$-events, consistent with a utility function of the form (7.5) or (7.21), will assign a higher uncertainty premium to the first of the two bets and will therefore strictly prefer the second one for any small Δ, positive or negative. The same would be true for any other 2×2 partition of the two sets of events, for example, $A = A_2$ and $B = B_1 \cup B_2$, which yields the following pair of bets, of which the second would again be strictly preferred:

TABLE 7.8: Test bets for a different 2×2 partition

Bet $\mathbf{x}_{A:B}(\Delta)$			Bet $\mathbf{x}_{B:A}(\Delta)$		
	$B_1 \cup B_2$	B_3		$B_1 \cup B_2$	B_3
A_2	$+3.33\Delta$	$+20\Delta$	A_2	$+3.33\Delta$	-20Δ
$A_1 \cup A_3$	-2.5Δ	-4Δ	$A_1 \cup A_3$	$+2.5\Delta$	-4Δ

A caveat here is that a systematic preference for B:A-bets over A:B-bets may be apparent only for small values of Δ if the situation is one in which the decision-maker is relatively much more uncertainty-averse than ambiguity averse (i.e., when the second-order premium term in part (d) of Theorem 7.1 is comparatively small) and risk-neutral probabilities have extreme values. For example, if event $A\overline{B}$ has a very low risk-neutral probability, then the payoffs of $\mathbf{x}_{A:B}(\Delta)$ and $\mathbf{x}_{B:A}(\Delta)$ in event $A\overline{B}$ will be very large in magnitude with opposite signs, as they are in the last pair of payoff tables ($\pm 20\Delta$). A decision-maker who is highly uncerainty-averse will wish to avoid a large negative payoff and therefore might prefer $\mathbf{x}_{A:B}(\Delta)$ if Δ is positive and large enough to invalidate the quadratic approximation, obscuring the fact that she may be modestly ambiguity-averse as well.

7.5   The second-order-uncertainty smooth model

The functional form (7.18) for the utility function can be given another interpretation, in which the decision-maker has second-order uncertainty about her first-order probabilities and/or utilities, and she is averse to that as well as to first-order uncertainty about payoffs. In this interpretation, the $\mathcal{B}$-events are first-order events that are observable, while the $\mathcal{A}$-events are second-order events that are credal states of the decision-maker, each characterized by its own first-order subjective probability distribution over the first-order events and its own state-dependent and perhaps even credal-state-dependent Bernoulli utility function representing first-order aversion to uncertainty. First-order acts and second-order acts are defined as acts whose payoffs depend only on first-order events or second-order events, respectively. Second-order acts are not really available. They are just tools of thought that the decision-maker or her theorist may use to construct her first-order preferences. To facilitate comparison with the two-source model, let the set of second-order states be henceforth denoted by $\mathcal{C}$ and indexed by i ($= 1, ..., I$), and let the first-order states have a two-way partition $\mathcal{A} \times \mathcal{B}$ with elements indexed by jk. The general state-dependent version of the utility function is then

$$U(\mathbf{f}) = \sum_{i=1}^{I} u_i \left(\sum_{j=1}^{J} \sum_{k=1}^{K} v_{ijk}(f_{jk}) \right), \qquad (7.19)$$

and the particular case that is obtained when the first- and second-order utility functions are entirely state-independent is

$$U(\mathbf{f}) = \sum_{i=1}^{I} \mu_i u \left(\sum_{j=1}^{J} \sum_{k=1}^{K} p_{jk|i} v(f_{jk}) \right), \qquad (7.20)$$

where μ_i is the probability of credal state i and $p_{jk|i}$ is the probability of first-order event jk in credal state i. This is the discrete form of the KMM model, which was axiomatized in a Savage-type framework[5] with the set of consequences being an interval of real numbers (like amounts of money). Insofar as it has the same functional form as the two-source model, it could also be axiomatized in the state-preference framework, although (as in the KMM model) this would require the explicit articulation of preferences over second-order acts, and here it would also require a corresponding reinterpretation of the compound application of the independence axiom.

---

[5]Some questions about the axiomatization of second-order beliefs in this framework have been raised by Epstein (2010), with a response from Klibanoff et al. (2011a).

Notwithstanding those issues, the results of the previous section concerning uncertainty premia will be recast here, and the risk-neutral probabilities and conditional expectations will be indexed by i as well as j and k as if bets on the credal state were possible, but the status quo wealth vector $\mathbf{f}$ and the bet $\mathbf{x}$ will be indexed only by j and k so as to reflect only first-order payoffs, i.e., acts that are concrete.

The local risk-neutral distribution determined by the state-independent version of the second-order uncertainty model is a joint distribution over the first-order and second-order states, and it satisfies

$$\pi_{ijk}(\mathbf{f}) \propto (\mu_i u_i')(p_{jk|i} v_{jk}'), \tag{7.21}$$

where u_i' is shorthand for $u'\left(\sum_{j=1}^{J}\sum_{k=1}^{K} p_{jk|i} v(f_{jk})\right)$ and v_{jk}' is shorthand for $v'(f_{jk})$, again suppressing the dependence of all terms on $\mathbf{f}$. Let u_i'' and v_{jk}'' similarly stand for $u''\left(\sum_{j=1}^{J}\sum_{k=1}^{K} p_{jk|i} v_{jk}(f_{jk})\right)$ and $v''(f_{jk})$, and also let

$$v_i' := \sum_{j=1}^{J}\sum_{k=1}^{K} p_{jk|i} v_{jk}' \tag{7.22}$$

be the *expected marginal utility of money* in credal state i.[6] Next, define vectors $\mathbf{s}(\mathbf{f})$ and $\mathbf{t}(\mathbf{f})$ exactly as before:

$$s_i := -(u_i''/u_i')v_i' \tag{7.23}$$

$$t_{jk} := -v_{jk}''/v_{jk}' \tag{7.24}$$

Again, t_{jk} is interpretable as a first-order state-dependent Arrow-Pratt measure for aversion to uncertainty about payoffs, but now s_i is a second-order state-dependent Arrow-Pratt measure for expected utilities in credal states, which measures aversion to uncertainty about the credal state. t_{jk} is not also subscripted by i (unlike the analogous dependence in the two-source model) because neither v nor $\mathbf{x}$ depends on the credal state when only first-order acts are considered.

Now let $\pi_i := \sum_{j=1}^{J}\sum_{k=1}^{K} \pi_{ijk}$ denote the risk-neutral probability of credal state C_i, let $\pi_{jk|i} := \pi_{ijk}/\pi_i$ denote the conditional risk-neutral probability of $A_j B_k$ given C_i, and let $\pi_{jk} := \sum_i \pi_{ijk}$ denote the unconditional risk-neutral probability of $A_j B_k$. (Note that π_{jk} is observable through preferences for real bets, while π_i, π_{ijk}, and $\pi_{jk|i}$ are not. They are imaginary betting rates that characterize the credal states.) The conditional risk-neutral expected payoff of a bet $\mathbf{x}$ in credal state i is

$$\overline{x}_i := \sum_{j=1}^{J}\sum_{k=1}^{K} \pi_{jk|i} x_{jk}, \tag{7.25}$$

---

[6]The first-order marginal utility of money, v_{jk}', is independent of the credal state, but its expected value varies with i insofar as the probabilities of first-order events depend on i.

in terms of which we have the following analog of Theorem 3:

Theorem 7.3   Under the second-order uncertainty model, the uncertainty premium of a neutral bet **x** has the local quadratic approximation:

$$\widehat{b}(\mathbf{x}; \mathbf{f}) = \frac{1}{2} \sum_{i=1}^{I} \pi_i s_i \overline{x}_i^2 + \frac{1}{2} \sum_{j=1}^{J} \sum_{k=1}^{K} \pi_{jk} t_{jk} x_{jk}^2. \qquad (7.26)$$

In this expression, which looks almost exactly like the one in the previous theorem, the summation on the right is the quadratic approximation of the uncertainty premium that would be assessed by a decision-maker with state-dependent expected utility preferences whose risk-neutral probabilities are $\{\pi_{jk}\}$ and whose state-dependent Arrow-Pratt uncertainty aversion measures are $\{t_{jk}\}$. The summation on the left is the quadratic approximation of an ambiguity premium assessed by a decision-maker who is also averse to the uncertainty about her credal state, as represented by a concave second-order utility function u. The terms in that summation are not directly observable, but the ambiguity premium can be backed out if π_{jk}, t_{jk}, and $\widehat{b}(\mathbf{x}; \mathbf{f})$ can be assessed separately.

An important difference between this model and the previous one is that it has a lot of free parameters to think about, not merely a measure of greater aversion to betting on $\mathcal{A}$-events. A constructive application of it would require an explicit enumeration of credal states and an assignment of both first-order and second-order probabilities rather than just the assessment of a second-order utility function to represent greater aversion to uncertainties that are more ambiguous. Even Ellsberg's two-urn problem offers a rich set of possibilities. The decision-maker might believe that the ambiguous urn contains either all red balls or all black balls, modeled by a binary distribution. Or she might believe that all proportions are equally likely, modeled by a uniform distribution. She might think there is some chance that the two urns are identical after all. Her second-order beliefs could be a mixture over these scenarios and many others, and it would be hard to determine the parameters of such a representation from observations of behavior. It is more appropriate to think of this model as one that is attributed to an idealized decision-maker for purposes of exposition, say, for deriving qualitative testable implications or comparative statics.

7.6 Discussion

State-preference theory provides a convenient starting point for modeling ambiguity-averse preferences insofar as it doesn't give a privileged role to the independence axiom in the first place. Its most basic assumptions are those of consumer theory, in which the additivity property of utility that follows from the independence axiom is quite often behaviorally unrealistic. The intuition it brings to the modeling of problems such as Ellsberg's paradox is that consequences received in mutually exclusive events may be complementary goods under some conditions, and the decision-maker's appetite for some types of uncertainty may be satiated more quickly than others. These are familiar sentiments when choosing among other types of consumption bundles. It is well-suited to modeling decisions that take place in financial markets, because it gives a very privileged role to money, and the rationality principles that do or do not apply to the individual are largely the same as those that do or do not apply to groups of investors or the market as a whole. Models of smooth ambiguity-averse preferences are readily constructed in this framework, and two such models—the source-dependent-utility model and the second-order-uncertainty (KMM-type) model—allow a decomposition of an overall uncertainty premium into distinct components for uncertainty and ambiguity via formulas that nicely generalize Pratt's result.

In KMM's second-order-uncertainty model, a focal objective is to obtain a clean separation between the characterizations of uncertainty and ambiguity and also a clean separation between attitudes toward them, which is important from a theoretical modeling perspective. In their setup, the first-order probability distributions represent perceptions of uncertainty about which state will happen, and the second-order distribution represents ambiguity about which perception is correct. The first-order utility function models aversion to uncertainty, and the second-order utility function models aversion to ambiguity. These separate parameters are uniquely identified, although this requires the articulation of preferences on sets of counterfactual acts even stranger than those in Savage's model.[7]

Denti and Pomatto (2022) have recently developed an alternative axiomatization of the second-order-uncertainty smooth model which does not involve unobservable credal states, building off an interpretation suggested by KMM and further

---

[7]The second-order-uncertainty model is also sensitive to background risk in a way that can obscure the presence of ambiguity aversion. (Nau 2006a) In particular, if the decision-maker's exposure to ambiguity has been hedged while her exposure to risk has been amplified through correlated prior investments, she may exhibit less aversion to neutral bets on ambiguous events than to neutral bets on unambiguous ones.

explored/implemented by Klibanoff et al. (2014, 2022). It separates events into two categories: those that satisfy the sure-thing principle and those that don't, as reflected in the decision-maker's preferences. The key new axiom states that an act **f** is unambiguously preferred to another act **g** if **f** is preferred to **g** conditional on every event that satisfies the sure-thing principle. This allows the set of ambiguous events to be determined endogenously from preferences without the introduction of second-order acts. It leads to a version of the smooth model in which the first-order probability distribution over payoff-relevant events is identifiable from the state that occurs, as if one element of the state vector is an indicator for the probability law that applies to the others. The second-order distribution applies to that element of the state vector, which is assumed to be observable ex post. In its application to Ellsberg's 2-urn problem, the model thus assumes that the composition of the ambiguous urn will eventually be revealed.

The source-dependent version of the model provides a parsimonious approach to restricting preferences in a way that allows ambiguity aversion to be separated from uncertainty aversion in simple two-source examples such as Ellsberg's. The local curvature of the decision-maker's indifference curves reveals her attitude toward different sources of uncertainty regardless of the status quo, and this can be measured through observations of preferences for small neutral bets as illustrated in the Alice-Bob-Carol-David example in the previous chapter. It needs as little as one additional parameter to represent attitudes toward ambiguity in the compositions of Ellsberg's two urns. The decision-maker may choose a convenient parametric form for a second-order utility function, such as an exponential or power function, without also facing the high-dimensional problem of selecting a personal set of second-order probability distributions. (What distributions would *you* choose or advise others to choose to represent ambiguous beliefs in the urn problems?). In the general model of Theorem 7.1 there is no attempt to separate probability from utility and no explicit representation of ambiguity in terms of sets of probabilities. Rather, there are just higher degrees of aversion to some sources of uncertainty, which could have many explanations. In the more restrictive model of Theorem 7.2, probabilities are separated from utility via additional assumptions of tradeoff consistency, but the probabilities themselves need not be ambiguous. Again, there are just situation-dependent differences in attitudes toward uncertainty. In neither model is there any reference to acts that are intrinsically counterfactual.

The use of the state-preference framework for the models of ambiguity aversion described above is lacking in generality insofar as it deals only with consequences that are units of real money, and it depends on the payoffs on any given dimension (corresponding to a particular source of uncertainty) being large enough to reveal curvature of the utility function along that dimension but not so large as to invalidate a quadratic approximation. However, this is the most elementary setting for modeling attitudes toward uncertainty and ambiguity in a way that applies to behavior outside the laboratory, such as financial interactions between agents and transactions in stock markets and sports betting markets, while working in units that are subject to precise numerical measurement. State-preference theory also directly addresses fundamental measurement issues such as the fact that utility for money (or anything else) may be state-dependent due to background uncertainty or quality-of-life issues that are inherent in the definitions of states. Where events of personal importance are concerned, there is no such thing as a riskless financial position or a riskless state-of-being in the status quo, so these imaginary landmarks ought not to have special significance. Any modeling framework that aspires to greater generality still needs to address these issues.

7.7 Some history of non-expected-utility theory

Over the last 50 years an immense literature on the topic of ambiguity has emerged, as part of the broader field of non-expected-utility. This section provides a brief review.

In the 1940s and 1950s, in parallel with early developments in expected utility theory, experimental psychologists were already exploring issues such as non-linear transformation of objective probabilities and differing attitudes toward gain versus loss probabilities (e.g., Preston and Baratta (1948), Edwards (1953, 1954, 1955, 1962)). Unlike economists of the time, they were not so tied to the rationality/normative justification of expected utility. Meanwhile, Allais (1953) and Ellsberg (1961) devised paradoxical thought-experiments which showed that not only do decision-makers empirically deviate from the prescriptions of expected utility and subjective expected utility theory: they do it deliberately in some situations. They may be unduly averse to very small probabilities of loss and averse to ambiguity in the values of probabilities. These ideas gradually gained traction and seem to have first entered the economic theory literature in a paper by Handa (1977) which proposed a "new theory of cardinal utility" that used transformations of probabilities in the evaluation of risky prospects. It had serious flaws which provoked a number of critical letters to the editor of *Journal of Political Economy*, whose writers included Peter Fishburn (1978), Mark Machina, and John Quiggin.[8]

---

[8] Only Fishburn's letter was published.

Subsequently, economists began working on more rigorous approaches to generalizing the expected-utility axioms.

Another watershed event was the 1979 publication of Kahneman and Tversky's paper on "prospect theory." It is a descriptive model of robust patterns of behavior that are typically seen in experiments. One of its key ideas is that preferences for risky acts (prospects) are based on the gains and losses that they yield relative to a reference point (often the status quo), not their absolute coordinates in the terminal payoff space. Utility functions are downward kinked at the status quo and bend in different directions on either side, concave for gains and convex for losses. Another of the key ideas is that "decision weights" that are used in evaluating uncertainty are systematically distorted relative to probabilities by a reverse-S-shaped "probability weighting function." This leads subjects to act as if they overestimate very small probabilities and underestimate moderate to large probabilities. Kahneman and Tversky showed that these assumptions are able to explain a variety of well-documented preference anomalies, including the Allais paradox.[9,10] Whereas every single ingredient of prospect theory had been described before, prospect theory was the first to put it all together in an operational model. Thus, it was the first "rational model of irrational behavior," something thought impossible up to that point, and it established a new behavioral approach in decision theory.

The same stylized facts were addressed by new and more general axiomatic models that economists and decision theorists were developing. In some of them the key is to relax the independence axiom in a way that allows preferences to depend nonlinearly on probabilities (Machina (1982), Chew (1983)). For example, in a situation involving 3-outcome gambles with objective probabilities, indifference curves in the probability simplex do not need to be parallel lines: they could "fan out." Some models seek to retain some features of subjective probabilistic beliefs while generalizing other aspects of preferences (Machina and Schmeidler (1992), Machina (2004, 2005)). Other models allow for systematic distortions of cumulative probabilities in addition to systematic distortions of payoffs (Quiggin (1982), Yaari (1987), Segal (1987), Schmeidler (1989), Diecidue and Wakker (2001)). In some models indifference curves are allowed to be kinked at riskless points in payoff space as discussed below. Incompleteness of preferences and state-dependence of utility have also been

---

[9]The original version of prospect theory had a problem: the distortion of probabilities could lead to behavior that violates first-order stochastic dominance. The problem was later corrected in Tversky and Kahneman (1992) and Wakker and Tversky (1993), who incorporated Quiggin's (1982) systematic distortion of cumulative rather than separate-outcome probabilities. Cumulative prospect theory has become the most widely studied non-expected utility model. A definitive treatment of the topic is given by Wakker (2010).

[10]Source-dependent uncertainty attitude in the context of prospect theory has been studied by Tversky and Fox (1995) and Abdellaoui et al. (2011).

widely studied as described in Chapter 5. Some of these generalized models are for decision under risk, where they aim at explaining phenomena such as Allais' paradox. Others are for decision under uncertainty and ambiguity where they seek to explain phenomena such as Ellsberg's paradox. Surveys of the literature on ambiguity can be found in Etner et al. (2012), Trautmann and van de Kuilen (2015), and Gilboa and Marinacci (2016).

The incomplete-preference approach to the modeling of ambiguous beliefs, which relaxes the completeness axiom rather than the independence axiom of SEU theory, is historically the oldest. It originated in the fields of statistics and philosophy (Koopman (1940), Smith (1961), Levi (1974)) in the context of Bayesian inference with imprecise prior probabilities, which has become its own subfield since the publication of Walley's (1991) book. Even de Finetti's "Fundamental Theorem of Probability" (1974) is stated in terms of lower and upper bounds on subjective probabilities. A theory of expected utility without the completeness axiom was first proposed by Aumann (1962), and it leads to a representation of preferences by convex sets of utility functions as shown in Theorem 4.3. Relaxation of the completeness axiom in the context of subjective expected utility, leading to joint imprecision in probabilities and utilities, has been explored by Seidenfeld et al. (1990a, 1995), Nau (2006b), Ok et al. (2012), Galaabaatar and Karni (2012, 2013), and many others, as discussed in Sections 5.4 and 5.5. A model of preferences that are incomplete due to indeterminacy of probabilities was introduced into microeconomics by Bewley (1986), who referred to it as "Knightian" uncertainty.[11] Its implications for phenomena such as general equilibrium have been elaborated by Rigotti and Shannon (2005) and others. A decision-maker with incomplete preferences does not have fixed indifference curves, and she may exhibit indecision or inertia when called upon to make a choice under ambiguous conditions. This model provides a natural explanation for incompleteness of markets. Indeed, an agent with incomplete preferences is a microcosm of an incomplete market.

The kinked-preference approach originated in the work of Schmeidler (1989) and Gilboa and Schmeidler (1989), who proposed models in which the decision-maker's beliefs are represented by a Choquet capacity (extending Quiggin's 1982 model to uncertainty) or else by a set of priors accompanied by a decision rule (maximization of minimum expected utility) that completes the preference relation despite the imprecision of beliefs. A key property of Schmeidler's (1989) model is that the scope of the independence axiom is limited to comparisons of acts that are "comonotonic," i.e., whose outcomes determine the same rank-ordering of states,

---

[11]The term is based on a distinction between risk and uncertainty proposed by Keynes (1921) and Knight (1921), but that was prior to the development of subjective probability theory and its integration with expected utility theory in microeconomics and finance.

thus not directly hedging each other's risk. The preferences of a Choquet-expected-utility decision-maker have an indifference curve representation in which the curves have kinks at points where the same outcome is received in two or more states, at which she exhibits first-order uncertainty aversion rather than second-order uncertainty aversion. That is, her uncertainty premia decrease linearly rather than quadratically as payoffs go to zero. The maxmin expected utility model may also yield kinked indifference curves: kinks occur at places where a jump to a different prior occurs when computing minimum expected utility over the set. Local preferences still have an expected-utility representation within regions where the same prior applies.

Many extensions and applications of such models have been explored. Diecidue and Wakker (2002) showed that when the scope of the no-arbitrage criterion is restricted to sets of monetary gambles that are comonotonic, this leads to the Choquet-expected-value model (Choquet expected utility with linear utility for money). Tversky & Kahneman (1992) developed an ambiguity model that generalizes Schmeidler (1989) by incorporating reference dependence and loss aversion. Sarin and Wakker (1998) showed that the multiple priors model satisfies consistency conditions that allow it to be used in dynamic choice situations. The same model was proposed independently by Epstein and Schneider (2003) and has been applied to intertemporal choice in asset markets.

The smooth-preference approach for ambiguity has been developed more recently (Nau (2001, 2006a, 2011b), Klibanoff et al. (2005, 2011ab, 2014, 2022), Chew and Sagi (2008), Ergin and Gul (2009), Seo (2009), Denti and Pomatto (2022)), and it explains ambiguity aversion as a form of ordinary second-order uncertainty aversion in which indifference curves bend in a way that is incompatible with the additive-utility effect of the independence axiom. This representation could have various psychological explanations.

Most of the models in the three aforementioned general categories adhere to the spirit of the SEU model insofar as they resolutely try to separate beliefs from tastes. That is, they typically have one set of parameters (maybe a very big set of parameters) for modeling beliefs about events and another set of parameters for modeling preferences for outcomes. In contrast, the model presented in this chapter does not require the decision-maker to have probabilities that represent beliefs in isolation. In the presence of issues such as background uncertainty and state-dependence of utility, probabilities and utilities cannot be separated from the viewpoint of an observer anyway, as emphasized throughout this book.

Chapter 8

Noncooperative games

8.1 Introduction

Game theory occupies the great middle ground in the spectrum of models of rational choice: "the theory of 2, 3, 4... participants" in the words of von Neumann and Morgenstern. In fulfillment of their vision, game theory has become a universal wrench for modeling interactions among small numbers of self-interested agents ("players") who may be individual humans or groups thereof or animals or plants or even computer programs. As it is conventionally presented, game theory invokes expected utility or subjective expected utility theory to model the objectives of any one player, and to those models it adds new principles of 2, 3, 4...-player rationality: *solution concepts*. These typically take the form of *equilibrium* conditions in which each player's strategy is one that maximizes her expected utility, taking the strategies of the others as given. Most commonly, the solution concept is that of *Nash equilibrium* in which the players choose among singular moves or independently randomized moves. In general there could be multiple equilibria, and in some such cases additional constraints, so-called *refinements*, may be invoked. In other cases it may be desired to impose weaker constraints on solutions, so-called *coarsenings* of

Nash equilibrium, of which the most often used are *correlated equilibrium, rational-izability*[1] and *correlated rationalizability* In games of *incomplete information* the players may have reciprocal uncertainties about each others' utilities and/or there may be exogenous events that are payoff-relevant. In such settings the uncertainties are often modeled by supposing that the players have a *common prior distribution* over them.

Correlated equilibrium (Aumann 1974, 1987) is a simple generalization of Nash equilibrium in which randomized or subjectively uncertain strategies are permitted to be correlated among players, and it will play a starring role in the story that follows. It is usually defined in this way: a correlated equilibrium distribution is a commonly-known prior probability distribution on the game's outcomes which has the property that, if it is used by a mediator to randomly generate a recommended outcome, and each player is privately informed (only) of her personal strategy in that outcome, then it is optimal for the player to follow that recommendation on the assumption that other players will do likewise. This means that when each player performs Bayesian updating on the common prior distribution to obtain a personal posterior distribution on the strategies that will be played by others, it is expected-utility-maximizing for her to comply with the recommendation she received. The story of a mediator and a common prior distribution is not to be taken literally, however. It is an as-if condition that (by duality) is a requirement of ex-post-coherence (no-ex-post-arbitrage) for bets that may be placed on the game's outcome, just as the existence of a subjective probability distribution which assigns positive probability to the state that occurs is an as-if requirement of ex-post-coherent betting on states of nature. (This will be proved in Theorem 8.2) The bets that are postulated have a special structure, which makes them incentive-compatible for the players to accept a priori, regardless of their beliefs about what other players will do. The possibility of betting on the game's outcome is itself only hypothetical (usually), but it is concrete and provides a commonly-understood language for thinking and communicating in quantitative terms.

The usual methods for modeling behavior in games raise many practical and philosophical questions. How much do players need to know about each others' incentives in the game and their methods of reasoning, and how might they come to know it? If the expected utility model of preferences applies to everyone, how do we know *that,* and how much can one player know about the utility function of another? What does it mean for rationality (as defined by having expected-utility

---

[1]Rationalizability and correlated rationalizability will also be discussed later, but briefly, a strategy of a player is rationalizable if it survives a process of iteratively deleting strategies that are not best responses under some system of beliefs regarding the behavior of opponents, with or without or without allowing for correlation.

preferences) to be *common knowledge*, and what does it mean for *utilities* to be common knowledge?[2] And given the expected utility model for individual decision-making, why should the players' joint decisions be predicted or guided by a solution concept such as Nash equilibrium? What if it yields an inefficient result? What if there is more than one such equilibrium? How would the players find their way to the right probabilities when randomizing their strategies? (The set of Nash equilibria can be quite ugly and hard to compute, and randomized strategies could involve probabilities that are irrational numbers.)   If the players do choose randomized stagies, why shouldn't they be allowed to use correlated randomization, yielding a correlated equilibrium? (Such an equilibrium is easily computable and could be a more equitable solution or even a dominant solution.) Why should probabilities that represent reciprocal uncertainty be objective, as if generated by a random device, whether or not they are independent among players? (In most decisions of any importance, probabilities are personal and subjective.) In a game of incomplete information, why should there be a common prior, and how would *its* parameters come to be commonly known? *Where do all the numbers come from?* A player's payoff matrix encodes two different types of data: *personal incentives* and *external effects.*[3] Only the personal incentives data is relevant to the determination of equilibria, while the efficiency of those equilibria may depend on the external effects, as illustrated by the games discussed in Section 8.4 below. The two types of data may require separate methods of elicitation (unilateral vs. multilateral contracts) and they have different kinds of impacts on preferences. The entire payoff matrix cannot necessarily be assumed into existence all at once as an object of common knowledge (as is conventionally done). Lastly, if the players' actions are not consistent with an equilibrium model, particularly in a game that is novel and will be played only once, *so what?* In what sense may we say they have acted irrationally?

---

[2] In conventional models of common knowledge in game theory (e.g., Brandenburger and Dekel (1989)) the payoff structure of the game (the set of joint strategies and the players' utilities attached to them) is generally regarded as data that is known to everyone and attention is focused on the questions of how to model common knowledge of events and common knowledge of rationality and how to endogenously determine rational beliefs. There is no simple answer to the question of how the payoffs themselves become common knowledge in an elementary example such as a 2×2 game that is not repeated or not drawn from a larger set of possible games. At the very end of their 1991 textbook, Fudenberg and Tirole analyze the implications of common knowledge in a couple of simple games (pp. 546-548), and they include this caveat: "Throughout our discussion, we assume that the structure of the game is common knowledge in an informal sense. Applying formal definitions of common knowledge to the structure of the game leads to technical and philosophical problems that we prefer not to address."

[3] Personal incentives are changes in a player's payoffs that are determined by changes in her own move, holding the moves of other players fixed. External effects are changes in a player's payoffs that are determined by changes in *other* players' moves and/or nature's moves, holding her *own* move fixed.

There has long been tension between Bayesian decision theory and noncooperative game theory in regard to these issues.[4] From the viewpoint of Bayesian decision theory, beliefs are subjective and they are the personal property of the decision-maker, not to be constrained a priori except by considerations of coherence. Whereas, from the viewpoint of game theory, beliefs ought to be determined endogenously from commonly known parameters of interlocking decision problems faced by the players.

This chapter will present an approach to noncooperative game theory at its most basic level (strategic-form games[5]) that seeks to reconcile these conflicting views by merely expanding the scope of no-arbitrage standard of rationality to cover the players as a group: "joint" coherence. As such it provides a direct integration of game theory with the models of individual decision-making that have been covered in earlier chapters. Mathematically it is dual to conventional approaches to game theory, in the sense of the duality theorem of linear programming. It gives a special role to money and emphasizes the viewpoint of an observer, and as such it provides the simplest and most tractable setting in which to define and explore the concepts of common knowledge and strategic rationality. In particular:

- Events whose occurrence is common knowledge (i.e., "public" events) can be defined as those upon which it is possible to place gambles.[6]

- Common knowledge of the rules of the game (which events are controlled by which players and what are the personal-incentive elements of their payoff functions) can be achieved through the acceptance of small gambles with a special structure (so-called preference gambles).

- Common knowledge of rationality can be defined by the criterion of joint coherence (no-ex-post-arbitrage), i.e., avoiding sure losses as a group. The existence of a common prior distribution emerges as a theorem, not a primitive assumption.

It will be shown that *these assumptions lead straight to the conclusion that rational outcomes of the game are those which have positive support in correlated equilibria.* Nash equilibrium goes farther than needed to protect the players from arbitrage

---

[4]See the paper by Kadane and Larkey (1982a), which questions whether game theory is consistent with a subjective view of probability. See also the reply by Harsanyi (1982a), the ensuing rejoinders by Kadane and Larkey (1982b) and Harsanyi (1982b), and the subsequent paper by Kadane and Larkey (1983).

[5]A strategic-form game (also known as a normal-form game) is a simultaneous-move game among a finite number of players with finite strategy sets.

[6]Throughout this chapter the terms "gamble" and "bet" will be used interchangeably.

losses, and rationalizability and correlated rationalizability sometimes do not go far enough. A solution of the game is in general a set of correlated equilibria, or a risk-neutral generalization thereof, not a singleton. This method of analysis, which can be applied only in simple environments, is not intended as a set of tools for applications. Rather, it is intended to highlight foundational questions that arise more broadly in game theory in regard to the nature of interactive beliefs, the precision with which they be modeled, and the constraints that they do or don't impose on rational behavior.

8.2   Solution of a 1-player game by no-arbitrage

To set the stage for the solution of a game by the gambling-based method of analysis described above, first consider the more basic problem of a decision under uncertainty: a "game against nature" in which a single player must choose among a finite number of courses of action whose consequences depend on a finite set of states. Suppose that she is also willing to accept monetary side gambles on the game's outcome and that she uses such gambles to communicate with those around her in regard to her beliefs and preferences. For the time being, let's assume that the player acts as if she has state-independent linear utility for money with respect to the side gambles. This assumption is not critical and its relaxation will be explored later. That merely leads to models in which the parameters of revealed beliefs are risk-neutral probabilities.[7]

As a simple example, suppose that Alice faces a decision of whether to carry an umbrella as she leaves her apartment to go to work, and the event that is under nature's control is whether it will be raining when she returns in the evening. A friend is going to give her a ride to work, but she must take the bus on the return trip, and the bus stop is a block away from her apartment. Carrying the umbrella is costly—a nuisance—so she would rather leave it at home if it won't be needed. But getting wet in the rain is also costly. It would be best to have the umbrella with her if and only if it rains. Suppose that the Alice reflects on her preferences by the following method of decision analysis: she asks herself what monetary gambles ought to be acceptable to her in the same state of mind in which she would take the umbrella. And further, suppose that she is willing to publicly reveal this information in the language of conditional gambles which have a special structure: *the gamble's*

---

[7]The method of analysis presented in this section is discussed by Nau and McCardle (1991), Nau (1995a), and Smith and Nau (1995).

payoff depends on the state of nature which occurs, and it is conditioned on the action she chooses.

Suppose that Alice is carrying out her decision analysis in the presence of two neighbors, Bob and Carol. They're having coffee together in Carol's apartment, and in a few minutes Alice is going to go back to her own apartment and gather her belongings to take to work. Bob is a weather expert, and Carol is someone who likes to gamble. After reflecting on her preferences, Alice says the following:

> Alice: "I will take my umbrella only if I believe that the probability of rain is at least 50%, and I won't take it only if I believe that the probability of rain is no more than 50%."

Let's assume that Alice is willing to accept small monetary gambles consistent with her reported probabilities, at the discretion of anyone else in the room. (More concretely, we may *define* her personal probabilities according to the gambles she is willing to accept.) When she says "I will do x only if I believe that the probability of event E is at least [no more than] p," this means that for some arbitrary small positive number α chosen by Carol she will accept a gamble in which, in the event that she does x, she will win [lose] $\$(1 - p)\alpha$ if E occurs and she will lose [win] $\$p\alpha$ if it doesn't. This is equivalent to setting a unit price p at which she is willing to buy [sell] arbitrary small numbers of lottery tickets which yield a payoff of \$1 if E occurs and nothing otherwise, *given* that she does x. Thus, Alice's statement above means that she is willing to accept the following monetary gambles in small non-negative multiples that may be chosen by Carol[8]:

TABLE 8.1: Payoff vectors of Alice's acceptable gambles on rain conditional on taking or not taking her umbrella

	Umbrella Rain	Umbrella No Rain	No Umbrella Rain	No Umbrella No Rain
Gamble #1	1	−1		
Gamble #2			−1	1

Gambles of this nature will be referred to as "preference gambles." They depend on Alice's choice (only) as a conditioning event, and they do not reveal any information about her probabilities for states of nature nor do they assign probabilities to her own actions. They merely reveal information about her comparative utilities in different outcomes of the situation. It is as if her utility function has the following form:

---

[8]Here and elsewhere, blank cells denote zeros.

TABLE 8.2: A possible utility function for Alice

	Rain	No Rain
Umbrella	1	-1
No Umbrella	-1	1

A person who has this utility function for real outcomes and who also has state-independent marginal utility for money *and* who is willing to admit this in public would accept the gambles of Table 8.1 in arbitrary small multiples. However, as with any utility function, the values in Table 8.2 are not uniquely determined. The origin and scale are arbitrary, and an arbitrary constant may also be added to each column without affecting comparisons of expected payoffs between alternatives. For example, an equivalent utility function that better represents the effort of carrying an umbrella and the displeasure of getting rained upon is:

TABLE 8.3: Another possible utility function for Alice

	Rain	No Rain
Umbrella	0	1
No Umbrella	-2	3

It has been obtained from the earlier one by subtracting 1 from Alice's utility in outcomes where it rains and adding 2 in outcomes where it doesn't rain. Thus, it makes different assumptions about external effects that the weather has on Alice. The modified utility function yields an intuitive strict preference ordering of outcomes ({No Umbrella, No Rain} $\succ$ {Umbrella, No Rain} $\succ$ {Umbrella, Rain) $\succ$ {No Umbrella, Rain}), whereas the former one does not, yet they are strategically equivalent in terms of their implications for conditional gambles that Alice would take. This underscores the point that communication and observation which take place under "noncooperative" conditions do not reveal everything about a player's preferences. The parameters of Table 8.1 are observable and actionable, hence they are plausibly common knowledge, while those of Tables 8.2 and 8.3 are not. You may ask: why can't Alice carry out a textbook-style assessment of her utilities (Howard (1968), Raiffa (1968)) and announce the results in public? The answer is that this would entail the contemplation of hypothetical—even counterfactual—acts such as objective lotteries over the 4 outcomes. Such an exercise might help Alice to clarify her preferences in her own mind, but it would have no material implications for anyone else, hence the additional parameters of utility would not be verifiable.

We need not assume, a priori, that Alice is an expected-utility maximizer. She may have any reasons whatever for accepting the gambles summarized in Table 8.1. The supporting utility functions of Tables 8.2 and 8.3 are merely suggestive interpretations of her thinking. But if Alice does think in the fashion of expected-utility analysis, and if she doesn't mind who knows her utilities (or if they are already common knowledge), then it is in her own interest to accept the gambles of Table 8.1 if she has state-independent marginal utility for money. Regardless of the probability that she may assign to the event of rain prior to making her choice, the gambles in Table 8.1 cannot decrease her total expected utility. Indeed, unless she ends up perfectly indifferent between carrying the umbrella or not, they can only increase her total expected utility. The defining property of the gambles in Table 8.1, under an expected-utility interpretation, is that they amplify whatever differences in expected utility Alice perceives between her two alternatives. For example, if she later concludes that carrying the umbrella has higher expected utility than not carrying it because her probability of rain turns out to be greater than 50%, then she will carry the umbrella and gamble #1 will be in force, and gamble #1 yields positive marginal utility precisely in the case that her probability of rain is greater than 50%.

Now suppose that Bob reveals his expert beliefs with the following assertion:

> Bob: "I believe that the probability of rain is at least 75% whether or not you carry your umbrella."

This means that Bob is willing to accept either or both of the gambles in Table 8.4 below, in arbitrary non-negative multiples, at the discretion of Carol. Gambles of this nature will be referred to as "belief gambles." Bob's statement reveals information about his beliefs regarding the event of rain, not his preferences in regard to the consequences of Alice carrying the umbrella. Assuming that money is equally valuable to him whether or not it rains, his own personal probability of rain is evidently at least 75%, and he does not regard Alice's behavior as informative.

TABLE 8.4: Payoff vectors for Bob's acceptable gambles on rain, independent of Alice's decision

	Umbrella Rain	Umbrella No Rain	No Umbrella Rain	No Umbrella No Rain
Gamble #3	1	−3		
Gamble #4			1	−3

The parameters of all four gambles are de facto common knowledge among Alice, Bob, and Carol in the specular sense of that term. Everyone can see them on the table, and everyone understands their material implications in the same way because they are in units of money, and the information they contain can be aggregated in those terms. We are now in a position to predict (or perhaps prescribe) what Alice should do, namely: *she should carry the umbrella, because otherwise there is ex-post-arbitrage.* If Carol takes gamble #2 with Alice (scaling it up by a factor of two) and gamble #4 with Bob, the result is:

TABLE 8.5: Payoff vector of combined acceptable gambles

	Umbrella Rain	Umbrella No Rain	No Umbrella Rain	No Umbrella No Rain
Gamble #2 for Alice (×2)			−2	2
Gamble #4 for Bob (×1)			1	−3
Total payoff to players			−1	−1

Thus, Carol earns a riskless profit in the event that Alice fails to carry her umbrella.

Why should Alice's behavior be constrained in any way by Bob's beliefs? Perhaps she believes that the probability of rain is less than 50% even after hearing Bob's statement, in which case, based on the utility parameters implied by her own previous testimony, she should not carry the umbrella. But if this is so, she should not just be sitting there. She ought to show an interest in betting with Bob, because taking the other side of gamble #4 would have positive expected value for her. And the same for Bob with respect to Alice: taking the other side of gamble #2 will have positive expected value for him if she fails to carry the umbrella. Their revealed beliefs are inconsistent if Alice fails to carry the umbrella, which presents a riskless profit opportunity for Carol, as well as positive-expected-value betting opportunities for themselves. And here is the core principle that links the minds of strategically rational players: *to an observer it doesn't matter whether the gambles on the table have been offered by one agent or by separate agents. The gambles are in units of the same money for everyone.* Bob's statement can just as well be made by Alice herself (perhaps after conferring with him or visiting weather.com), in which case the conclusion is the same: she should take the umbrella, otherwise she might be throwing money away.

Decision analysis has been carried out in a novel fashion in this example. Rather than asking the player to articulate a probability distribution and utility function and then advising her to choose the alternative with the highest expected utility, we asked the player—and perhaps also those around her—to articulate some gambles they were willing to accept and then advised her to choose an alternative that avoids ex-post-arbitrage.

8.3   Solution of a 2-player game by no-arbitrage

The passage from a single-player game against nature to a multi-player game of strategy entails no additional modeling assumptions or rationality concepts. As an illustration, consider a modification of the Alice-Bob-Carol problem in which the uncertainty for Alice is not about whether it will rain but about whether Bob will dump a bucket of water on her as she walks into the building on the way home. He lives in a 3rd-floor apartment directly above the front steps and he has a malicious desire to see her get wet. If she doesn't have the umbrella, he will get pleasure from dumping the water, otherwise he will have wasted his time in filling the bucket and waiting for her to arrive. Alice knows this and she will take her umbrella to work if she thinks it is sufficiently likely that Bob will dump the bucket of water. He can't see whether she has the umbrella as she approaches the building. She keeps it under her coat and can quickly pull it out and raise it if she sees Bob starting to dump the water She enjoys thwarting Bob in this way, but it is inconvenient to take the umbrella to work if she is not going to need it. Thus, each is uncertain about the strategy that will be used by the other and they have opposing interests. Suppose that they both reveal some limited information about their payoffs in this game by offering to accept gambles with Carol that are of the same kind that Alice used in the previous example.

> Alice: "I will take my umbrella only if I believe there's at least a 50% probability that Bob will dump a bucket of water on me when I return, and I won't take it only if I believe that the probability is no more than 50%."

This means she will accept the following gambles:

TABLE 8.6: Payoff vectors of Alice's acceptable gambles on Bob's move, conditional on her own move

	Umbrella Bucket	Umbrella No Bucket	No Umbrella Bucket	No Umbrella No Bucket
Gamble #1	1	−1		
Gamble #2			−1	1

Alice's implied utility function is the same as before (Tables 8.2 and 8.3), with the event label "Rain" merely replaced by "Bucket." Bob is now cast in the role of rainmaker rather than weatherman, and, unlike nature, he wants to see Alice get wet, as indicated by the following claim:

> Bob: "I will dump a bucket of water out the window as Alice returns only if I believe there's at least a 50% probability that she isn't carrying her umbrella, and I won't dump it only if I believe it is no more than 50%."

This means that he will accept the following gambles:

TABLE 8.7: Payoff vectors of Bob's acceptable gambles on Alice's move, conditional on his own move

	Umbrella Bucket	Umbrella No Bucket	No Umbrella Bucket	No Umbrella No Bucket
Gamble #3	−1		1	
Gamble #4		1		−1

He is now revealing information about his own relative utilities for outcomes, not his beliefs. It is as if his payoff function is of the form:

TABLE 8.8: A possible payoff function for Bob

	Bucket	No Bucket
Umbrella	−1	1
No Umbrella	1	−1

because (only) someone with a utility function equivalent to this one would accept the gambles that Bob has accepted (assuming constant marginal utility for money). Putting Alice's and Bob's apparent utility functions together, we find it is as if they are players in a noncooperative game with the payoff matrix:

TABLE 8.9: Payoff functions for both players ("true rules of the game")

	Bucket	No Bucket
Umbrella	1, −1	−1, 1
No Umbrella	−1, 1	1, −1

where the numbers in the cells are the utilities for Alice and Bob respectively.[9] This is the game of "matching pennies," and it has a unique Nash equilibrium in which both players randomly choose among their two alternatives with equal probabilities.

What rational solution(s) of the game can be derived from the information about the rules of the game that has been openly revealed? The set of all acceptable gambles is now as follows:

TABLE 8.10: Payoff vectors of acceptable gambles for both players ("revealed rules of the game")

	Umbrella Bucket	Umbrella No Bucket	No Umbrella Bucket	No Umbrella No Bucket
Gamble #1 for Alice	1	−1		
Gamble #2 for Alice			−1	1
Gamble #3 for Bob	−1		1	
Gamble #4 for Bob		1		−1

As it happens, ex-post-arbitrage is not possible in any outcome, so Alice and Bob may do whatever they please. However, *from Carol's perspective, Alice and Bob appear to believe that they are implementing the Nash equilibrium solution.* That is, they appear to believe that all four outcomes are equally likely. To see this, note that Carol can rescale and add up the gambles in the following way:

TABLE 8.11: Combination of gambles equivalent to betting *on* {Umbrella, Bucket} at odds of 3 to 1 against

	Umbrella Bucket	Umbrella No Bucket	No Umbrella Bucket	No Umbrella No Bucket
Gamble #1 Alice (×4)	4	−4		
Gamble #2 Alice (×2)			−2	2
Gamble #3 Bob (×1)	−1		1	
Gamble #4 Bob (×3)		3		−3
Total payoff to players	3	−1	−1	−1

The bottom line is equivalent to Alice and Bob jointly betting on the outcome {Umbrella, Bucket} at odds of 3-to-1 against—i.e., betting as if the probability of this outcome is at least 25%. Alternatively, the gambles can be combined this way:

---

[9]Each of the utility functions could also be modified w.l.o.g. through manipulation of external effects, in the fashion of Table 8.3.

TABLE 8.12: Combination of gambles equivalent to betting *against* {Umbrella, Bucket} at odds of 3 to 1 against

	Umbrella Bucket	Umbrella No Bucket	No Umbrella Bucket	No Umbrella No Bucket
Gamble #1 Alice (×0)				
Gamble #2 Alice (×2)			−2	2
Gamble #3 Bob (×3)	−3		3	
Gamble #4 Bob (×1)		1		−1
Total payoff to players	−3	1	1	1

Here the bottom line is equivalent to betting as if the probability of {Umbrella, Bucket} is no more than 25%. Hence, between them, Alice and Bob appear to believe the probability of {Umbrella, Bucket} is *exactly* 25%, and by symmetry the same trick can be played with all the other outcomes. *The game already has been solved from the viewpoint of an observer.*

The remarkable feature of this example is that neither Bob nor Alice has revealed any information whatever about his or her present beliefs. They have merely revealed some partial information about their utilities via acceptable conditional gambles. Yet this turns out to be operationally equivalent to asserting beliefs that correspond to a mixed-strategy Nash equilibrium. Why did this happen? It happened because it is a generalization of our earlier no-arbitrage theorems to the case of a game of strategy. By the ex-post fundamental theorem of subjective probability (Section 3.7), we know that the players behave rationally (avoid ex-post-arbitrage) if and only if there is a supporting probability distribution that assigns non-negative expected value to every gamble accepted by every player and assigns strictly positive probability to the outcome that occurs. The supporting probability distribution (or set thereof) can be interpreted to represent the commonly-held (and possibly imprecise) beliefs of the players, notwithstanding any distortions that might be introduced by state-dependent marginal utility for money. If the situation happens to be a game of strategy and the players accept gambles which reveal some information about their relative utilities for its outcomes, in the manner illustrated above, then the supporting probability distribution must be a correlated equilibrium of the game defined by those utilities. (This result will be proved below as Theorem 8.2.) The game between Alice and Bob, which is strategically equivalent to matching pennies, happens to have a unique correlated equilibrium, which is also the unique Nash equilibrium. Hence, as soon as the rules of the game are revealed through acceptable gambles, the players' apparent beliefs about its outcome are uniquely determined.

8.4   Games of coordination: chicken, battle of the sexes, and stag hunt

The matching-pennies game that was analyzed above is a strictly competitive game (in fact, a zero-sum game) with a unique correlated equilibrium, which (by its uniqueness) is also necessarily a Nash equilibrium, giving the appearance that the players are independently randomizing their strategies. It is not necessary for them to be doing this explictly, but that's the way the situation looks to an observer. A few other canonical examples will be analyzed in this section: the well-known games of "chicken" and "battle-of-the-sexes" and "stag hunt." These are games in which coordination may be advantageous, and they are exemplified by the following payoff matrices:

TABLE 8.13a: Chicken

	Straight (B_1)	Swerve (B_2)
Swerve (A_1)	2, 5	4, 4
Straight (A_2)	0, 0	5, 2

TABLE 8.13b: Battle-of-the-sexes

	Ballet (B_1)	Boxing (B_2)
Ballet (A_1)	4, 3	1, 1
Boxing (A_2)	0, 0	3, 4

TABLE 8.13c: Stag hunt

	Stag (B_1)	Hare (B_2)
Stag (A_1)	5, 4	0, 3
Hare (A_2)	3, 0	1, 2

The stories that go with the games are as follows:

- **Chicken:** Alice and Bob are driving cars toward each other at high speed on a one-lane road, and each has the option to go straight or to swerve when they meet. If they both go straight, they will crash and die, which has a utility of 0 for both of them. If one swerves and the other doesn't, the one who swerves gets a utility of 2 for being considered as a chicken and the other gets a utility of 5 for enjoying a moment of excitement and living to tell about it. If both swerve, they each get a utility of 4 for being equally considerate.

• **Battle-of-the-sexes:** Alice and Bob must each choose whether to go to a ballet or to a boxing match for their evening's entertainment. Alice prefers ballet and Bob prefers boxing, but they strongly prefer to go somewhere together rather than separately. Alice gets a utility of 4 if both go to the ballet and a utility of 3 if both go to the boxing match, and vice versa for Bob. If each goes to his or her preferred venue separately, they each get a utility of 1, and if each goes to the *other's* preferred venue separately, they each get a utility of 0.

• **Stag hunt (asymmetric):** In the woods are 3 hares and 1 stag which Alice and Bob may choose to hunt. Both of them are needed to kill the stag, while only one is needed to kill one or more hares. The stag is worth 9 hares, and if both choose to hunt the stag, they will share it in proportions of 5 to 4 hare-equivalents. If only one hunts the stag, that player will get nothing while the other will get all 3 hares. If both hunt the hares, Alice will get 1 and Bob will get 2.

The games are superficially quite different in terms of payoff functions, but they are identical in terms of the constraints they place on the collective beliefs that would be attributed to the players by an observer, because *they have the same revealed-rules matrix*.

Henceforth the revealed-rules matrix of a game will be generically represented by $\mathbf{G}$, with rows denoted by $\{\mathbf{g}_{njk}\}$, where n is an index for a player, j is an index for a strategy of that player, and k is an index for an alternative strategy of the same player. The columns of $\mathbf{G}$ are indexed by outcomes of the game, denoted by s. Row $\mathbf{g}_{njk}$ is the payoff vector for player n of a gamble that is constructed as follows. If s is an outcome in which she chooses strategy j, the value of $\mathbf{g}_{njk}(s)$ is equal to the actual payoff she receives there *minus* the payoff she would have gotten by playing strategy k instead, other things (i.e., other players' strategies) being equal. If s is not an outcome in which player n chooses strategy j, then $\mathbf{g}_{njk}(s) = 0$. (The revealed-rules matrix for the matching-pennies game was shown in Table 8.10.)

The logic of this construction is as follows. In the event that player n is observed to play strategy j, she evidently weakly prefers the vector of payoffs yielded by j over the vector of payoffs yielded by k, in which case she ought to be willing to accept a *gamble* whose payoffs are proportional to those of strategy j minus those of strategy k. Such a gamble has non-negative expected value precisely when strategy j yields an expected payoff greater than or equal to that of strategy k, which is the only situation in which j would be chosen. The gamble is conditioned on the event that she chooses strategy j, so its elements are zeroed-out in outcomes where other strategies are played. If the player has linear utility for money, she cannot be worse off by accepting such a gamble, and she could be strictly better off if the expected payoff of j turns out to be strictly greater.than that of k, according to whatever her beliefs may be at the time she makes her move. (She need not have formed her beliefs yet.) Therefore, on the margin, she ought to be willing to accept an additional payoff of αg_{njk} for some small positive multiplier α that may be chosen by an observer. We do not assume that the observer knows the full details of the player's payoff function. Rather, the player publicly reveals only the gambles $\{g_{njk}\}$ that are acceptable to her, because is in her interest to do so regardless of her present beliefs.

It is possible that, at the very beginning of interactions among the players, before very much is known, the act of offering gambles could give rise to its own strategizing and adjustment of beliefs and perturbations of payoffs, but we assume that the situation as seen by an observer is one in which the players have already, by whatever means, exhausted their incentives for making further gambles with each other. Some acceptable gambles are in view, and their parameters are stable, and everyone knows this (and knows that, etc.). This assumption is without loss of generality relative to what is conventionally assumed about common knowledge of the rules of a noncooperative game.

The three games above have the same revealed-rules matrix, which looks like this (up to scaling):

TABLE 8.14: Common revealed-rules matrix (**G**) for the three games

	A_1B_1	A_1B_2	A_2B_1	A_2B_2
g_{112}	2	-1		
g_{121}			-2	1
g_{212}	1		-2	
g_{221}		-1		2

For example, $g_{112} = (2, -1, 0, 0)$ because, in the chicken game, the payoff of this gamble for player 1 in outcome A_1B_1 is 2 minus 0, its payoff in outcome A_1B_2 is 4 minus 5, and it is zeroed-out if strategy 1 is not played by player 1. Similarly, $g_{212} = (1, 0, -2, 0)$ because its payoff for player 2 in outcome A_1B_1 is 5 minus 4, its payoff in outcome A_2B_1 is 0 minus 2, and it is zeroed-out if strategy 1 is not played by player 2. Identical values for the **G** matrix elements are obtained for stag hunt, and proportional values are obtained for battle-of-the-sexes

The differences among the games lie in their external effects. Each of their payoff matrices can be derived from the others by adding or subtracting amounts from each player's payoff function which depend on the move made by the other, possibly with some linear rescaling. For example, Alice's payoff function in stag hunt is obtained from her payoff function in chicken by adding 3 to her payoff if Bob plays B_1 and subtracting 4 if he plays B_2. Alice's payoff function in battle-of-the sexes is obtained from her payoff function in chicken by scaling the latter up by a factor of 2 and then subtracting 7 if Bob plays B_2. Adjustments of this nature do not affect the directions of vectors of *differences* in payoffs between strategies of a given player (the rows of **G**), because they are constants that get subtracted out.

The rows of **G** also have another (dual) property that was introduced in the matching-pennies example: they are the vectors of coefficients in the linear constraints that determine the set of correlated equilibria of the game. A correlated equilibrium is a probability distribution π on the set of outcomes which satisfies $G\pi \geq 0$, meaning that for every njk, player n's expected payoff for playing strategy j is greater than or equal to her expected payoff for playing stategy k, conditional on the event that she plays strategy j. The set of all correlated equilibrium distributions determined by these constraints is a convex polytope.

The correlated equilibrium polytope for these three games is a hexahedron with 5 vertices, two of which are pure Nash equilibria (A_1B_1 and A_2B_2), one of which is a completely mixed Nash equilibrium that assigns 2/3 probability each to A_1 and B_2, and two of which are correlated equilibria that place zero probability on one of the off-diagonal strategies.[10] They are summarized here along with their expected payoffs to the players:

TABLE 8.15: Extremal correlated equilibria (π^1, π^2, π^3 = Nash; * = efficient)

	Correlated equilibria				Expected payoffs		
	A_1B_1	A_1B_2	A_2B_1	A_2B_2	Chicken	B-O-S	Stag hunt
π^1	1	0	0	0	*2,5	*4,3	*5,4
π^2	0	0	0	1	*5,2	*3,4	1,2
π^3	2/9	4/9	1/9	2/9	3.33,3.33	2,2	1.67,2.67
π^4	1/4	1/2	0	1/4	*3.75,3.75	2.25,2.25	1.50,3
π^5	2/5	0	1/5	2/5	2.8,2.8	2.8,2.8	3,2.4

---

[10]The various possible shapes of the correlated equilibrium polytope for 2×2 games are discussed by Calvo-Armengól (2003).

The mixed Nash equilibrium (π^3) is strongly dominated by convex combinations of other Nash equilibria in all three games. One of the non-Nash correlated equilibria (π^4) is on the efficient frontier in the chicken game but not in the battle-of-sexes or stag hunt games, which illustrates that the orientation of the efficient frontier depends on external effects that are not represented in **G**.

Figure 8.1 shows the geometry of the set of correlated equilibria of these games. The tetrahedron is the set of all probability distributions on outcomes of the game. The saddle-shaped surface is the set of probability distributions that are independent between players. The hexahedron is the set of correlated equilibria, and the Nash equilibria are its 3 points of contact with the saddle. An obvious equitable solution in the chicken and battle-of-the-sexes games is for the players to flip a coin or follow a taking-turns convention to choose between A_1B_1 and A_2B_2, which corresponds to the large dot in the middle of the long edge and is not a Nash equilibrium. In the chicken game the correlated equilibrium π^4 is actually a bit better than a coin flip: it yields an expected payoff of 3.75 to both players versus 3.5 for the coin flip. This is the vertex facing the viewer in the figure. It assigns 50% probability to the swerve-swerve outcome and 25% each to swerve-straight and straight-swerve. In the stag hunt game there is no benefit to deliberate randomization: the pure strategy profile A_1B_1 (stag-stag) is strictly dominant. However, hare-hunting provides assurance of avoiding a payoff of zero if there are mistakes or failures of communication, and it is not irrational.

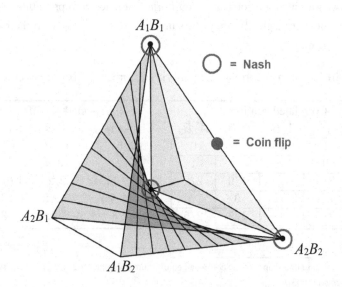

FIGURE 8.1: The set of correlated equilibria of all 3 games

The coin-flip correlated equilibrium is often not mentioned in discussions of the battle-of-sexes game that appear in surveys of rational choice theory, such as Binmore (2009), Gilboa (2010) and Peterson (2017). Binmore's book concludes with a detailed analysis of this very example, including the following passage that reflects the conventional view of the primacy of Nash equilibrium: "What is the solution of the Battle of the Sexes? If this question has an answer, it must be one of the game's Nash equilibria. If we ask for a solution that depends only on the strategic structure of the game, then the solution of a symmetric game like Battle of the Sexes must be a symmetric Nash equilibrium." (Binmore's solution is to use a "muddling box" that allows a wider range of symmetric independent strategies.) Yet, if Alice and Bob have such a lack of foresight that they wind up in a situation where their best option is to act independently and symmetrically, one might say that they have already made a mistake. A solution that involves correlated beliefs does not necessarily presume the explicit use of a correlated randomization device, although that would be an obvious strategy for Alice and Bob to employ here. (Another might be a taking-turns convention over the long run.) In general, the correlation in a correlated equilibrium could merely represent the nature of subjective uncertainty that exists in the mind of an observer, whatever its source. An observer of this game might reasonably expect Alice and Bob to have coordinated their choices by some means, even if he doesn't know which way. In constructing his own beliefs, he would probably assign a probability greater than one-half to the event that they will spend the evening together.

Figure 8.1 illustrates a general property of Nash and correlated equilibria that is true for games with any number of players and any number of discrete strategies. The set of correlated equilibria is a convex polytope, and it must touch the saddle of independent distributions at one or more points, and those are the Nash equilibria. However, the saddle cannot pass through the interior of the polytope. *The Nash equilibria always lie on the surface (boundary or relative boundary) of the polytope*, not inside it, although they need not be vertices as shown here. For example, in higher dimensions, some Nash equilibria could lie along curves in faces of the polytope. Or they could be points in the middle of an edge, and those could have irrational coordinates. Examples are presented below. The extreme points of the polytope always have rational coordinates insofar as they are solutions to sets of linear equations with rational coefficients. The result is formalized in:

Theorem 8.1: Geometry of Nash and correlated equilibria (Nau et al. (2004))

In any finite non-trivial game, the Nash equilibria are on the boundary of the correlated equilibrium polytope. If the polytope is of full dimension, the Nash equilibria are on its relative boundary.[11,12]

Proof: If a Nash equilibrium is not completely mixed, it assigns zero probability to one or more joint strategies, hence it satisfies at least one of the non-negativity constraints ($\pi \geq 0$) with equality. If it is completely mixed, a Nash equilibrium renders every player indifferent among all of her own strategies, hence it satisfies all of the incentive constraints ($\mathbf{G}\pi \geq 0$) with equality, at least one of which is non-trivial if the game is non-trivial. Hence every Nash equilibrium satisfies at least one non-negativity constraint or non-trivial incentive constraint with equality, and that constraint (together with the total probability constraint) determines a face of C whose dimension is less than $N - 1$.

It is possible for the number of vertices of the polytope to be enormous even in a modest-sized game, although the number is typically small. In a 2×2 game it cannot be larger than 5. Nau et al. (2004) carried out a simulation in which vertex enumeration was performed on equilibrium polytopes of 250 randomly generated 4×4 games with positively correlated payoffs (These were determined by a system of 25 constraints in 16 variables.) Positive correlation of payoffs was used to make them more likely to be games in which correlation of strategies would be desirable. It was found that around half of these games had polytopes with 5 or fewer vertices, although one had more than 100,000.

The gambles that are the rows in the revealed-rules matrix have the property that their acceptability is independent of the players' beliefs about the game's outcome and independent of cooperative agreements among them, if any. As such they yield a solution of the game that consists of the entire set of correlated equilibria. It is also

---

[11]A game is *non-trivial* if at least one player has a payoff function that is not constant. The correlated equilibrium polytope is of *full dimension* if its dimension is $|S| - 1$. π is on the *boundary* of the polytope if it lies in a face whose dimension is less than $|S| - 1$, which is true if and only if it lies on a supporting hyperplane whose normal vector is non-constant, i.e., linearly independent from the total probability constraint. If the polytope is of less than full dimension, then all of its points are boundary, but if it is not a singleton it still has a relative interior whose border is the relative boundary.

[12]A Nash equilibrium may exist in the relative interior of a correlated equilibrium polytope of *less than full dimension* only if the following special and rather uninteresting conditions are satisfied: (i) the Nash equilibrium assigns positive probability to every coherent strategy of every player, and (ii) in every correlated equilibrium, the incentive constraints for defecting from one coherent strategy to another coherent strategy are all satisfied with equality. See Nau et al. (2004) for details.

possible that the players could reveal more information about the outcome(s) of the game that they expect to occur, based on private communication or history of play or other information not directly visible to an observer. For example, Alice could make one of the following statements in the context of playing the game of chicken:

> *a* :"I am sure that at least one of us will swerve."
> *b*: "I am sure that exactly one of us will swerve."
> *c*: "I am going to go straight and I am sure that Bob will not."
> *d*: "I am going to swerve and I'm sure that Bob will not."

By saying "I am sure that x will occur" she means that she will accept a non-positive gamble in which she will gain nothing if x occurs and will lose a finite amount of money if it does not. By saying "I am going to do y" she means that she will accept a non-positive gamble in which she will gain nothing if she does y and will lose a finite amount of money if she does not. Statements a, b, c, and d then translate into the following acceptable gambles that refine the set of equilibria in various directions:

TABLE 8.16: Additional gambles whose acceptance refines the solution

	A_1B_1	A_1B_2	A_2B_1	A_2B_2	CE allowed:
g_a			-1		π^1,π^2,π^4,π^5
g_b		-1	-1		π^1,π^2
g_c	-1	-1	-1		π^2
g_d		-1	-1	-1	π^1

Each of the statements rules out one or more correlated equilibria as a description of the players' beliefs by rendering one or more game outcomes to be jointly incoherent. The players might also offer to accept gambles that directly determine the probabilities of outcomes, such as the following:

> *e*: "The probability that both of us will swerve is $1/2$ and the probabilities that only I will swerve or only Bob will swerve are both $1/4$."
> *f*: "The probabilities that only I will swerve or only Bob will swerve are both equal to $1/2$."

Statement e describes a situation in which the players intend to implement the specific correlated equilibrium π^4 but have not yet performed the randomization (say, via a mediator) and will not do so until after the observer has had a chance to place gambles. Statement f similarly describes the correlated equilibrium in which a coin is going to be flipped at some future moment to decide who will swerve.

8.5   An overview of correlated equilibrium and its properties

The modeling framework sketched above provides the basis for a unifying theory of games in environments where small monetary gambles are available as a tool of communication in numbers that are commonly understood and which have real implications for the players. This is the best-case scenario for modeling the strategic aspects of *economic* behavior in precise quantitative terms, and as such it establishes some limits on what is possible more generally. Whatever is knowable by an observer is something that is knowable in common by the players, and this provides an operational, specular definition of common knowledge of a game's parameters. Any outcome of the game that an observer would consider to be irrational, given what he knows, is one that the players should consider to be jointly irrational. So, how might some of the numerical parameters of the game come to be known by an observer, and on what basis would he consider an outcome to be irrational?

Our approach, which was illustrated in the preceding two sections, is a direct application of the ex-post fundamental theorem of subjective probability (Theorem 3.6) to a situation in which there are multiple actors in the same scene and some of the gambles that they offer to accept may refer to events that are under their own control. Outcomes of the game that are considered irrational are those that lead to an ex-post-arbitrage profit for an observer: they are *jointly incoherent*. Outcomes of the game that do *not* lead to ex-post-arbitrage (of which there must be at least one) are jointly *coherent*. It as if the players have revealed mutually consistent and potentially *positive* subjective probabilities for them, which may or may not be uniquely determined. From an observer's perspective, the set of players is equivalent to a single player who may have imprecise or incompletely revealed beliefs, represented by a convex set of probability distributions as discussed in Section 3.7. In that context the rationality criterion was also referred to as ex-post-coherence.

The gambles that are made on the game's outcome have a special structure in which a player's own moves are used (only) as conditioning events for gambles she will accept. They reveal beliefs that *she would need to have* in order to support the move that she made. For example, she might say something like: "Conditional on my choosing to play Top rather than Bottom in this game, I will accept an even-odds gamble on my opponent playing Left rather than Right." In other words, she would take the even-odds gamble on Left in the same state of mind in which she would choose to play Top. Such a statement indirectly reveals information about her utility function without revealing anything about her unconditional beliefs and her intentions. When gambles of this nature are offered as a way of revealing information about the rules of the game, the criterion of no-ex-post-arbitrage leads immediately, via linear programming duality, to the result that the only rational outcomes of the game are ones that could have arisen from implementation of a correlated equilib- rium. (Theorem 8.2) This means that the reciprocal beliefs of the players, as seen by an observer (or a game theorist?), must lie in the set of correlated equilibria. And, if the players have nonlinear or otherwise state-dependent utility for money, and/or the game also has important non-monetary consequences, the parameters of the correlated equilibria merely become risk-neutral probabilities as in the model of risk-averse agents developed in Chapter 6. (Theorem 8.4)

From an observer's perspective, the players are acting as one unit and whether or not they are rational ought to be judged by the same standard that would apply to an individual. An observer cannot see inside the players' heads to distinguish their separate personalities and find numbers that represent the true values that they as- sign to outcomes, measured on whatever scale. However, *gambles* that the players are willing to accept, with payoffs in units of money, *are* visible from the outside. The use of monetary gambles to elicit the rules of a noncooperative game is admit- tedly stylized and artificial, but it's the most elementary special case inasmuch as monetary transactions and pricing are the very basis of quantitative reasoning and quantitative communication in economics. Any theoretical issues that arise here in regard to numerical modeling of behavior in games are not going to go away in more general applications that do not involve money.

Here is a more detailed outline of the stylized facts and claims about Nash and correlated equilibrium, as they apply to games with finite strategy sets, that will be discussed in the remainder of this chapter:

1. Aumann (1974, 1987) is correct in asserting that correlated equilibrium is the expression of Bayesian rationality in noncooperative games. It is the natural solution concept under a subjective view of probability. The late discovery of this fact is an unfortunate historical accident. It should be viewed as a broader concept than Nash equilibrium, not a special case of it.

2. A correlated equilibrium is not necessarily a solution that is deliberately im- plemented via correlated randomization, although that is one possibility. More generally, a correlated equilibrium (or set thereof) is a representation of collec- tive subjective beliefs of the players as seen by an observer who communicates with them in the language of betting.

3. The study of correlated equilibrium (as carried out here) starts from an oper- ational definition of *common knowledge of the rules of the game* in terms of acceptable monetary gambles of a special form: each player offers to accept

conditional gambles on moves of other players, where the conditioning events are her own moves and where the gamble payoffs are proportional to differences between the game payoffs of the chosen strategy and those of some other strategy. Such gambles are incentive-compatible and they determine linear constraints on the probabilistic beliefs that the player would need to have in order to support the choice of a given move over some other available move. (They also reveal which events she regards as her own moves, thus addressing the issue of how those too might become common knowledge.) The parameters of the gambles are de facto common knowledge insofar as the gambles are offered in public in a common currency. The matrix whose rows are the payoff vectors of these gambles will be called the "revealed-rules" matrix, and it constitutes the real rules-of-the-game for purposes of noncooperative analysis. It contains the information needed to compute the Nash equilibria and correlated equilibria of the game, and no more than that. (It does not reveal information about external effects, as discussed in point #20 below.).

4. An outcome of the game is consistent with *common knowledge of rationality*, in the Bayesian sense, if it is jointly coherent, i.e., does not lead to an ex-post-arbitrage profit for an observer who takes the other side of the gambles. *This condition holds if and only if the outcome is one that has positive probability in a correlated equilibrium*, which follows from application of the ex-post version of de Finetti's fundamental theorem of subjective probability. (Theorem 8.2)

5. The proof of this result is based on a separating hyperplane theorem (Lemma 2.2 and Theorem 3.6 here) that is a variant of the theorem-of-the-alternative that appears in von Neumann and Morgenstern's 1944 book, where it is applied to the narrower problem of finding a minimax solution to a 2-player zero-sum game.

6. If a forecaster joins the betting, there is no ex-ante arbitrage opportunity (a *sure* win) for the observer if and only if the forecaster's revealed subjective probability distribution is a correlated equilibrium. (Corollary 8.1)

7. The existence of a supporting common prior distribution is a requirement of the no-arbitrage condition rather than a primitive assumption. In this respect the method of analysis presented here is "dual" to that of Aumann (1987), which begins with the common prior assumption.

8. The independence constraint that Nash equilibrium imposes is hard to express or enforce as a rationality condition to be satisfied ex-ante by subjective beliefs, particularly when those beliefs may not be represented by a unique

probability distribution. It is hard for an observer to verify that such a constraint is satisfied except with respect to an already-determined probability distribution, and a violation of this constraint per se does not expose the players to exploitation at the hands of the observer.[13]

9. The set of all correlated equilibria of a finite game is a convex polytope whose extreme points have rational coordinates. (See Figure 8.1 or the book's cover for a 2×2 game example.) It is determined by a system of linear incentive constraints whose coefficients are parameters of the conditional gambles that the players accept as a way of unilaterally revealing some (but not all) information about their payoff functions.

10. Extremal correlated equilibria (vertices of the polytope) can be found by solving linear programs in which an observer attempts to find opportunities for arbitrage among the gambles that the players have offered to accept. The primal problem is to solve for a correlated equilibrium in which a particular outcome has positive probability, and the dual problem is to look for an ex-post-arbitrage profit when the game has that outcome.

11. The set of Nash equilibria is the intersection of the correlated equilibrium polytope with the set of distributions that are independent between players, and solving for a Nash equilibrium in a general game is a nonconvex nonlinear optimization problem.

12. Nash equilibria lie on the surface (boundary or relative boundary) of the correlated equilibrium polytope, although not necessarily at vertices (Theorem 8.1). The set of Nash equilibria may be nonconvex and disconnected. In a 3-player game it is possible to have a unique Nash equilibrium with irrational coordinates, located in the interior of a line segment on the surface of the polytope. (See Table 8.22 for an example.)

13. Proof of existence of a correlated equilibrium is very simple (almost obvious) and it follows from the existence of a stationary distribution of a Markov chain in which players iteratively adjust their separate strategies to defeat a would-be arbitrageur (Theorem 8.3). The existence proof for a Nash equilibrium applies the fixed point theorem, a powerful tool of topology.

14. Proof of convergence is another issue. There is no universal mechanism for driving the players' reciprocal beliefs to a Nash equilibrium, particularly for a game that is played only once and/or which has more than one equilibrium.

---

[13] For example, in a battle-of-sexes game, how can an observer verify that the players have not engaged in private communication or followed some convention for coordinating their moves?

Whereas, the mere revelation of the rules of the game (via betting) forces revealed beliefs to "converge" to the set of correlated equilibria. (In fact, they begin there.)

15. Imprecision of beliefs or inexact knowledge of game payoffs are more problematic for Nash equilibrium than for correlated equilibrium insofar as Nash equilibria are determined by equality constraints while correlated equilibria are determined by inequality constraints that yield convex sets of solutions in general.

16. The fact that the set of correlated equilibria may be larger than the convex hull of the set of Nash equilibria is similar in principle to what happens in the SEU model when the completeness assumption is dropped, as discussed in Chapter 5 (Theorem 5.3). There the convex set of state-dependent utilities may be larger than the convex hull of the set of agreeing state-independent utilities.

17. The alternative solution concepts of rationalizability and correlated rationalizability do not necessarily protect the players against ex-post-arbitrage, i.e., they may allow outcomes that are jointly incoherent.

18. In the more general case where the players are risk-averse with respect to gambles on the game's outcome, as represented by nonlinear utility for money, the parameters of the correlated equilibria become risk-neutral probabilities, a result which yields a unification of noncooperative game theory and asset pricing theory. (Section 8.9)

19. When the same modeling technique is applied to games of incomplete information, the result is a correlated generalization of Harsanyi's concept of Bayesian equilibrium. (Section 8.11)

20. An important qualification of this entire method of analysis (which applies to both Nash and correlated equilibrium and noncooperative thinking in general) is that it does not take external effects into account, i.e., the way in which one player's payoff is affected by the strategies of others, holding her own strategy fixed. These determine the possible benefits of cooperation and the orientation of the Pareto frontier of the set of correlated equilibria if it is not a singleton.[14] The problem is that external effects cannot be revealed through unilateral contracts ("I agree to do X at your pleasure," where X might be the acceptance of a gamble) as opposed to multilateral contracts ("I agree to

---

[14] For example, the games of chicken, stag hunt, and battle-of-the-sexes have identical sets of correlated equilibria, but they differ in their interpersonal elements and their sets of efficient equilibria, as shown in Table 8.15. Another example is given in Tables 8.19 and 8.20.

do X if you agree to do Y," where X and Y might both be "don't confess" in the prisoner's dilemma). External-effect details of the payoff functions are subtracted out when the revealed-rules matrix is constructed as the (sole) basis for computing Nash and correlated equilibria, as carried out for a 2 × 2 game in Tables 8.13 and 8.14 and illustrated in Tables 8.25-8.27 for the prisoner's dilemma.

8.6 The fundamental theorem of noncooperative games

Consider a strategic-form noncooperative game in which there are N players. Suppose that player n has a finite set of strategies denoted by S_n and let $S := S_1 \times ... \times S_N$ denote the set of all joint strategies, i.e., outcomes of the game. A generic outcome is denoted by $s := (s_1, s_2, .., s_N)$ where $s_n \in S_n$. The jth strategy of player n is denoted by s_{nj}. Let f_n denote player n's payoff function, with $f_n(s)$ being her payoff in outcome s. Let $\mathcal{G}$ denote the set of parameters of these payoff functions (an array of N-tuples of payoffs). It represents the "true" rules of the game. In a game of "complete information," the values of those parameters are assumed to be common knowledge, although what it might mean for players to have common knowledge of the numerical values of each others' payoffs is not usually addressed. That issue will be directly confronted here.

For the purposes of this section, it will be assumed that payoffs are amounts of money and players behave as if they have state-independent linear utility for money. These assumptions are not essential, and their relaxation will be considered in a later section. What is essential is that the players should have the opportunity to make small monetary *side gambles* that depend on the game's outcome in the fashion of those illustrated in Sections 8.3 and 8.4. Such side gambles will be used as the language in which the players converse in quantitative terms with each other and with an outside observer. The parameters of side gambles that are publicly offered are de facto common knowledge by virtue of the opportunities that they create for reciprocal monetary transactions.

Let $f_{nj}(s)$ denote player n's payoff when she chooses s_{nj} and the rest of the players choose their own respective parts of s:

$$f_{nj}(s) := f_n(s_{nj}, s_{-n}) \tag{8.1}$$

where the notation (s_{nj}, s_{-n}) means $(s_1, s_2, ..., s_{n-1}, s_{nj}, s_{n+1}, .., s_N)$. Let $\mathbf{1}_{nj}$ denote an indicator vector for the event in which player n chooses s_{nj}:

$$\mathbf{1}_{nj}(s) \quad : \quad = 1 \text{ if } s_n = s_{nj} \tag{8.2}$$
$$: \quad = 0 \text{ otherwise.}$$

In the event where player n chooses s_{nj} in preference to some other strategy s_{nk}, others can infer that s_{nj} has a greater-or-equal expected payoff than s_{nk}, given her beliefs at that moment. In that case a small side gamble with payoffs proportional to $f_n(s_{nj}, s_{-n}) - f_n(s_{nk}s_{-n})$ will have non-negative expected value, and therefore she ought to be willing to accept it if someone offers to take the other side. Let $\mathbf{g}_{njk}$ denote the vector of payoffs of this gamble. Its value in state s is

$$g_{njk}(s) := 1_{nj}(s)(f_{nj}(s) - f_{nk}(s)) \tag{8.3}$$

Suppose that each player is willing to *publicly* reveal the gambles of this nature that are acceptable to her, if they are not already public knowledge. She does this by announcing (or else affirming what is already known) "I'm willing to accept a small gamble with payoffs $\alpha_{njk}\mathbf{g}_{njk}(s)$ where α_{njk} is positive number chosen by someone else." This is an ordinary gamble which just happens to have the property that it is conditioned on an event that is under the control of the player who offers it, as in the examples of Sections 8.3 and 8.4. From an observer's perspective this distinction is irrelevant: *an event is an event and a gamble is a gamble.*

Let $\mathbf{G}$ denote the matrix whose rows are these payoff vectors of acceptable gambles $\{\mathbf{g}_{njk}\}$. $\mathbf{G}$ has $|S|$ columns, one for every outcome of the game, and for player n it has $|S_n| \times (|S_n| - 1)$ rows, one for every alternative strategy of any given strategy of her own. $\mathbf{G}$ constitutes the observable rules of the game and will henceforth be called the "revealed-rules matrix."[15] It determines the sets of correlated and Nash equilibria, as will be shown below.

Here's what the construction looks like in the 2×2 case, as in the examples of Sections 8.3 and 8.4. For visual clarity, the symbols A_j and B_k rather than s_{1j} and s_{2k} are used as labels for strategies of the players, and a_{jk} and b_{jk} rather than $f_1(s_{1j}, s_{2k})$ and $f_2(s_{1j}, s_{2k})$ are used to denote the payoffs of the players in outcome jk.

TABLE 8.17: Table of unobservable real payoffs (true rules $\mathcal{G}$)

		Player #2:	
		B_1	B_2
Player #1:	A_1	a_{11}, b_{11}	a_{12}, b_{12}
	A_2	a_{21}, b_{21}	a_{22}, b_{22}

[15] $\mathbf{G}$ corresponds to $\mathbf{A}^T$ in the notation originally used by Nau and McCardle (1990), and thus $\alpha \cdot \mathbf{G}$ here corresponds to $\mathbf{A}\alpha$ there.

TABLE 8.18: Matrix of observable payoffs of acceptable gambles (revealed rules **G**)

	A_1B_1	A_1B_2	A_2B_1	A_2B_2
g_{112}	$a_{11} - a_{21}$	$a_{12} - a_{22}$		
g_{121}			$a_{21} - a_{11}$	$a_{22} - a_{12}$
g_{212}	$b_{11} - b_{12}$		$b_{21} - b_{22}$	
g_{221}		$b_{12} - b_{11}$		$b_{22} - b_{21}$

An observer of the game may choose a vector of small non-negative weights α to be applied to the acceptable gambles, thus getting the payoff vector $-\mathbf{G} \cdot \alpha$, and everyone knows this and knows that everyone knows it, etc., because they can all share the observer's viewpoint. Notice that $\mathbf{G}$ contains some *but not all* information about the situation faced by the players. It encodes only personal incentives, not external effects. The latter information determines the benefits of cooperation, which cannot be communicated through unilateral offers to accept gambles. It requires the use of other forms of shared information such as cheap talk (e.g., "notice that we will both be better off if I play A_1 and you play B_1") or bilateral contracts that change the rules of the game (e.g., "I will agree to pay you \$100 to play B_1 if you will agree to pay me \$100 to play A_1)." The data that is needed to assess the benefits of cooperation is subtracted out when the $\mathbf{G}$ matrix is constructed, which is why the payoff functions for Alice in Tables 8.2 and 8.3 are strategically the same. Some famous "dilemmas" of game theory arise from this fact.

Suppose that the observer looks for a riskless profit opportunity, namely a vector of non-negative weights α that satisfies $\mathbf{G} \cdot \alpha \leq 0$ and $[\mathbf{G} \cdot \alpha](s) < 0$ in some game outcome s. If another set of weights α' yields a riskless profit opportunity in some other outcome s', the observer can obtain a riskless profit in both outcomes by using the weight vector $\alpha + \alpha'$, and so on. Overall there is a maximal set of outcomes of the game (if any) in which the observer can obtain a riskless profit with a well-chosen set of weights applied to the acceptable gambles. If one of those outcomes occurs, the players have jointly violated the same ex-post-coherence standard of rationality that applies to subjective probability assessments by an individual.

> **Definition:** Outcome s of the game is *jointly coherent* if and only if there is no ex-post-arbitrage opportunity when s occurs, i.e., if and only if there is no vector of non-negative weights α such that $\mathbf{G} \cdot \alpha \leq \mathbf{0}$ and $[\mathbf{G} \cdot \alpha](s) < 0$.

Now consider the following story about how a solution of the game might be implemented through the help of a mediator. (This is not to be taken literally. It's just a dramatization.) The mediator will use a random device to generate a recommended outcome of the game. Let π denote the prior probability distribution on S that is used inside the device, and suppose that π is common knowledge, which is to say, everyone knows how the device works. π could be a degenerate distribution, assigning probability 1 to some outcome, or it could assign strategies randomly to some or all players, and randomization could be either independent or correlated among players. Suppose that after selecting s in this way, the mediator privately tells each player only what her own recommended strategy is. Player n hears s_n and she is able to perform Bayesian updating to obtain a personal posterior distribution over the strategy recommendations received by the other players. Should she comply with her own strategy recommendation, assuming that others do likewise, or should she defect?

The unconditional probability that player n receives a recommendation to play s_{nj} is $\mathbf{1}_{nj} \cdot \pi$. If she and everyone else is going to comply, then the unconditional expected payoff of the gamble $\mathbf{g}_{njk}$ is $\mathbf{g}_{njk} \cdot \pi$, and its conditional payoff, given that she has received the recommendation s_{nj}, is:

$$E_\pi(\mathbf{g}_{njk}|s_n = s_{nj}) := (\mathbf{g}_{njk} \cdot \pi)/(\mathbf{1}_{nj} \cdot \pi) = E_\pi(f_{nj} - f_{nk}|s_n = s_{nj}) \quad (8.4)$$

The last term on the right is the difference between the expected payoff of strategy s_{nj} and that of strategy s_{nk} conditional on receiving recommendation s_{nj}, and it is optimal for player n to stick to s_{nj} rather than defect to s_{nk} if and only if this quantity is non-negative. This yields the system of inequalities $\mathbf{G}\pi \geq 0$ as the condition for a probability distribution π to be an equilibrium, and insofar as it allows recommended strategies to be correlated among players it is called a correlated equilibrium:

> **Definition:** A probability distribution π on outcomes of the game is a *correlated equilibrium* if it has the property that if it were used by a mediator to generate possibly-random private recommended strategies for all the players, it would be optimal for each player to comply with her own strategy recommendation if all other players do likewise. This means that π satisfies the following conditions with respect to the revealed-rules matrix $\mathbf{G}$:

$$\begin{aligned} \mathbf{G}\pi &\geq 0 \\ \pi &\geq 0 \\ \mathbf{1} \cdot \pi &= 1 \end{aligned} \quad (8.5)$$

A Nash equilibrium is a special case of a correlated equilibrium in which randomization (or as-if-randomization) of strategies is statistically independent among players, which rarely occurs outside of actual games of chance. Even where events are causally independent, they may be probabilistically dependent in the minds of observers if the parameters of the joint probability distribution are uncertain, as in a case where there are multiple equilibria.

We now come to the main theorem, which may be appropriately called "fundamental" insofar as it fits seamlessly into the spectrum of the other fundamental theorems that are applications of the separating hyperplane theorem in the context of a no-arbitrage standard of rational public behavior.

Theorem 8.2: Fundamental theorem of noncooperative games (Nau and McCardle (1990))

An outcome of a finite noncooperative game is jointly coherent if and only if there is a correlated equilibrium in which it has positive probability.

Proof: By Lemma 2.2, a version of von Neumann and Morgenstern's (1944) theorem-of-the-alternative, the system of equations $\alpha \geq 0$, $\mathbf{G} \cdot \alpha \leq 0$ and $[\mathbf{G} \cdot \alpha](s) < 0$ has no solution, i.e., there is no ex-post-arbitrage opportunity in outcome s, if and only if the system $\mathbf{G}\pi \geq 0$, $\pi \geq 0$, $\pi(s) > 0$, has a solution, i.e., there is a correlated equilibrium assigning positive probability to outcome s.

From the viewpoint of an observer who knows only $\mathbf{G}$, the solution of the game is the entire set of correlated equilibria, which is a convex polytope of probability distributions that satisfy the system of inequalities $\mathbf{G}\pi \geq 0$. The fact that the polytope may contain correlated strategies does not mean that a correlated randomization device needs to have been used. It could be merely a reflection of limits on what can be observed prior to the play of the game via (only) noncooperative communication. The players' beliefs may be entirely personal and subjective, and they may also be incomplete.

Here is another way to view the situation. Consider an analyst of the game—a forecaster—who knows the revealed-rules matrix $\mathbf{G}$. Suppose that after some study the forecaster is asked to reveal his own subjective probability distribution over outcomes of the game, also in the language of acceptable gambles. Now consider an observer of the entire scene. Application of the fundamental theorem of subjective probability to everyone's acceptable gambles yields the following:

Corollary 8.1: There is no ex-ante (unconditional) arbitrage opportunity for the observer if and only if the forecaster's revealed subjective probability distribution lies in the set of correlated equilibria.

There is no comparably simple, self-justifying argument for a priori restricting the forecaster's revealed beliefs to the set of Nash equilibria in an arbitrary one-shot game. *A forecaster therefore ought to subscribe to the view that correlated equilibrium is the natural expression of Bayesian rationality in noncooperative games, as asserted by Aumann (1987).* Yet, Nash equilibrium is almost universally portrayed as the more fundamental solution concept. Authoritative expositions of game theory (e.g., Laraki et al. (2019), Maschler et al. (2020)) typically define a correlated equilibrium of a game as a Nash equilibrium of the extended game that is obtained by introducing an actual mediator who makes correlated recommendations. That approach takes the mediator story literally rather than merely an as-if condition. The more fundamental rationality principle is no-arbitrage, not equilibrium for its own sake.

The proof of existence of a correlated equilibrium is simple and intuitive, almost obvious. If there is no correlated equilibrium, then all outcomes of the game are jointly incoherent, which means that there is a sure win opportunity for the opponent. Life would be very unfair if this were true. The players have merely revealed some facts about the rules of the game into which they have been thrust. They haven't necessarily formulated any strategies nor speculated about what others will do. They ought not to be doomed to have money stolen from them just because they have been honest about the facts that everyone is assumed to know. If it were possible for them to be doomed in this way, that would be a famous paradox. Because life cannot be this unfair, and there is no such famous paradox, a correlated equilibrium must exist, QED. Of course, the existence of a Nash equilibrium proves the existence of a correlated equilibrium, because the former is a special case of the latter. However, we might hope that there is a more direct way to prove the existence of a correlated equilibrium, given that it is defined by a system of linear inequalities rather than nonlinear equations, and indeed there is. The trick is to show that for any α chosen by the observer there is a way for each player to choose a pure or mixed strategy that yields her an expected loss of zero, by using a process of iterative adjustment.[16]

Theorem 8.3: Existence of correlated equilibria (Nau and McCardle (1990))

There is at least one jointly coherent outcome of the game and hence there exists a correlated equilibrium.

Proof: Suppose it is possible for an observer to choose a vector of non-negative weights α that yields a sure loss to the players, i.e., $\mathbf{G} \cdot \alpha < \mathbf{0}$. Then any pure or randomized strategies of the players must have strictly

---

[16]Hartand Schmeidler (1989) prove this result by a different and somewhat more technical method.

negative expected value for them as a group. We will demonstrate a contradiction, namely that for any α there exists a pure or independently randomized strategy for each player that yields zero expected value to the observer. Consider the situation that α presents to player n. Each strategy available to her is a lottery over the uncertain actions of her opponents, and α_{njk} is the multiple of the strategy s_{nk} lottery that the observer will ask her to give up in return for the same multiple of the strategy s_{nj} lottery if she chooses s_{nj} in the game. Without loss of generality, assume that $\sum_{k:k\neq j} \alpha_{njk} < 1$ for all j, and define $\alpha_{njj} := 1 - \sum_{k:k\neq j} \alpha_{njk}$, whence $\sum_{k=1}^{K} \alpha_{njk} = 1$ and $\alpha_{njj} > 0$ for all j. (The exchange of a lottery for itself yields no net transaction so α_{njj} is arbitrary.) Suppose now that player n employs an independently randomized strategy with probability distribution $\mathbf{p}$ over her own actions. Then the expected multiple of the strategy s_{nk} lottery that she will give to the observer is $\sum_{j=1}^{J} \alpha_{njk} p_j$, and the expected multiple of it that she will receive is $p_k \ (= \sum_{j=1}^{J} \alpha_{nkj} p_k)$ regardless of the actions of her opponents. (Note the reversal of k and j subscripts on α in the latter summation.) The player's net expected gain or loss in multiples of the strategy s_{nk} lottery will be zero if these are equal. Therefore, if $\mathbf{p}$ satisfies

$$\sum_{j=1}^{J} \alpha_{njk} p_j = p_k \ \forall k, \tag{8.6}$$

then the player's net expected transaction in every strategy will be zero, and hence her overall expected gain will be zero regardless of the strategies chosen by her opponents. It remains to show that such a distribution $\mathbf{p}$ always exists. Let $\mathbf{W}$ be the square matrix whose element in row k and column j is $w_{kj} := \alpha_{njk}$ (Here too, note the order of subscripts.). $\mathbf{W}$ is a stochastic matrix (i.e., non-negative with columns summing to 1) with positive diagonal elements, hence it may be considered as the transition matrix for a finite aperiodic Markov chain. The system of equations above can now be written as $\mathbf{W}\mathbf{p} = \mathbf{p}$, which is the equation that defines $\mathbf{p}$ as a stationary distribution of that Markov chain. Every finite aperiodic Markov chain has a stationary distribution, which can be constructed as $\mathbf{p} = \lim_{m\to\infty} \mathbf{W}^m \mathbf{p}^*$ for an arbitrary initial distribution $\mathbf{p}^*$, and that is a zero-expected-gain randomized strategy for player n.

The stationary strategy for each player's response to the observer's choice of α is trivially computable to arbitrary precision by iteration of her own Markov chain. And given that a correlated equilibrium is known to exist, one can be found by solving a linear program: maximize $\pi \cdot \beta$ over all π such that $\mathbf{G}\pi \geq \mathbf{0}$, $\pi \geq \mathbf{0}$, $\mathbf{1} \cdot \pi = 1$, for any vector of weights β in the objective function. The objective function could also be a weighted sum of expected utilities. The (extremal) correlated equilibrium probabilities found in this way will always be rational numbers. By contrast, the proof of existence of a Nash equilibrium needs to invoke a fixed point theorem, a much more powerful tool, and the computational complexity of finding them is also much greater. For example, the problem of solving for a Nash equilibrium that maximizes a weighted sum of expected payoffs or achieves a specified minimum payoff is NP-complete (Conitzer and Sandholm (2008)), and a game may have a unique Nash equilibrium in irrational numbers as shown in an example below.

8.7   Examples of Nash and correlated equilibria

In the examples of Section 8.4 (chicken, etc.), there are 3 Nash equilibria of which at least one is efficient within the set of correlated equilibria. It is possible to have a game in which there are no efficient Nash equilibria, such as the following, due to Aumann (1987):

TABLE 8.19: 4×4 game whose unique Nash equilibrium is dominated in expectation by a correlated equilibrium

	B_1	B_2	B_3	B_4
A_1	0, 0	10, 5	5, 10	6, 6
A_2	5, 10	0, 0	10, 5	6, 6
A_3	10, 5	5, 10	0, 0	6, 6
A_4	6, 6	6, 6	6, 6	7, 7

This game has a unique Nash equilibrium, namely the pure strategy A_4B_4. Whereas, all outcomes of the game are jointly coherent, and there is a correlated equilibrium that assigns a probability of 1/6 to each of the outcomes whose payoffs are 5 and 10. The latter equilibrium yields an expected payoff of 7.5 to both players, which dominates the Nash equilibrium payoffs of 7 to both players. This correlated equilibrium might even be regarded as focal in this example if the full details of the players' payoff functions are publicly known.[17] An observer who knows only what can be revealed by unilateral gambles would not have this much information, though.

---

[17]This correlated equilibrium is not a vertex of the polytope. It is at the center of a hexagon formed

A caveat in regard to this example—and games with multiple equilibria in general—is that by adding to each player's payoffs a function of the other player's move we obtain a strategically equivalent game in which the Nash and correlated equilibria are the same but may be differently ordered by their expected payoffs. For example, consider this variation on the 4×4 game in Table 8.19:

TABLE 8.20: Strategically equivalent 4×4 game with a different efficient frontier

	B_1	B_2	B_3	B_4
A_1	$0, 0$	$10, 5$	$5, 10$	$99, 6$
A_2	$5, 10$	$0, 0$	$10, 5$	$99, 6$
A_3	$10, 5$	$5, 10$	$0, 0$	$99, 6$
A_4	$6, 99$	$6, 99$	$6, 99$	$100, 100$

Each player now receives a large bonus if the opposing player chooses his or her 4th strategy. This game has the same Nash and correlated equilibria as the earlier one, but the Nash equilibrium strongly dominates the non-Nash correlated equilibria. It is now the focal outcome of the game from the perspective of someone who knows everything. It's what should be expected if the players are able to engage in cheap talk or otherwise have some shared awareness of the benefits of cooperation. This is the same phenomenon that was illustrated by the three 2×2 games discussed in Section 8.4, which have different orderings of outcomes in terms of total expected utility despite being strategically equivalent. Where they differ is in their external effects.

In the game of Table 8.19, a correlated strategy yields payoffs that are greater in expectation than those of the only Nash equilibrium, but not greater for sure. Here's an example of a 3-player game in which a correlated equilibrium yields a higher sure payoff than any Nash equilibrium.[18]

TABLE 8.21: $2 \times 2 \times 2$ game whose unique Nash equilibrium is dominated with certainty by a correlated equilibrium

| | C_1 | | C_2 | | C_3 | |
	B_1	B_2	B_1	B_2	B_1	B_2
A_1	$0, 1, 3$	$0, 0, 1$	$2, 2, 2$	$0, 0, 0$	$0, 1, 0$	$0, 0, 1$
A_2	$1, 1, 1$	$1, 0, 0$	$0, 0, 0$	$2, 2, 2$	$1, 1, 1$	$1, 0, 3$

---

by 6 vertices that place unequal weights on the 6 cells with payoffs of 5 and 10 and which all yield an expected payoff of 7.5 to both players.

[18]This example appeared in Moulin and Vial (1978) and is widely used.

The row, column and matrix players are Alice, Bob and Carol. Here the only Nash equilibria are those in which Alice and Bob jointly choose A_2B_1 and Carol chooses either C_1 or C_3 or some mixture thereof, yielding a sure payoff of 1 to each player. Whereas, there is a set of correlated equilibria in which Carol chooses C_2 and Alice and Bob jointly choose A_1B_1 or A_2B_2 with probabilities ≈ between 1/3 and 2/3. A solution of that kind yields a sure payoff of 2 to each player, and it would require only a limited amount of coordination to implement. The probabilities need not be precise or objective. For example, Carol might say to Alice and Bob: "I am going to play C_2, and you two should go off and flip a coin to choose between A_1B_1 and A_2B_2." Or, Alice and Bob could say to Carol: "the two of us are going to flip a coin to choose between A_1B_1 and A_2B_2 so you should play C_2." It is not necessary for Alice and Bob to carry out any explicit correlated randomization. It suffices that Carol should believe that private coordination of Alice and Bob's moves is possible. This sort of cheap talk would convey information about external effects that is relevant to an efficient solution.

The vertices of the polytope (extremal correlated equilibria) must have rational coordinates because they are determined by systems of linear constraints. Nash equilibria must satisfy some additional nonlinear constraints and they could have irrational coordinates. For example, a game could have a unique Nash equilibrium with irrational coordinates lying in the middle of an edge of the polytope, as in the following 3-player game, similar to one devised by Shapley and discussed in Nash's 1951 paper:

TABLE 8.22: 3-player game whose unique Nash equilibrium has irrational parameters

	C_1		C_2	
	B_1	B_2	B_1	B_2
A_1	3, 0, 2	0, 2, 0	1, 0, 0	0, 1, 0
A_2	0, 1, 0	1, 0, 0	0, 3, 0	2, 0, 3

This game has a unique Nash equilibrium in mixed strategies in which the probabilities are all irrational numbers:

$$\pi(B_1) = x := (-13 + \sqrt{601})/24 \approx 0.480 \qquad (8.7)$$
$$\pi(A_1) = (9x - 1)/(7x + 2) \approx 0.0.619$$
$$\pi(C_1) = (-3x + 2)/(x + 1) \approx 0.0.379$$

It yields expected payoffs {0.764..., 0.843..., 0.854...}. The correlated equilibrium polytope is 7-dimensional (full dimension) with 33 vertices, and the Nash equilibrium lies in the interior of a line segment connecting two of them. (Details are given in Nau et al. (2004).) How should this game be solved? Clearly the players ought to avoid the four outcomes in which one player gets 1 and the others get zero. Thus, Alice and Carol should jointly choose between $A_1 C_1$ and $A_2 C_2$, in which case Bob ought to independently randomize his choice between B_1 and B_2 so that Alice and Carol can't outguess him. Here's an example of a Pareto efficient correlated equilibrium in which Alice and Carol choose between $A_1 C_1$ and $A_2 C_2$ with probabilities 3/5 and 2/5 and Bob chooses between B_1 and B_2 with probabilities 2/3 and 1/3:

TABLE 8.23: An efficient correlated equilibrium of that game

	C_1		C_2	
	B_1	B_2	B_1	B_2
A_1	6/15	3/15		
A_2			4/15	2/15

This solution can be obtained by solving a linear program to find the correlated equilibrium that maximizes the sum of the expected payoffs. The individual expected payoffs are $\{1.467, 1.2, 1.2\}$, much better than the Nash expected payoffs. *The same solution can also be obtained by hand.* If Bob is randomizing his strategy independently from Alice and Carol, then effectively it is a 2-player game in which Bob chooses from $\{B_1, B_2\}$ while Alice and Carol jointly choose from $\{A_1 C_1, A_2 C_2\}$. In equilibrium, Alice and Carol must use a mixed strategy that renders Bob indifferent between B_1 and B_2. Let p denote the probability of $A_1 C_1$ in this mixture. The condition it must satisfy is $p \times 0 + (1 - p) \times 3 = p \times 2 + (1 - p) \times 0$, which yields $p = 3/5$. Now let q denote the probability of B_1. As a condition of Alice not wanting to defect from A_1 to A_2 when $A_1 C_1$ is supposed to be played, q must satisfy $q \times 3 + (1 - q) \times 0 \geq q \times 0 + (1 - q) \times 1$, which implies $q \geq 1/4$ As a condition of her not wanting to defect from A_2 to A_1 when $A_2 C_2$ is supposed to be played, q must satisfy $q \times 0 + (1 - q) \times 2 \geq q \times 3 + (1 - q) \times 0$, which implies $q \leq 2/3$. (The corresponding non-defection constraints on Carol are also satisfied under these conditions.) The solution above is the extremal one that is obtained with $q = 2/3$. Setting $q = 1/4$ we get the other extremal solution:

Its individual expected payoffs are $\{1.05, 1.2, 1.2\}$, which are weakly dominated by the other solution. This probability distribution is also a vertex of the polytope and it is connected to the other correlated equilibrium by the line segment along which q

TABLE 8.24: Another correlated equilibrium on the same edge of the polytope

	C_1		C_2	
	B_1	B_2	B_1	B_2
A_1	3/20	9/20		
A_2			2/20	6/20

varies from $1/4$ to $2/3$.[19] Thus we have found an edge of the polytope along which $\pi(A_1C_1) = 3/5$, $\pi(A_2C_2) = 2/5$, and (independently) $1/4 \leq \pi(B_1) \leq 2/3$. That was a good deal easier than solving for the Nash equilibrium and it yielded a solution that is intuitive and simple to implement.

Last but not least, consider the 2×2 game known as the "prisoner's dilemma." Discussions of game theory often begin with this example, and for critics of the field they often end there.[20] The payoff matrix looks like this (or something strategically equivalent):

TABLE 8.25: The prisoner's dilemma

	B_1	B_2
A_1	10, 10	0, 11
A_2	11, 0	1, 1

Assume that the game is played for money and the payoffs are in units of dollars. For each player, strategy 2 strictly dominates strategy 1, so there is a unique Nash and correlated equilibrium, A_2B_2, which yields payoffs of \$1 to each. This outcome is Pareto inferior to A_1B_1, which yields \$10 each, hence the dilemma. Here is another 2×2 game which might be called a "no-brainer":

TABLE 8.26: The no-brainer

	B_1	B_2
A_1	0, 0	10, 1
A_2	1, 10	11, 11

It is obtained from the prisoner's dilemma game by decreasing Alice's payoff by

---

[19]Both 8-variable solutions satisfy 8 linear constraints with equality: 3 incentive constraints, 4 non-negativity constraints, and the total probability constraint.

[20]See Poundstone (1992) for a detailed history. The game was originally studied by Melvin Dresher and Merrill Flood at RAND Corporation in 1950 as a metaphor for strategic conflict, and it was given its colorful name by Albert Tucker. An equivalent game appears in Nash's (1950a) paper as the second of his "simple examples" (p. 291).

$10 if Bob plays B_1 and increasing it by $10 if Bob plays B_2, and similarly for Bob's payoffs. These adjustments are (merely) manipulations of external effects. The unique equilibrium solution is (again) A_2B_2, but now it is Pareto *superior* to A_1B_1. Yet, these two games are the same from the viewpoint of Nash and correlated equilibrium. In both of them, the revealed-rules-of-the-game matrix $\mathbf{G}$ is:

TABLE 8.27: Common revealed-rules matrix ($\mathbf{G}$) for the two games

	A_1B_1	A_1B_2	A_2B_1	A_2B_2
g_{112}	-1	-1		
g_{121}			1	1
g_{212}	-1		-1	
g_{221}		1		1

The sum of acceptable gambles g_{112} and g_{212} yields a payoff vector of $(-2, -1, -1, 0)$ to the players, which is an ex-post-arbitrage profit opportunity for the observer if any outcome other than A_2B_2 occurs. The problem is that $\mathbf{G}$ does not include any information about benefits of cooperation: it has been subtracted out. It is as if the players have engaged (only) in the following dialog:

Dialog 1 (both games):

Alice: "Regardless of what you do, I'm $1 better off if I play A_2 rather than A_1."

Bob: "Regardless of what you do, I'm $1 better off if I play B_2 rather than B_1."

Based only on this information, the outcome A_2B_2 ought to be expected, and its inferiority in the prisoner's dilemma is not apparent. The missing information would be revealed by a second dialog, which differs between the two games:

Dialog 2 (prisoner's dilemma):

Alice: "Regardless of what I do, I'm $10 better off if you play B_1 rather than B_2."

Bob: "Regardless of what I do, I'm $10 better off if you play A_1 rather than A_2."

Dialog 2 (no-brainer):

Alice: "Regardless of what I do, I'm $10 better off if you play B_2 rather than B_1."

Bob: "Regardless of what I do, I'm $10 better off if you play A_2 rather than A_1."

Now it is apparent that A_2B_2 is Pareto superior in the no-brainer game while it is Pareto inferior in the prisoner's dilemma. The problem is that only the Dialog 1 information—personal incentives—can be revealed by unilateral contracts such as offers to take conditional bets on events. The Dialog 2 information—external effects—requires some other form of communication or contracting or convention in order to be credibly revealed. For example, each player could promise to pay the other for playing a particular strategy if there is a way for such promises to be enforced through bilateral contracts or handshake agreements. Or, if the game is repeated, some sort of signaling or reciprocation could take place, such as tit-for-tat (always match the other player's previous move). Or, if the game has a social context, norms could have evolved to turn dilemmas into no-brainers. Thousands of papers have been written on these aspects of the game.

The various games discussed above illustrate that the reasonableness of a Nash (or correlated) equilibrium as a solution may depend on parameters of the game that it does not consider, namely the external effects that players' strategies have upon each other. The literature of the field is filled with examples of behavioral experiments which were designed to test whether Nash equilibrium or one of its refinements would predict the behavior of real subjects, never mind that it ignores important information that might be focal to their attention and their desires. Experiments that look for paradoxes exploit this very fact. By taking correlated equilibrium rather than Nash equilibrium as the fundamental solution concept, we at least allow for some creative enlargement of the efficient frontier in situations where correlation may be mutually advantageous and/or where independence among randomized strategies cannot be verified by an observer and/or the observer may be subjectively uncertain about which of several equilibria has been implemented and/or computation of equilibria may be difficult.

8.8   Correlated equilibrium vs. Nash equilibrium and rationalizability

Nash equilibrium is a less general solution concept than correlated equilibrium insofar as it imposes an additional constraint on the solution, yet it is typically portrayed as a more general concept: a correlated equilibrium is viewed as Nash equilibrium of a game that is a priori augmented by a correlated randomization device with respect to which each player's assigned move is "do what the device says." (E.g., see the textbooks by Laraki et al. (2019) and Maschler et al. (2020).) Which concept is really the fundamental one? That depends on the vantage point(s), the modeling objective(s), the definition of "rational" individual behavior and joint behavior, and (last but not least) on assumptions about what the players know in common about the rules of the game and how they know it. The latter knowledge provides the data from which solutions will ultimately be computed. In applications of Nash equilibrium, the vantage points are those of individual players and the goal is to determine a strategy for each player which can be implemented independently of what others are doing and which solves some problem of prediction or prescription or mechanism design. What should they be expected to do, or what should they be advised to do, or how can they be motivated to do what is desired of them within a given set of rules? Individual rationality is defined by adherence to the EU or SEU model, noncooperative is defined to mean that they care only about how their own payoffs are affected by changes in their own strategies, and joint rationality is defined by an equilibrium condition in which randomization, if any, is performed independently. Common knowledge usually means that they are all reading as-if from the same sheet music, where everyone's full payoff matrix is written down in units of money or utility.

A correlated equilibrium can be viewed from that same perspective, merely with the possibility of correlated randomization thrown in, but that's not only way, nor the most appropriate way, in which to view it. More broadly, a correlated equilibrium is a description of uncertainty about the game's outcome as seen by an observer who knows exactly as much as the players commonly know. From the shared vantage point of the observer, that uncertainty is represented by a convex set of subjective probability distributions determined by linear inequalities as in the theory of lower and upper probabilities. The coefficient vectors in the inequalities are payoff vectors of gambles that are incentive-compatible for the players to accept and which serve as a medium of communication through which (some of) the game's parameters become common knowledge. The set of probabilities that satisfy these linear constraints is the set of correlated equilibria of the game, which also determines the Nash equilibria. The players as a group should satisfy the same rationality standard as an individual, namely, that they should not expose themselves to arbitrage via their acceptable gambles. The fundamental theorem states that there is no ex-post-arbitrage if and only if the outcome of the game has positive probability in one of the correlated equilibria. There is no essential distinction between individual and strategic rationality, just as an individual bettor is a microcosm of a financial market.

The Nash requirement that the players' moves should be independent is sometimes defended with analogies like this: "It is helpful to think of players' strategies as corresponding to various 'buttons' on a computer keyboard. The players are thought of as being in separate rooms, and being asked to choose a button without

communicating with each other." (Fudenberg and Tirole (1991))   Laboratory experiments sometimes do exactly that. In real games players live in a messy environment
where their understanding of the situation may be imprecise, they may have relevant personal experience, their actions may be informed and supported by others,
they may be bound by social norms or customs, and there may be opportunities for
cheap talk. The very assumption that the numerical parameters of the game are common knowledge presumes some shared information. If such knowledge reveals that
coordination would be mutually beneficial, we ought to expect that rational players would consider it. Coordination does not necessarily require explicit correlated
randomization, as in the 3-player game of Table 8.21.

Another issue is that of what to do when a game has multiple Nash equilibria.
From the perspective of an observer who expects the players to implement one of
several Nash equilibria but is subjectively uncertain about which one it will be, the
game's solution is a convex combination of the Nash equilibria, which is a correlated
equilibrium. This situation might well arise in the battle-of-sexes game. There may
also be attractive correlated equilibria that are outside the convex hull of the Nash
equilibria, as in the game of chicken, whose solutions (and those of battle-of-the-
sexes and stag hunt) are pictured in Figure 8.1 and on the cover of this book. The
main point, though, is that correlated equilibrium should not be treated as a special-
ized tool that applies only when a game is extended with a correlated randomization
device. Rather, it is the fundamental solution concept that follows from a material
definition of common knowledge and a material test for irrationality, and correlated
uncertainty in a game's solution may be only a reflection of limits of reciprocal
knowledge, which could have many subjective sources. It could also arise from uncertainty about which of several Nash equilibria will be played if there is more than
one. All that being said, the independence constraint that Nash equilibrium imposes
is useful for applied theory due to its parsimony and tractability and its support for
comparative statics, high dimensional modeling, and mechanism design.

Another widely studied generalization of Nash equilibrium is *rationalizability*
(Bernheim (1984), Pearce (1984)), which applies to profiles of pure strategies. A
pure strategy of a player is (independently) rationalizable if it survives a process of
iterative deletion of strategies that are not best replies to any pure or independently
randomized strategies of other players. This concept is weaker than Nash equilibrium but neither weaker nor stronger than correlated equilibrium. Any strategy that
has positive probability in a Nash equilibrium is rationalizable, although a rationalizable pure strategy may not be a pure strategy Nash equilibrium. For example, every
pure strategy is rationalizable in the matching-pennies game of Table 8.9, but Nash
equilibrium requires 50-50 randomization. Rationalizability also may rule out strategies that are supported by correlated equilibria but not Nash equilibria. For example,

in the 3-player game of Table 8.21, the only rationalizable pure strategy profiles are the two pure-strategy Nash equilibria that yield payoffs of 1 to each player, whereas there is a strictly dominant correlated equilibrium that yields 2 to each player. The more general concept of *correlated rationalizability* deals with cases like this. In games with more than 2 players, it admits strategies that are best replies to correlated strategies of other players. In the game of Table 8.21, Carol's strategy C_2 is a best reply to correlated strategies of Alice and Bob in which they flip a coin to choose between A_1B_1 and A_2B_2, and this is also a correlated equilibrium as discussed earlier.

Rationalizability and correlated rationalizability are often taken to be the very embodiments of common knowledge of rationality.[21] Yet it is possible to have strategies that are rationalizable but do not have positive probability in any correlated equilibrium, which is to say, via the fundamental theorem, they expose the players to a collective arbitrage loss, and this is common knowledge. In the following example (Bernheim (1984), Nau and McCardle (1990)), only A_2B_2 is jointly coherent, yet all strategies are rationalizable. Alice would play A_1 if she thought that Bob would play B_3, which he would play if he thought that Alice would play A_3, which she would play if she thought that Bob would play B_1, which he would play if he thought that Alice would play A_1. Since each of these 4 strategies is a best response to one of the others, none can be the first eliminated on the grounds of not being a best reply to a possible strategy of the other player. Yet the play of any of these 4 strategies leads to an arbitrage loss for the players as a group if the commonly known facts of the game are revealed or affirmed through acceptance of gambles.[22]

TABLE 8.28: A 3×3 game with rationalizable strategies that are jointly incoherent

	B_1	B_2	B_3
A_1	0, 7	2, 5	7, 0
A_2	5, 2	3, 3	5, 2
A_3	7, 0	2, 5	0, 7

Aumann (1974) also proposed an even more general solution concept: *subjec-*

---

[21]Brandenburger and Dekel (1989) state: "Our starting point is that rationalizability is the solution concept implied by common knowledge of rationality." This view is widely shared (e.g., Maschler et al. (2020), p. 187).

[22]This example has been tuned-up to make the A_2B_2 payoffs inferior to the average payoffs in the other cells (3 vs. 3.5) but this is meaningless as far as a comparison of noncooperative solution concepts is concerned. External effects could be manipulated to turn A_2B_2 into a Pareto superior solution by adding 5 to each player's payoff if the other plays his or her 2nd strategy, and this would have no bearing on the rationalizability or joint coherence of any strategies. The revealed-rules matrix and dominance relationships would be unaffected. The same point is illustrated by the games in Tables 8.19 and 8.20.

tive correlated equilibrium. In a subjective correlated equilibrium, the common prior assumption is not invoked. Players are allowed to have heterogeneous prior distributions on a source of information to which their strategy choices may be pegged, e.g., a correlated randomization device. Such a device would be welcome for playing a zero-sum game or the game of Table 8.28. It would provide a basis for implementing correlated strategies in which each player thinks that she will be the winner. However, if the players have state-independent marginal utility for money, such differences in priors will create arbitrage opportunities. This can be avoided if their differences in priors are compensated by differences in state-dependent marginal utility, so that they have common prior risk neutral probabilities. That situation is treated in the following section.

8.9   Risk aversion and risk-neutral equilibria

In the models and results discussed so far, it has been assumed that players are not risk-averse[23] and they have state-independent marginal utility for money and no unobservable prior stakes in outcomes of the game. Let's now consider a more general setting in which the players *are* risk-averse and they *do* have unobserved prior stakes in the game's outcome and/or they have *state-dependent* marginal utility for money. Under these conditions their lower and upper previsions determined by offers to accept small gambles must be interpreted as lower and upper expectations with respect to convex sets of *risk-neutral probabilities*, rather than true subjective probabilities, as discussed in previous chapters. The same consideration applies to the analysis of games. A game's own payoffs are a source of background uncertainty with respect to gambles on its outcome, and if the players are sufficiently risk-averse, this will give rise to distortions when the rules of the game are revealed through betting. The result will be that a rational solution of the game is characterized by a convex set of correlated equilibria whose parameters are risk-neutral probabilities.

Suppose that each player has strictly risk-averse subjective-expected-utility preferences with respect to profiles of monetary payoffs in the game, and let U_n denote the strictly concave von Neumann-Morgenstern utility function of player n. Then the payoff profiles $\{f_n\}$ translate into utility profiles $\{U_n(f_n)\}$. Let $\mathcal{G}^*$ denote the true game that is determined by the utility profiles, and let $\mathbf{u}_n$ denote the utility profile vector for player n, whose value in outcome s is $u_n(s) := U_n(f_n(s))$. The payoff matrix of $\mathcal{G}^*$ in units of utility looks like this in the 2×2 case in *A-B* notation:

---

[23]Throughout this section the term "risk-averse" will be used in place of "uncertainty-averse" in order to match up with use of the term "risk-neutral probability." However, there is no objective risk, just subjective uncertainty.

TABLE 8.29: Unobservable payoffs of 2×2 game in units of utility

	B_1	B_2
A_1	$u_A(a_{11}), u_B(b_{11})$	$u_A(a_{12}), u_B(b_{12})$
A_2	$u_A(a_{21}), u_B(b_{21})$	$u_A(a_{22}), u_B(b_{22})$

If U'_n denotes the first derivative of U_n, strict concavity requires that $U'_n(x) < U'_n(y)$ whenever $x > y$. Let $\mathbf{u}'_n$ denote the corresponding marginal utility vector whose value in outcome s is $U'_n(f_n(s))$. Also, let $\mathbf{u}_{nj}$ denote the vector constructed from $\mathbf{u}_n$ in the same way that f_{nj} was constructed from f_j in Section 8.4 above, namely $u_{nj}(s) := U_n(f_{nj}(s))$. In other words, $u_{nj}(s)$ is the utility that player n would receive by playing her strategy s_{nj} when all others play according to s. Let $\mathbf{u}'_{nj}$ denote the corresponding vector of marginal utilities for money, i.e., $u'_{nj}(s) := U'_n(f_{nj}(s))$.[24] By the same argument as in the risk-neutral case, player n will choose strategy s_{nj} in preference to strategy s_{nk} only if her beliefs are such that she would be willing to exchange the utility profile $\mathbf{u}_{nk}$ for the utility profile $\mathbf{u}_{nj}$, hence a small monetary gamble yielding a profile of *changes in marginal utility* that is proportional to $\mathbf{u}_{nj} - \mathbf{u}_{nk}$ should be acceptable if the event $s_n = s_{nj}$ is observed to occur. When strategy j is chosen, the player's profile of marginal utilities for money is $\mathbf{u}'_{nj}$, and a monetary gamble that yields a profile of marginal utilities proportional to $\mathbf{u}_{nj} - \mathbf{u}_{nk}$ can be obtained by dividing the utility differences by the appropriate marginal utilities. Therefore player n should be willing to accept a small gamble whose monetary payoffs are proportional to $(\mathbf{u}_{nj} - \mathbf{u}_{nk})/\mathbf{u}'_{nj}$ conditional on $s_n = s_{nj}$. Such a gamble has an *unconditional* payoff vector of $((\mathbf{u}_{nj} - \mathbf{u}_{nk})/\mathbf{u}'_{nj})\mathbf{1}_{nj}$ in units of money.[25]

Let $\mathbf{G}^*$ now denote the matrix whose rows are indexed by njk and whose columns are indexed by s and whose njk^{th} row is the vector $\mathbf{g}^*_{njk} := ((\mathbf{u}_{nj} - \mathbf{u}_{nk})/$

---

[24] It is assumed here that the macroscopic outcomes of the game for player n are significant amounts of money $\{f_n(s)\}$ to which a strictly concave smooth utility function U_n is applied. This restriction to monetary outcomes is not necessary except in the context of Theorem 8.5. All that is needed otherwise is for money to be *one attribute of utility* and for its marginal utility to be a smooth decreasing function of total utility. We could dispense with f_n and U_n and work directly with the utility vector u_n and marginal-utility-for-money vector u'_n subject only to the condition that $u_n(s) > u_n(s^*)$ implies $u'_n(s) < u'_n(s^*)$.

[25] When the utility functions of the players are strictly concave rather than linear, the bet with payoff vector $((u_{nj} - u_{nk})/u'_{nj})\mathbf{1}_{nj}$ is technically only acceptable to player n on-the-margin, so a bet with an aggregate payoff vector of $\alpha \cdot G^*$ may not be quite acceptable to the players for finite α. In such a case the observer may need to make a small side payment ϵ to the players to get them to agree to the deal, which makes the observer's position not entirely riskless. However, if $\alpha \cdot G^* \leq 0$ and $[\alpha \cdot G^*](s) < 0$, then by choosing α sufficiently small, the magnitude of the required side payment can be made arbitrarily small in relative terms in comparison to the aggregate loss the players will suffer if they play s, which will be considered here as sufficient grounds for not playing s. This could be made precise by using the concept of ϵ-acceptable bets (Nau 1995a), but it will not be pursued here in the interest of brevity.

$u'_{nj})1_{nj}$. This is the revealed-rules matrix for the game, representing all the information about it that can be made common knowledge through unilateral offers to accept small gambles when the players are risk-averse. Here's a summary of the steps in its construction:

$f_n(s)$ = player n's monetary payoff in outcome s (not necessarily observable)

$f_{nj}(s) := f_n(s_{nj}, s_{-n})$: player n's payoff when she chooses her strategy s_{nj} and others play according to s

$U_n(x)$: player n's utility for monetary payoff x

$u_n(s) := U(f_n(s))$: player n's utility for the monetary payoff in outcome s (not observable)

$u'_n(s) := U_n'(f_n(s))$: player n's marginal utility for money in outcome s

$u_{nj}(s) := U_n(f_{nj}(s))$: player n's utility for the payoff obtained when she chooses her strategy s_{nj} and others play according to s

$u'_{nj}(s) := U_n'(f_{nj}(s))$: player n's local marginal utility for money when she chooses her strategy s_{nj} and others play according to s

$1_{nj}(s) :=$ 0-1 indicator for the event $s_n = s_{nj}$

$g^*_{njk}(s) := ((u_{nj}(s) - u_{nk}(s))/u'_{nj}(s))1_{nj}(s)$: observable payoff in outcome s of revealed-rules gamble for player n choosing strategy s_{nj} over strategy s_{nk}.

$\mathbf{u}_n,\ \mathbf{u}'_n,\ \mathbf{u}_{nj},\ \mathbf{u}'_{nj},\ \mathbf{1}_{nj},\ \mathbf{g}^*_{njk}$ = the corresponding vectors with elements indexed by s

$\mathbf{G}^*$: revealed-rules matrix whose njk^{th} row is $\mathbf{g}^*_{njk}$

The matrix $\mathbf{G}^*$ looks like this in the 2×2 case in A-B notation:

TABLE 8.30: Observable payoffs of revealed-rules gambles ($\mathbf{G}^*$ matrix)

	$A_1 B_1$	$A_1 B_2$	$A_2 B_1$	$A_2 B_2$
g^*_{112}	$\dfrac{u_A(a_{11}) - u_A(a_{21})}{u'_A(a_{11})}$	$\dfrac{u_A(a_{12}) - u_A(a_{22})}{u'_A(a_{12})}$		
g^*_{121}			$\dfrac{u_A(a_{21}) - u_A(a_{11})}{u'_A(a_{21})}$	$\dfrac{u_A(a_{21}) - u_A(a_{11})}{u'_A(a_{21})}$
g^*_{212}	$\dfrac{u_B(b_{11}) - u_B(b_{12})}{u'_B(b_{11})}$		$\dfrac{u_B(b_{21}) - u_B(b_{22})}{u'_B(b_{21})}$	
g^*_{221}		$\dfrac{u_B(b_{12}) - u_B(b_{11})}{u'_B(b_{12})}$		$\dfrac{u_B(b_{22}) - u_B(b_{21})}{u'_B(b_{22})}$

If an observer chooses a small non-negative vector α of multipliers for the revealed-rules gambles, the players as a group will receive the vector of payoffs $\mathbf{G}^* \cdot \alpha$ and the observer will receive the opposite payoffs.

The same rationality criterion that was applied in the risk-neutral case also applies here: an outcome s is jointly coherent if there is no $\alpha \geq \mathbf{0}$ such that $\mathbf{G}^* \cdot \alpha \leq \mathbf{0}$ and $[\mathbf{G}^* \cdot \alpha](s) < 0$. The definition of correlated equilibrium and the fundamental theorem of games can now be generalized accordingly. The proof is the same.

> **Definition:** A probability distribution π on outcomes of a game among risk-averse players with revealed-rules matrix $\mathbf{G}^*$ is a *risk-neutral equilibrium* if and only if $\mathbf{G}^* \pi \geq \mathbf{0}$, which means that for every player n and every strategy s_{nj} and alternative strategy s_{nk} of that player, either $P_\pi(s_n = s_{nj}) = 0$ or else $E_\pi(g^*_{njk}) \geq 0$

> **Theorem 8.4: Fundamental theorem of noncooperative games among risk-averse players (Nau (2011a, 2015))**

> An outcome of a game among risk-averse players is jointly coherent if and only if there is a risk-neutral equilibrium in which it has positive probability.

> **Proof:** Same as Theorem 8.2 with $\mathbf{G}$ replaced by $\mathbf{G}^*$.

A risk-neutral equilibrium is a special case of a subjective correlated equilibrium (Aumann (1974, 1987)), which is an equilibrium that can be implemented through a mediator who uses a randomizing device about whose properties the players may hold differing beliefs. Such a device would be welcome in playing a zero-sum game: all players might believe their expected payoffs to be positive! Aumann (1987) remarks that such a result depends on "a conceptual inconsistency between the players." By permitting such inconsistencies, subjective correlated equilibrium places only weak restrictions on solutions of many games. A risk-neutral equilibrium adds the nontrivial restriction that the players' risk-neutral prior probabilities should be mutually consistent, as in an equilibrium of a financial market. Whenever players are risk-averse with significant prior investments in events, their true probabilities are not revealed by their preferences among financial assets, and inconsistencies among them are neither surprising nor problematic. This is the norm in financial markets.

As in the risk-neutral case, there is more to be said about the rational solution of the game than to identify the outcomes that are jointly coherent. It is also possible to place bounds on risk-neutral probabilities of events and on risk-neutral expectations of financial assets that depend on the outcome of the game, namely whatever bounds are determined by the system of inequalities $\mathbf{G}^*\boldsymbol{\pi} \geq 0$ that defines the convex polytope of risk-neutral equilibria. These bounds are bid-ask spreads for assets that the players are jointly offering to the observer through their gambles that reveal information about the rules of the game.

A simple example of the concept of risk-neutral equilibrium is provided by the zero-sum game of matching pennies, which was discussed earlier. Its payoffs in units of money are:

TABLE 8.31: Monetary payoffs in matching-pennies game

	B_1	B_2
A_1	$1, -1$	$-1, 1$
A_2	$-1, 1$	$1, -1$

When played by risk-neutral players the revealed-rules matrix $\mathbf{G}$ scaled to a maximum value of 1 is:

TABLE 8.32: Revealed-rules matrix ($\mathbf{G}$) for risk-neutral players

	A_1B_1	A_1B_2	A_2B_1	A_2B_2
g_{112}	1	-1		
g_{121}			-1	1
g_{212}	-1		1	
g_{221}		1		-1

This game has a unique correlated/Nash equilibrium in which it is as if the players use independent 50-50 randomization, so the graph of the set of equilibria consists of the single point $(\frac{1}{4}, \frac{1}{4}, \frac{1}{4}, \frac{1}{4})$ in the center of the saddle-shaped set of distributions that are independent between $\{A_1, A_2\}$ and $\{B_1, B_2\}$.

Now suppose that both players are risk-averse, and in particular assume that they have identical exponential utility functions, $U(x) = 1 - \exp(-rx)$, in which the risk aversion coefficient is $r = \ln(\sqrt{2})$. In units of utility, the payoff matrix of the matching-pennies game is then:

TABLE 8.33: Utility payoffs for risk-averse players in matching-pennies game

	B_1	B_2
A_1	a, b	b, a
A_2	b, a	a, b

where $a = 1 - \sqrt{1/2} \approx 0.293$ and $b = 1 - \sqrt{2} \approx 0.414$. The corresponding marginal utilities of money in the vicinity of the payoffs a and b are 0.245 and 0.490, respectively, which conveniently differ by a factor of exactly 2.

This game is constant-sum and strategically equivalent to the original one, having the same unique correlated/Nash equilibrium that uses independent 50-50 randomization. However, the revealed-rules matrix is not equivalent because of the distortions of nonlinear utility for money. When each of its rows is scaled to a maximum value of 1, it is:

TABLE 8.34: Revealed-rules matrix ($\mathbf{G}^*$) for risk-averse players

	A_1B_1	A_1B_2	A_2B_1	A_2B_2
g^*_{112}	1	$-1/2$		
g^*_{121}			$-1/2$	1
g^*_{212}	$-1/2$		1	
g^*_{221}		1		$-1/2$

The polytope of risk-neutral equilibria determined by the inequalities $\mathbf{G}^*\pi \geq 0$ is no longer a single point. Rather, it expands into a tetrahedron with these vertices:

TABLE 8.35: Extremal risk-neutral equilibrium distributions

	A_1B_1	A_1B_2	A_2B_1	A_2B_2	Gamble with positive expected value
Vertex #1	8/15	1/15	4/15	2/15	g_{112} (#1 chooses A_1 over A_2)
Vertex #1	2/15	4/15	1/15	8/15	g_{121} (#1 chooses A_2 over A_1)
Vertex #3	1/15	2/15	8/15	4/15	g_{212} (#2 chooses B_1 over B_2)
Vertex #4	4/15	8/15	2/15	1/15	g_{221} (#2 chooses B_2 over B_1)

None of them is independent between players, so none is a Nash equilibrium of a game with these strategy sets. Each of these probability distributions satisfies 3 out of the 4 incentive constraints with equality, i.e., it assigns an expected value of zero to 3 out of the 4 rows of $\mathbf{G}^*$. The label of the row whose expected value is positive is shown in the rightmost column of the table. The graph of this solution is shown in Figure 8.2. The tetrahedron of risk-neutral equilibria is suspended in the middle of the probability simplex, and the saddle of independent distributions cuts through its interior, a situation that would be impossible for a set of correlated equilibria of a game among risk-neutral players.

When players are risk-averse, the small side gambles they are willing to accept do not fully reveal the between-strategy differences in utility profiles that they face in the game, so the set of risk-neutral equilibria of the revealed game is larger than

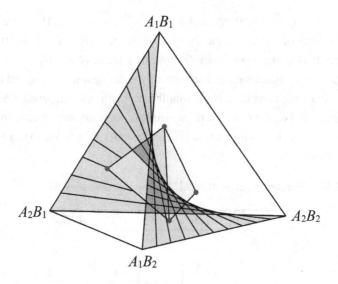

FIGURE 8.2: Risk neutral equilibria of matching-pennies for risk-averse players.

the set of correlated equilibria of the true game. For example, the uniform distribution that is the unique equilibrium of the game when the true utility differences of the players are common knowledge lies in the strict interior of the risk-neutral equilibrium polytope. This is true in general, as formalized here:

Theorem 8.5: Geometry of the set of risk-neutral equilibria (Nau (2015))

The set of correlated equilibria of a game with monetary payoffs played by risk-neutral players is a subset of the set of risk-neutral equilibria of the same game played by risk-averse players.

Proof: If player n is risk-neutral, she will accept a gamble with payoff vector $(\mathbf{u}_{nj} - \mathbf{u}_{nk})\mathbf{1}_{nj}$, while if she is risk-averse, she will accept a gamble with payoff vector $((\mathbf{u}_{nj} - \mathbf{u}_{nk})/\mathbf{u}'_{nj})\mathbf{1}_{nj}$. The multiplier $\mathbf{1}_{nj}$ can be ignored because it zeroes-out the same elements of both vectors. By the subgradient inequality, $U(z) < U(y) - U'(y)(y - z)$, because the value of a strictly concave function U at z must lie below the tangent to its graph at any other point y. Plugging in $y = f_{nj}(s)$ and $z = f_{nk}(s)$ yields $u_{nk}(s) \leq u_{nj}(s) - u'_{nj}(s)(f_{nj}(s) - f_{nk}(s))$, which rearranges to $(u_{nj}(s) - u_{nk}(s))/u'_{nj}(s) \geq f_{nj}(s) - f_{nk}(s)$, with strict inequality if $f_{nj}(s) \neq f_{nk}(s)$. This means that $\mathbf{G}^* \geq \mathbf{G}$ pointwise, i.e., the gamble that player n is willing to accept when she chooses strategy j in preference to k if she is risk-neutral is weakly dominated by the gamble she will accept in the same game if she is risk-averse. It follows that $\mathbf{G}\boldsymbol{\pi} \geq \mathbf{0}$ implies $\mathbf{G}^*\boldsymbol{\pi} \geq \mathbf{0}$ for any probability distribution $\boldsymbol{\pi}$, so if $\boldsymbol{\pi}$ is a correlated equilibrium of the game played by risk-neutral players, then it is a risk-neutral equilibrium of the same game when it is played by risk-averse players.

Hence, risk aversion introduces even more imprecision into the probabilistic solutions of noncooperative games when their rules must be revealed through credible gambles.

In the story just told, the outcome of the game for player n is a given amount of money $f_n(s)$, her utility for it is determined by applying a given von Neumann/Morgenstern utility function U_n, and her marginal utilities for money are obtained by differentiation of U_n. However, there is nothing in this setup that actually requires the outcomes of the game to consist only amounts of money nor for any calculations to be carried out in units of utility. It suffices to have money as merely one attribute of the game's real outcome. The observable parameters of the game are the rows of the matrix $\mathbf{G}^*$ which can be *interpreted* to be computed as $\mathbf{g}_{njk}^* := ((\mathbf{u}_{nj} - \mathbf{u}_{nk})/\mathbf{u}_{nj}')\mathbf{1}_{nj}$. However, the elements of this matrix are (minimally) all that the observer needs to know and all that the players can reveal by using money as a language of communication under noncooperative conditions. The gambles that are the rows of $\mathbf{G}^*$ do not need to be decomposed into utility terms, although the players might choose to do it that way in their own minds if they are trained in decision analysis. So, the theory that has been sketched here is not limited to games that are played only for money. A player merely needs to be willing and able to reveal *some* amount of information about her preferences, namely the parameters of conditional *side gambles* that she is willing to accept, which can be used to indirectly determine whether the game's outcome is one that is rational.

8.10   Playing a new game

In situations where the acceptable gambles that reveal the rules of the game do not determine a unique probability distribution over its outcomes, the players could choose to accept additional gambles in order to reveal more precise information about their joint beliefs. This was illustrated for the game of chicken in Table 8.16. Or, to turn the argument around again, if their joint beliefs are somehow already commonly known to lie in a smaller set, then it is as if they have offered to accept some additional gambles. If they are risk-neutral and have in fact implemented a Nash or correlated equilibrium, which induces a common prior distribution over outcomes of the game, they cannot both be made strictly better off through gambles with each other. When players are risk-averse, this is not necessarily true, and the matching-pennies game provides a good example. When played by risk-averse players, it is a negative-sum game in units of utility, and for both players the unique Nash equilibrium has an expected utility that is below their status quo utility. *Risk-averse players would rather not play this game at all.* Furthermore, player 1's marginal utility of money is greater in outcomes A_1B_2 and A_2B_1 (her losing outcomes) than in the other two, and vice versa for player 2. The Nash equilibrium is therefore not a competitive equilibrium of a financial market in which it is possible for the players to make additional gambles that reveal their solution of the game in addition to the gambles that reveal the rules of the game (the latter being the rows of $\mathbf{G}^*$). In the context of the Nash equilibrium, it is desirable to both players to make a gamble with each other in which player 1 receives x from player 2 if A_1B_2 or A_2B_1 occurs and player 2 receives x from player 1 if A_1B_1 or A_2B_2 occurs, for any positive $x \leq 1$. Such a gamble is a bilateral contract which changes the rules of the game in a substantive way, but coin-flipping remains a Nash equilibrium. By choosing $x = 1$ they can even zero-out their payoffs, dissolving the game altogether.[26] If they do not gamble with each other in this fashion, but instead gamble separately with an observer, there is an arbitrage opportunity for the observer that arises from the fact that, at the outset, the players' risk-neutral probabilities do not agree if their true probability distributions are uniform.

8.11 Games of incomplete information

The games studied in the preceding sections are so-called games of "complete information," in which the players have exact knowledge of the differences among payoffs that they will receive when given moves are made by themselves and others, and rational beliefs are endogenously determined (perhaps imprecisely) by no-arbitrage requirements and dual equilibrium conditions. For much broader applicability, the model can be generalized to one of "incomplete information," following Harsanyi (1967). In such a game there may be exogenous uncertainty about payoff functions and players may be in possession of private information at the time when their moves are made, thus allowing their beliefs and their moves to be conditioned on their information states. Harsanyi proposed that the reciprocal uncertainty about

[26]"If the players in this game are so smart, why are they playing this silly game? Why don't they change the rules and play a game where they can do better?" (Abba Schwarz, quoted in Kreps (1990))

game parameters should be modeled in the following way. Each player may have one or more "types," and different types of a player may have different payoff functions and/or states of information with regard to types of other players. Before planning their moves, the players have a common prior distribution over the set of all combinations of types and it is as if they do not yet know their own type. (They are behind a so-called "veil of ignorance.") Each player is endowed with an information partition that represents the knowledge about other players' types that she will have upon learning her own type. Upon receiving that information the player is able to perform Bayesian updating of the common prior in order to obtain a personal posterior distribution over the types of the other players. The common prior assumption ensures that players' *posterior* beliefs are mutually consistent, given their information.[27] The solution of the game, a so-called *Bayesian equilibrium*, is a Nash equilibrium of the game played among the types. In such an equilibrium, each type of each player is assigned a pure or independently randomized strategy and she has no incentive to defect from it if her other types and those of other players do not defect either.

In the context of games of complete information, the natural solution concept that emerges from a no-arbitrage characterization of rationality is that of correlated equilibrium, not Nash equilibrium, as shown in the preceding sections. In the more general context of games of incomplete information, the solution concept that emerges is a similar generalization of Harsanyi's in which types may play a correlated equilibrium. Again, the requirement that (as-if) randomized strategies should satisfy an independence condition is not supported by the rationality standard of avoiding a sure loss to a betting opponent. This result is described in detail in Nau (1992a) and will be illustrated more informally here.[28]

To fix ideas, consider a two-player incomplete-information game in which Alice and Bob both have 2 strategies, and Bob has 2 types, C_1 and C_2, which are determined by "chance" with prior probabilities of 0.5 each. The payoff matrix is shown below. Note that one submatrix is a 180-degree rotation of the other. The two subgames are strictly competitive and have unique Nash equilibria in mixed strategies

---

[27] The common prior assumption ought to be applied to the players' theoretically-observable risk neutral probabilities, where its violation would lead to arbitrage, as discussed earlier in previous two sections of this chapter. Nevertheless, it is conventional to apply it to imaginary true probabilities. That will also be done in this chapter for purposes of maintaining focus on the issue of correlation among randomized strategies, but these results immediately generalize to the case of risk neutral probabilities..

[28] Very similar notions of correlated equilibrium in games of incomplete information have been proposed by Forges (1993, 2006) and by Bergemann and Morris (2016) who gave it the name "Bayes correlated equilibrium." The important difference between our approach and theirs is that they treat the existence of a supporting prior distribution as a primitive, whereas here the existence of a supporting prior distribution is derived by applying the no-arbitrage condition to players' acceptable conditional bets on other players' types given their own types as well as acceptable conditional bets on other players' moves given their own moves.

where both players independently randomize with probabilities of 1/3 on one strategy and 2/3 on the other.

TABLE 8.36: Example of a game of incomplete information C_1 and C_2 are types of player 2 with equal prior probabilities

	C_1		C_2	
	B_1	B_2	B_1	B_2
A_1	4,4	2,5	4,0	0,2
A_2	0,2	4,0	2,5	4,4

Here are examples of Bayesian equilibrium and correlated Bayesian equilibrium for the full game:

TABLE 8.37: Bayesian and correlated Bayesian equilibria

	C_1		C_2			C_1		C_2	
	B_1	B_2	B_1	B_2		B_1	B_2	B_1	B_2
A_1	1/4			1/4	A_1	1/9	2/9	1/9	1/18
A_2	1/4			1/4	A_2	1/18	1/9	2/9	1/9

A Bayesian equilibrium	A correlated Bayesian equilibrium
Expected payoffs = 2, 3	Expected payoffs = 2.667, 3.333

In any Bayesian Nash equilibrium, Alice independently chooses A_1 with probability α for some α between 1/3 and 2/3. (The case of $\alpha = 1/2$ is shown above.) Bob independently chooses B_1 if his type is C_1 and B_2 if his type is C_2. This solution yields them expected payoffs of 2 and 3 respectively. However, there are many Pareto superior solutions in which Alice's and Bob's strategies are correlated. In some cases correlation might be implemented by a mediator or a designed mechanism, and in other cases it might be implemented by cheap talk. (More generally, correlation might exist in the eye of an observer of the game who can't detect whether communication or deliberate correlated randomization are possible.) In one correlated Bayesian equilibrium that is particularly interesting, Bob plays B_1 with probability 1/3 if his type is C_1 and plays it with probability 2/3 if his type is C_2 *and* he reveals his type to Alice, who plays A_1 with probability 2/3 if Bob's type is C_1 and with probability 1/3 if his type is C_2. This arrangement is incentive compatible and it yields an expected value of 2.667 to Alice and 3.333 to Bob.[29] 2.667 is the largest

---

[29]Observe that the Bayesian equilibrium places a total probability of 1/2 on outcomes $A_2 B_1 C_1$ and $A_1 B_2 C_2$, where Alice gets 0 (her lowest possible payoff) and Bob gets 2 (his second-lowest). These two outcomes have a total probability of only 1/9 in the correlated Bayesian equilibrium.

expected payoff that Alice can get in any equilibrium, and it wouldn't be surprising if she proposed this solution to Bob if the game's true parameters were known to both players. These two equilibria are shown in Table 8.37.

The case of general incomplete-information games with finite strategy and type sets will now be considered. As in the complete-information game model, suppose that player n has a set of strategies denoted by S_n and let $S := S_1 \times ... \times S_N$ denote the set of all joint strategies. Additionally, suppose that each player may be one of several types who may have different payoff functions and/or private information. Let T_n denote the set of types of player n and let $T := T_1 \times ... \times T_N$ denote the set of all joint assignments of types. These are interpretable as states of nature.[30] Let $s := (s_1, s_2, .., s_N)$ and $t := (t_1, t_2, .., t_N)$ denote elements of S and T, where $s_n \in S_n$ and $t_n \in T_n$. The jth strategy of player n is denoted by s_{nj}. f_n denotes player n's payoff function, with $f_n(s, t)$ being her payoff when the players implement the joint strategy s and their types are given by t.[31] Let $\mathcal{G}$ denote the array of parameters of these payoff functions. It represents the "true" rules of the game.

As in Section 8.4 it will be assumed for simplicity that payoffs are amounts of money and players behave as if they have state-independent linear utility for money, and they have the opportunity to make monetary side gambles that depend on the game's outcome (s, t), which is assumed to be fully revealed ex post.[32] Let $f_{nj}(s, t)$ denote player n's payoff when she chooses s_{nj} and the rest of the players choose their own respective parts of s, and their types are specified by t:

$$f_{nj}(s, t) := f_n(s_{nj}s_{-n}, t) \tag{8.8}$$

Let $\mathbf{1}_{njr}$ denote an indicator vector for the event in which player n chooses s_{nj} and her type is $r \in T_n$:

$$
\begin{aligned}
\mathbf{1}_{njr}(s, t) \quad &: \quad = 1 \text{ if } s_n = s_{nj} \text{ and } t_n = r \qquad (8.9)\\
&: \quad = 0 \text{ otherwise.}
\end{aligned}
$$

In the event where player n has type r and she chooses s_{nj} in preference to some other available strategy s_{nk}, others can infer that s_{nj} has a greater-or-equal expected payoff than s_{nk} when her beliefs and her payoffs are those of type r. In that case a

---

[30] Equivalently we could begin by assuming a set T of states of nature and then defining T_n as a partition of T that represents the characteristics and state of knowledge of player n.

[31] Note that a player's payoff may depend on another player's type as well as the other player's strategy. For example, player 1's payoff could depend on whether player 2 is "weak" or "strong." A player's type could also refer to her state of knowledge about another player's type.

[32] This assumption can be generalized to the case of nonlinear and/or state-dependent utility exactly as in Section 8.9, merely changing the units of analysis to risk-neutral probabilities.

small side gamble with payoffs proportional to $f_n(s_{nj}s_{-n},t) - f_n(s_{nk}s_{-n},t)$ will have non-negative expected value, and therefore she ought to be willing to accept it if someone offers to take the other side. Let $\mathbf{g}_{njkr}$ denote the vector of payoffs of this gamble. Its value in outcome (s,t) is

$$g_{njkr}(s,t) := 1_{njr}(s,t)(f_{nj}(s,t) - f_{nk}(s,t)) \qquad (8.10)$$

Gambles of this form are (again) called "preference gambles" because they reflect the player's conditional preferences among payoff vectors without revealing anything about her beliefs in regard to either strategies *or* types. As in the model of the complete-information game, assume that player n is willing to *publicly* reveal that she is willing to accept a small gamble with payoff vector $\alpha_{njkr}\mathbf{g}_{njkr}$ where α_{njkr} is a positive weight chosen by someone else, along with any other gambles that are acceptable to her. Let $\mathbf{G}$ denote the matrix whose rows are these payoff vectors of acceptable gambles, $\{\mathbf{g}_{njkr}\}$. $\mathbf{G}$ has $|S| \times |T|$ columns, one for every outcome of the game (strategy/type combinations for all players), and for player n it has $|S_n| \times (|S_n| - 1) \times |T|$ rows, one for every alternative strategy of any given strategy of a given type of herself. As before let $\boldsymbol{\alpha}$ denote the vector whose elements are the non-negative weights that the observer chooses for the gambles that are the rows of $\mathbf{G}$.

The description of the game is completed by introducing "belief gambles" which the players may accept as a way of revealing information about their subjective probabilities regarding the occurrence of types. In particular, *each player reveals (perhaps incompletely) her conditional probabilities for types of other players, given her own type*. Let I_n denote a set of index numbers for the belief gambles that are acceptable to player n. For all $i \in I_n$, let $\mathbf{E}_{ni}$ and $\mathbf{F}_{ni}$ denote names as well as indicator vectors (defined on $S \times T$) for events that are subsets of T_{-n} and T_n, respectively. Following de Finetti's method of defining and eliciting conditional subjective probabilities, suppose that for each such pair of events player n announces a *lower conditional probability for $\mathbf{E}_{ni}$ given $\mathbf{F}_{ni}$*, which will be denoted p^*_{ni}. This means that she is willing to accept small multiples of a gamble that yields a gain of $1 - p^*_{ni}$ if both $\mathbf{E}_{ni}$ and $\mathbf{F}_{ni}$ occur, a loss of p^*_{ni} if $\mathbf{F}_{ni}$ occurs but $\mathbf{E}_{ni}$ does not, and zero if $\mathbf{F}_{ni}$ does not occur, regardless of the strategies played. Let $\mathbf{h}_{ni}$ denote the vector of payoffs of this gamble. Its value in outcome (s,t) is

$$h_{ni}(s,t) := (E_{ni}(s,t) - p^*_{ni})F_{ni}(s,t) \qquad (8.11)$$

A probability distribution π on $S \times T$ is called a *supporting probability distribution for player n* if it assigns non-negative expected value to every belief gamble that she accepts, i.e., if it satisfies

$$\sum_{s,t} h_{ni}(s,t)\pi(s,t) \geq 0 \qquad (8.12)$$

for all $i \in I_n$. Let $\mathbf{H}$ denote the matrix whose rows are the payoff vectors of all the acceptable belief gambles. It has $|S| \times |T|$ columns and one row for every belief gamble accepted by some player. Let β_{ni} denote a non-negative weight chosen by the observer for the belief gamble whose payoff vector is $\mathbf{h}_{ni}$, the $(n,i)^{th}$ row of $\mathbf{H}$, and let β denote the vector of all such weights.

Definition: Outcome (s,t) of an incomplete-information game is *jointly coherent* if and only if there is no ex-post-arbitrage opportunity when it occurs, i.e., if and only if there are no vectors of non-negative weights α and β such that $(\mathbf{G} \cdot \alpha + \mathbf{H} \cdot \beta) \leq \mathbf{0}$ and $(\mathbf{G} \cdot \alpha + \mathbf{H} \cdot \beta)(s,t) < 0$.

Definition: A probability distribution π on $S \times T$ is a *correlated Bayesian equilibrium* if it has the property that if it were used by a mediator to select types and private recommended strategies for all the players, then (a) it would be optimal for each player to comply with her own strategy recommendation for each of her types if all other players do likewise, and (b) it is a supporting probability distribution for the belief gambles of all players. This means that π satisfies the following conditions with respect to the preference-gamble payoff matrix $\mathbf{G}$ and the belief-gamble payoff matrix $\mathbf{H}$:

$$\begin{aligned} \mathbf{G}\pi &\geq \mathbf{0} \qquad (8.13) \\ \mathbf{H}\pi &\geq \mathbf{0} \\ \pi &\geq \mathbf{0} \\ \pi &\neq \mathbf{0} \end{aligned}$$

Theorem 8.6: Fundamental theorem of games of incomplete information (Nau (1992a))

An outcome of a finite game of incomplete information is jointly coherent if and only if there is a correlated Bayesian equilibrium in which it has positive probability.

The statement of Theorem 8.6 does not draw a sharp distinction between S and T in terms of who or what is at fault if the outcome leads to an arbitrage loss. The players have chosen s and nature has chosen t, and evidently they do not all agree. From the perspective of a would-be arbitrageur, an event is an event, regardless of who or what controls it.

Here again is the payoff table for the game of incomplete information that was discussed above (Table 8.36). Bob (alone) gets to observe whether C_1 or C_2 has occurred prior to making his own move, so they are two types for him.

	C_1		C_2	
	B_1	B_2	B_1	B_2
A_1	4, 4	2, 5	4, 0	0, 2
A_2	0, 2	4, 0	2, 5	4, 4

The preference-gamble matrix **G** has 8 columns, one for each combination of types and moves, It has 2 rows for Alice's acceptable gambles, which are conditioned on her move, and 4 rows for Bob's acceptable gambles, which are conditioned on both his type and his move.[33]

TABLE 8.38: Preference gambles that reveal payoff information

	111	211	121	221	112	212	122	222
$A_1 \succsim A_2 \mid A_1$	4		−2		2		−4	
$A_2 \succsim A_1 \mid A_2$		−4		2		−2		4
$B_1 \succsim B_2 \mid B_1 C_1$	−1	2						
$B_2 \succsim B_1 \mid B_2 C_1$			−2	1				
$B_1 \succsim B_2 \mid B_1 C_2$					1	−2		
$B_2 \succsim B_1 \mid B_2 C_2$							2	−1

The belief-gamble matrix **H** represents the common prior. It has two rows which establish that 0.5 is both a lower and upper probability for C_1:

TABLE 8.39: Belief gambles that reveal prior probabilities of types

	111	211	121	221	112	212	122	222
$p(C_1) \geq 0.5$	0.5	0.5	0.5	0.5	−0.5	−0.5	−0.5	−0.5
$p(C_1) \leq 0.5$	−0.5	−0.5	−0.5	−0.5	0.5	0.5	0.5	0.5

As noted earlier this game has many correlated and uncorrelated Bayesian equilibria. Here are the probability distributions and expected payoffs of several of the most interesting ones:

---

[33]In Tables 8.38-8.40, the index ijk in the column heading is shorthand for $A_i B_j C_k$.

TABLE 8.40: Some correlated and uncorrelated Bayesian equilibria

	111	211	121	221	112	212	122	222	Alice	Bob
π^1	0.333	0.167					0.333	0.167	2.00	3.00
π^2	0.167	0.333					0.167	0.333	2.00	3.00
π^3	0.111	0.057	0.222	0.111	0.111	0.222	0.057	0.111	2.67	3.33
π^4	0.200	0.100	0.200			0.200	0.100	0.200	2.40	4.00

π^1 and π^2 are the endpoints of the interval of Bayesian Nash equilibria, and π^3 is the correlated Bayesian equilibrium mentioned earlier, which maximizes the expected payoff to Alice and could be implemented with cheap talk. π^4 is a correlated Bayesian equilibrium that maximizes Bob's expected payoff. The convex hull of the latter two is the efficient frontier. These equilibria can be found by using linear programming to solve for probability distributions that maximize or minimize the probabilities of particular outcomes or weighted sums of the expected payoffs.

These results are easily extended to settings in which the players may employ communication mechanisms to which they truthfully report their types and from which they receive recommended strategies. The behavioral variables for the players are "decisions" which consist of mappings from their information states (types) to inputs of the device, along with mappings from their information states and outputs of the device to their played strategies. The result is a generalization of Theorem 8.6: the decisions of the players are jointly coherent (avoid ex-post-arbitrage) if and only if they have positive probability in a communication equilibrium as defined by Forges (1986) and Myerson (1985, 1986). Details are given by Nau (1992a).

8.12 Discussion

This chapter has explored the simplest and most tractable case of a noncooperative game, one in which players have finite sets of strategies and outcomes are quantities of money. Several main objectives have been pursued. One objective has been to show that *Aumann is right: correlated equilibrium is the expression of Bayesian rationality in noncooperative games*. It fits seamlessly into a spectrum of theories of single-player and multi-player rationality which includes subjective probability (de Finetti's version) on the 1-player end and asset pricing on the n-player end for large n (to be discussed in the next chapter). The single- and multi-player theories are unified by the rationality standard of no-arbitrage (joint coherence), and their fundamental representation theorems are applications of the fundamental theorem of linear programming, which rests on the separating hyperplane theorem for convex sets. The mental parameters of rational agents (probabilities, utilities, equilibrium reasoning) are the primal variables in the linear programs. The actions of observers—would-be arbitrageurs—are the dual variables. Small and large groups of agents satisfy the same principles of rationality as individuals, as seen by others. It's odd that the connection between linear programming and the use of correlated strategies in noncooperative games was not noticed much earlier, especially in light of the interest in linear programming that already existed by 1950 (including von Neumann's own contributions to duality theory),[34] and even today correlated equilibrium tends to be treated as a minor topic which gets only brief coverage in the teaching of game theory, if any at all.

Nash's requirement that randomized (or as-if- randomized) strategies should be statistically independent is important for applications that seek to model the game from an individual player's perspective as parsimoniously as possible—which is to say, most applications in economics—but it is difficult to defend as a condition that ought to be satisfied by the beliefs of observers with respect to the players as a group. It is hard for an observer to prove a negative, namely that the players have not implicitly or explicitly coordinated their moves, particularly where there may be incentives for cooperation that are not represented in the revealed-rules matrix **G**. Players in real games generally have some amount of shared knowledge and experience that guide their moves. Also, Nash equilibrium lacks a good story or algorithm for converging to a fixed point, particularly in games that are not repeated or which have multiple equilibria, whereas the exploitation of arbitrage opportunities among side gambles would tend to drive the game toward an outcome that is supported by a correlated equilibrium. In light of the duality theorem, it is obvious why a game should have a correlated equilibrium (otherwise there would be no escape from arbitrage) but not so obvious why it should have a Nash equilibrium (that is the beauty of Nash's result). The very term "correlated equilibrium" is unfortunate, for it gives the impression of an equilibrium in which correlation has been implemented, whereas it really means that correlation is not forbidden a priori, and it arises naturally in the representation of subjective uncertainty in regard to

---

[34]Howard Raiffa came close in his 1951 PhD. thesis, where he observed that it could be beneficial for the players to privately correlate their randomized strategy choices before reporting them to an umpire. "This is a generalization of von Neumann's scheme of mixed strategies but in practice it fulfills the strict interpretation of the game, namely, that the game is resolved by the players giving the umpire a single pair of pure strategies... [I]t is a useful concept since it serves to convexify certain regions in the Euclidean plane."

multiple solutions. It's a property of imprecise probabilities in general. We ought to just refer to a correlated equilibrium as a "noncooperative equilibrium" (or perhaps an "Aumann equilibrium") with Nash equilibrium as the special case.

A second objective has been to emphasize, once again, the significance of the question of whether money is or isn't the unit of account in the situation being modeled, even hypothetically. It is subject to precise numerical measurement, in private or in public, it is a tool of quantitative thinking as well as an object of desire and a medium of exchange, and everyone can agree that if you are rational you shouldn't throw it away. These properties give a very clear meaning to assumptions of common-knowledge-of-payoffs and common-knowledge-of-rationality that are central to noncooperative game theory but which are tricky to define in terms of utilities. Expected utility was axiomatized by von Neumann and Morgenstern precisely so as to have a "single monetary commodity" (not necessarily actual money) to be used as a payoff scale for games in any setting. But real people do not quantify their personal experiences in units of utility, let alone do it with any precision, let alone communicate with others or observe the characteristics of others in those terms. In a payoff matrix for a hypothetical game, it matters whether the numbers are amounts of money or amounts of utility. In the latter case, assumptions in regard to the precision of common knowledge and sophistication of play should perhaps be somewhat less forceful.

A third objective has been to emphasize that even in this highly stylized and oversimplified case, there are still problems to be faced with respect to measurement issues. The Nash and correlated equilibria of a game are determined by a subset of its parameters, namely the elements of the matrix G which depend only on "internal" effects of changes in a player's strategy, taking the strategies of other players as fixed, thus deliberately ignoring the "external" effects that are imposed on her by the other player's strategies, taking her own strategy as fixed. No wonder that it is easy to construct pathological examples. There is no universally-applicable solution concept that deals systematically with the left-out game parameters. The as-if knowledge of those parameters would require more machinery such as written instructions or bilateral contracting or cheap talk or memory of history of play or social norms that are outside the rules of a single play of the game. Absent any such information, the solution of the game from an observer's perspective is the entire set of correlated equilibria, a convex polytope as in the theory of imprecise subjective probabilities.

A fourth objective has been to show that *risk-neutral probabilities* have a role to play in game theory as well as in the theories of single-agent risk aversion and many-agent asset pricing. The field persists in trying to uniquely separate personal probabilities and utilities in modeling strategic behavior (as well as individual behavior in the presence of risk and ambiguity), but it is not necessary to do this in the most general case and it is not consistent with the way individuals really think when they are juggling numbers in their heads. These issues are nicely finessed by changing the units of analysis to risk-neutral probabilities in situations where sizable amounts of money are at stake and players exhibit aversion to uncertainty or otherwise state-dependent marginal utility for money. An important stylized fact that emerges here is that the set of acceptable side gambles is smaller and the set of equilibria is larger than they would be for the same game played by risk-neutral players. Thus, the presence of risk aversion tends to reduce the precision of reciprocal knowledge of the rules of the game. Risk-averse players are inherently less candid than risk-neutral players in revealing information about their true payoff functions.

Chapter 9

Asset pricing

9.1 Introduction

Asset pricing is the field of application where rational choice theory finds its purest expression: vast numbers of people engage in quantitative economic reasoning on an ongoing basis in real time according to their knowledge and beliefs and attitudes towards risk and receipt of information and (often) the use of highly refined computer models. Even those who do not directly participate in financial markets have a stake in what goes on there (if only as an indicator of worldly events that may affect them) and up-to-the-minute values of stock indices and interest rates and exchange rates are displayed on news programs and web sites. In recent years the landscape has changed dramatically to make markets more accessible to everyone. It's now possible for individual investors to trade stocks instantly on their laptop computers and cell phones, and they are barraged with TV ads encouraging them to do exactly that. The field of sports has also become a financial market. Sports betting has been legalized nearly everywhere and TV sports coverage is permeated with quantitative financial news in the form of live odds and money lines. Online casinos offer complex prop bets ("proposition" bets) that allow wagers to be placed on fine details of what may happen in a game. Uncertainties are being converted to financial assets on many fronts.

The standard of economic rationality that naturally applies to asset prices is that there should be no arbitrage, and the terms of analysis are, in the most general case, risk-neutral probabilities. Indeed, the explicit use of risk-neutral probabilities

DOI: 10.1201/9781003527145-9

originated in the field of finance in the 1970s. (Cox and Ross (1976), Rubinstein (1976), Breeden and Litzenberger (1978), Ross (1978), Brennan (1979), Harrison and Kreps (1979), Stapleton and Subrahmanyam (1984)) This section will provide an overview of the two simplest examples of asset pricing: two-period markets in which asset price distributions are either discrete or normal. A hypothetical asset—a multivariate quadratic option—will be used to derive simple closed-form solutions for aggregation of the beliefs and risk preferences of heterogenous investors in a market with multiple primary assets. It will be used to illustrate (yet again) some fundamental issues concerning the measurement of subjective probabilities and utilities even under highly idealized circumstances.

9.2 Risk-neutral probabilities and the fundamental theorem

Consider a 2-period financial market in which there is a single representative investor (the market itself) and an arbitrageur, and there is a set of K risky assets that the arbitrageur may buy and sell in arbitrary amounts at time 0. The time-1 payoffs of the assets depend on a finite set of M states of nature. Let $P := (p_1, ...p_K)$ denote the vector of time-0 asset prices and let z_{km} denote the time-1 payoff of 1 unit of asset k in state m. (The states will be identified with realizations of payoffs of some of the assets at time 1.) Also assume that there is a riskless bond whose current price is p_0 and which yields a payoff of 1 at time 1, and let $r := 1/p_0$ denote the multiplicative return factor for the bond (1 plus the risk-free rate of return). The market is complete if the assets span the payoff space, i.e., if $K - 1$ of the asset payoff vectors are linearly independent of each other and the constant vector, and purchases and sales can be made at the same price. The purchase of 1 unit of asset k at time 0 for price p_k can be financed by borrowing at the risk-free rate and paying back $p_k r$ at time 1 while getting an asset payoff of z_{km}, for a net time-1 payoff of $z_{km} - p_k r$. The reverse transaction (a sale at time 0) is also possible.

The arbitrageur's problem can be formulated as follows: find a vector of portfolio weights $\{\alpha_1 ..., \alpha_K\}$, unrestricted in sign, that yield $\sum_{k=1}^{K} \alpha_k(z_{km} - p_k r) < 0$ in every state m. By invoking the separating hyperplane theorem (Lemma 2.1), we obtain the classic result:

Theorem 9.1: Fundamental theorem of asset pricing (for the case of a finite set of states and two time periods)

There is no arbitrage opportunity in a complete financial market if and only if there is a risk-neutral probability distribution with respect to which the present value of every asset's expected future payoff is equal to its current price, and correspondingly the expected return on any asset is equal to the risk-free rate of return.

Proof: Let $\mathbf{W}$ denote the $K \times M$ matrix whose km^{th} element is $(z_{km} - p_k r)$. Then by Lemma 2.1, there exists $\alpha \geq 0$ such that $\mathbf{W} \cdot \alpha < \mathbf{0}$ (there is an arbitrage opportunity) or else there exists $\pi \geq \mathbf{0}$, with $\pi \neq \mathbf{0}$, such that $\mathbf{W}\pi \geq \mathbf{0}$ (there is a risk-neutral distribution π according to which the expected return factor for every asset is at least r). If the market is complete, in which case the same prices hold for purchases and sales, the expected return factor for every asset must be exactly r.

This is merely a special case of the fundamental theorem of subjective probability[1] in which the individual bettor is the whole market, the acceptable bets have a special structure, and π is interpreted as a risk-neutral distribution.

For a simple example, suppose that there are 5 states and that the assets consist of a riskless bond, a stock, and call options[2] on the stock at three different strike prices. Let the possible values of the stock price at time 1 be 100, 110, 120, 130, and 140, respectively, in states 1 to 5, and let the strike prices of the available call options be 110, 120, and 130. The riskless bond, by definition, has a constant payoff of $1 in period 1. So, the time-1 payoffs of the assets are as follows.

TABLE 9.1: Time 1 payoff functions of bond, stock and European call options at 3 strike prices

State	Payoff at time 1				
	1	2	3	4	5
Bond	$1	$1	$1	$1	$1
Stock	$100	$110	$120	$130	$140
Call@110	$0	$0	$10	$20	$30
Call@120	$0	$0	$0	$10	$20
Call@130	$0	$0	$0	$0	$10

Note that the payoff of a call option is a piecewise linear function of the stock price. If the stock price is less than or equal to the strike price of the option on the date when it expires, the value of the option is zero: it is "out of the money." If the stock price is greater than the strike price, then the value of the call option is the difference

---

[1] The connection between the fundamental theorem of probability and the fundamental theorem of asset pricing is discussed by Schervish et al. (2008) with respect to the issue of unbounded payoffs.

[2] A call option is an option to buy a share of stock at given future date for a price (the so-called strike price) agreed upon today.

between the stock price and the strike price. Options with piecewise linear payoffs are the most common in practice, although we will see shortly that options with quadratic payoffs would be more convenient for theoretical purposes.

Suppose that the time-0 asset prices are the following:

TABLE 9.2: Prices of assets at time 0

	Bond	Stock	Call@110	Call@120	Call@130
Price	$0.950	$111.150	$8.075	$2.375	$0.475

The price of the bond determines the risk-free rate of interest, which in this case is $1.00/0.95 - 1 = 5.26\%$. Because the number of linearly independent assets equals the number of states, and we also assume no friction, the market is complete, which means that the available assets span the set of all possible payoff distributions at time 1.

An arbitrage opportunity exists if there is a linear combination of purchases and sales whose total payoff at time 0 is zero and whose total payoff at time 1 is strictly positive. Such an opportunity will exist if some assets are mis-priced relative to the others. From the fundamental theorem, we know that there is no arbitrage if and only if there exists a risk-neutral distribution such that the time-0 price of every asset is equal to the expected value of its time-1 payoff, discounted at the risk-free rate. It is a simple linear programming problem to find the risk-neutral distribution, if it exists, and in this case there is one, namely $\pi = (0.15, 0.25, 0.4, 0.15, 0.05)$. The fact that the risk-neutral distribution prices every asset according to its discounted expected future value is equivalent to saying that, according to the risk-neutral distribution, all assets have the same expected return, namely the risk-free rate. This can be verified by inspection of the following table which shows the state-by-state percentage returns of the assets:

TABLE 9.3: Returns on assets and unique risk-neutral distribution (π) that equates their expected returns

State	Percentage return					Mean	Var.
	1	2	3	4	5		
Bond	5.26%	5.26%	5.26%	5.26%	5.26%	5.26%	0
Stock	−10.03%	−1.03%	7.96%	16.96%	25.96%	5.26%	0.009
Call@110	−100%	−100%	23.84%	148%	272%	5.26%	1.116
Call@120	−100%	−100%	−100%	321%	742%	5.26%	5.097
Call@130	−100%	−100%	−100%	−100%	2005%	5.26%	21.05
π	0.15	0.25	0.40	0.15	0.05		

The assets have very different levels of risk, as measured by the variances of their returns, but nevertheless they all have the same expected return according to the risk-neutral distribution. This is the magical and counterintuitive property of risk-neutral probabilities. The variance was calculated here by using the risk-neutral distribution, since we have not specified a "true" distribution a priori, nor do we need to. *The risk-neutral variance is the best estimate of the true variance that is publicly observable at the present moment.* As will be shown below, the risk-neutral variance is a weighted average of the true subjective variances of the investors, with weights proportional to their risk tolerances which are measures of their relative market power. In this sense it is "unbiased."

Now suppose that you want to price a new, exotic derivative asset: a *quadratic option*. Its payoff at time 1 is the *square of the excess return on the stock price*. Thus, in state 1, the payoff of one unit of the quadratic option would be $(-10.03\% - 5.26\%)^2 = 0.02339$, and so on. The following figure illustrates the difference between the payoff function of a quadratic option and an ordinary European call option for this stock:

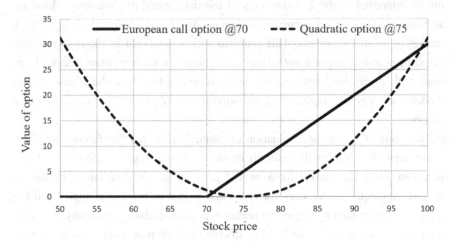

FIGURE 9.1: Payoff functions of European and quadratic options.

To determine the correct arbitrage-free price, all you need to do is compute the risk-neutral expected payoff, then discount it back to time 0 at the risk-free rate:

TABLE 9.4: Payoffs of a quadratic option and their risk neutral probabilities

State	1	2	3	4	5
Quad. option payoff	$0.02339	$0.00397	$0.00073	$0.01368	$0.04282
Risk-neutral prob. π	0.15	0.25	0.40	0.15	0.05

Here the undiscounted (future) price of the quadratic option is the risk-neutral expected value of its payoffs, which is $0.00899. The discounted (current) price is that value multiplied by 0.95, which is $0.00854. Any other strange new asset could be priced by the same method.

Here's the especially nice thing about quadratic options: suppose that the risk-neutral distribution is not yet fully known because the three call options have not yet been priced, but the market has already determined a price for the quadratic option. The undiscounted price of the quadratic option computed in Table 9.4 is precisely the *variance of the stock return* (in units of squared time-1 dollars) according to the risk-neutral distribution, as shown at the right of the second row in Table 9.3. Thus, the price of the quadratic option directly reveals the risk-neutral variance of the stock return, which is otherwise known as the "implied volatility" of the stock.

More generally, a quadratic option could be defined so that its payoff would be the *product* of the excess returns on *two different stocks*, and its price would then directly reveal the risk-neutral *covariance* between them. So, if you were primarily interested in the 2nd moments of the risk-neutral distribution—which are all that is needed for some purposes—you could elicit them directly by opening a market in quadratic options. This scenario will be analyzed in the next two sections. In practice, implied volatilities are typically estimated from prices of options with piecewise linear payoffs using formulas based on the Black-Scholes model that assumes a lognormal distribution for future prices (geometric Brownian motion).

Methods of risk-neutral valuation are sometimes called "preference free" because they do not directly refer to anyone's probabilities or utilities, but this is a misnomer. The market's risk-neutral probabilities are determined by the investors' aggregate preferences for risky assets, just as prices for apples and bananas are determined by aggregate preferences in an exchange economy. Investors do look at the empirical probability distributions of past asset returns, as well as recent trends and breaking news and private information, when constructing their preferences, but there no agreed-upon way to do this, and their hypothetical subjective probabilities are unobservable, as will be discussed in more detail below. The literature of "behavioral finance" that has emerged over the last few decades has documented various ways in which the behavior of asset prices departs from the predictions of models that assume perfectly rational investors armed with complete information and correct models of the structure of the economy.

9.3   The multivariate normal/exponential/quadratic model

Now let's consider a more general setting: a market for risky assets whose payoffs are continuously distributed. Suppose that investors have beliefs described by multivariate normal subjective probability distributions and risk preferences described by exponential (CARA) utility functions and that they hold portfolios of assets whose payoffs are quadratic functions of the prices as discussed above. Normal distributions and exponential utility functions and quadratic portfolios are "conjugate" to each other in the sense that they yield normal risk-neutral probability distributions. The normal/exponential assumption reasonable to use as an approximation on a small time scale and/or in situations where negative wealth positions are possible, and it will make it convenient to derive formulas for optimal portfolio selection and risk-neutral probabilities in equilibrium. This will lead to a subjective heterogeneous-expectations version of the capital asset pricing model (Nau and McCardle (1991)). The use of quadratic options in the context of the normal/exponential model has been widely studied by others, but mainly in contexts involving a single risky primary asset (e.g., Brennan and Cao (1996), Cao and Yang (2008), and Qin (2013)) Here the case of an arbitrary number of primary assets will be considered, and the pricing of covariance as well as variance will be addressed.

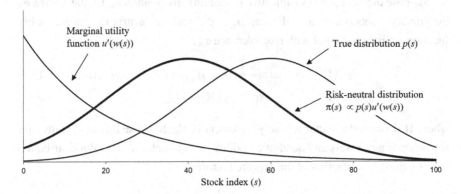

FIGURE 9.2: Construction of risk-neutral density function for a normal subjective probability distribution, exponential utility function, and linear wealth function.

The figure above illustrates the conjugate relationship between normal probability distributions and exponential utility functions. In general the investor's risk-neutral probability distribution for the future value of the stock price, $\pi(s)$, is the renormalized product of her true subjective probability distribution $p(s)$ and her marginal utility function $u'(w(s))$, where $u(w)$ is utility for wealth and $w(s)$ is wealth in state s. If her probability distribution is normal and her utility function is exponential and her wealth function is linear, the resulting risk-neutral distribution is also normal, its mean is left-shifted relative to that of the true distribution, and its variance remains the same. If the wealth function is quadratic, the variance will also change but the risk-neutral distribution will remain in the normal family. The multivariate version of this phenomenon is discussed in what follows.

Suppose the market contains K primary assets (stocks) as well as all pairwise quadratic options, and let R_j denote the multiplicative return factor for asset j (1 plus the percentage return). The payoff of 1 unit of a normalized quadratic option is

$$Q_{jk} := (R_j - r)(R_k - r), \tag{9.1}$$

i.e., the product of the excess returns on assets j and k relative to the risk-free return. Risk-neutral valuation of the quadratic option yields

$$P_{jk} = E_\pi[Q_{jk}]/r = E_\pi[(R_j - r)(R_k - r)]/r = \text{cov}(R_j, R_k)/r = \theta_{jk}/r \tag{9.2}$$

where P_{jk} denotes its price and θ_{jk} denotes the covariance between R_j and R_k under the risk-neutral distribution π. Hence, *prices of quadratic options directly reveal the risk-neutral variances and covariances of returns on primary assets.*[3]

Suppose that investor n's subjective probability distribution p_n for the returns on the primary assets is normal with mean $\boldsymbol{\mu}_n$ and covariance matrix Θ_n and her utility function u_n is exponential with risk tolerance τ_n:

$$p_n(\mathbf{R}) \quad \propto \quad \exp(-\frac{1}{2}(\mathbf{R} - \boldsymbol{\mu}_n) \cdot \Theta_n^{-1}(\mathbf{R} - \boldsymbol{\mu}_n)), \tag{9.3}$$
$$u_n(w_n(\mathbf{R})) \quad = \quad -\exp(-w_n(\mathbf{R})/\tau_n).$$

where $\mathbf{R}$ denotes the vector whose j^{th} element is R_j. Suppose further that investor n holds primary assets and quadratic options so as to achieve a wealth distribution that is a quadratic function of the expected returns:

$$w_n(\mathbf{R}) = a_n r + \mathbf{R} \cdot \mathbf{b}_n + \frac{1}{2}(\mathbf{R} - \mathbf{r}) \cdot \mathbf{C}_n(\mathbf{R} - \mathbf{r}) \tag{9.4}$$

for some constant a_n, vector $\mathbf{b}_n$, and *symmetric* matrix $\mathbf{C}_n$. $\mathbf{b}_n$ is the vector whose j^{th} element is the dollar value of asset j held by investor n at time 0, which may

---

[3]Notation for covariance matrices: Θ denotes the covariance matrix of the market's risk neutral distribution, whose elements are determined by quadratic option prices. Θ_n denotes the covariance matrix of investor n's true subjective probability distribution, and $\hat{\Theta}_n$ denotes the covariance matrix of her risk neutral distribution, whose difference from Θ_n reflects her holdings of quadratic options. In an arbitrage-free state, $\hat{\Theta}_n$ must be equal to Θ for all n.

be positive or negative (short selling is allowed), and $\mathbf{r}$ is the constant vector whose elements are all equal to r. $\mathbf{C}_n$ is the matrix whose jk^{th} element is the dollar value of the normalized quadratic option Q_{jk} held by investor n at time 0, which may also be positive or negative. Without loss of generality it will be assumed that $Q_{jk} = Q_{kj}$ for symmetrization of $\mathbf{C}_n$ insofar as they have identical payoffs (9.1).

Investor n's personal risk-neutral probability density is determined as usual by

$$\pi_n(\mathbf{R}) \propto p_n(\mathbf{R})u'_n(w_n(\mathbf{R})), \tag{9.5}$$

where u'_n denotes the first derivative of u_n. Ignoring multiplicative constants, this yields

$$\pi_n(\mathbf{R}) \quad \propto \quad \exp(-\frac{1}{2}(\mathbf{R} - \boldsymbol{\mu}_n) \cdot \Theta_n^{-1}(\mathbf{R} - \boldsymbol{\mu}_n)) \tag{9.6}$$

$$\times \exp(-(\mathbf{R} \cdot \mathbf{b}_n + \frac{1}{2}(\mathbf{R} - \mathbf{r}) \cdot \mathbf{C}_n(\mathbf{R} - \mathbf{r}))/\tau_n)$$

$$= \quad \exp[-\frac{1}{2}(\mathbf{R} - \boldsymbol{\mu}_n) \cdot \Theta_n^{-1}(\mathbf{R} - \boldsymbol{\mu}_n)) \tag{9.7}$$

$$-(\mathbf{R} \cdot \mathbf{b}_n + \frac{1}{2}(\mathbf{R} - \mathbf{r}) \cdot \mathbf{C}_n(\mathbf{R} - \mathbf{r}))/\tau_n)]$$

which is normal with mean $\widehat{\boldsymbol{\mu}}_n$ and covariance matrix $\widehat{\Theta}_n$ given by:

$$\widehat{\Theta}_n \quad = \quad (\Theta_n^{-1} + \mathbf{C}_n/\tau_n)^{-1} \tag{9.8}$$

$$\widehat{\boldsymbol{\mu}}_n \quad = \quad \mathbf{r} + \widehat{\Theta}_n\Theta_n^{-1}(\boldsymbol{\mu}_n - \mathbf{r}) - \widehat{\Theta}_n\mathbf{b}_n/\tau_n. \tag{9.9}$$

These equations can be rewritten as:

$$\widehat{\Theta}_n^{-1} \quad = \quad \Theta_n^{-1} + \mathbf{C}_n/\tau_n \tag{9.10}$$

$$\widehat{\Theta}_n^{-1}(\widehat{\boldsymbol{\mu}}_n - \mathbf{r}) \quad = \quad \Theta_n^{-1}(\boldsymbol{\mu}_n - \mathbf{r}) - \mathbf{b}_n/\tau_n \tag{9.11}$$

and also

$$\mathbf{C}_n \quad = \quad \tau_n(\widehat{\Theta}_n^{-1} - \Theta_n^{-1}) \tag{9.12}$$

$$\mathbf{b}_n \quad = \quad \tau_n(\widehat{\Theta}_n^{-1}(\widehat{\boldsymbol{\mu}}_n - \mathbf{r}) - \Theta_n^{-1}(\boldsymbol{\mu}_n - \mathbf{r})). \tag{9.13}$$

Thus, *the effect of linear+quadratic investments is to shift both the mean and covariance of the investor's risk-neutral distribution relative to those of her true distribution, while remaining in the normal family.* The *changes* in the dimensionless terms $\Theta_n^{-1}(\boldsymbol{\mu}_n - \mathbf{r})$ and Θ_n^{-1}, multiplied by her personal risk tolerance τ_n (in units of money), are additive constants that are equal to $\mathbf{b}_n$ and $\mathbf{C}_n$, her holdings of primary and quadratic assets (in units of money).

A first-order condition for an optimal solution of the investor's problem is that it must equate the mean vector of her risk-neutral distribution with that of the market, which is the constant vector determined by the risk-free interest rate: $\widehat{\boldsymbol{\mu}}_n = \mathbf{r}$. Plugging this requirement into the previous equation yields

$$\mathbf{b}_n = \tau_n \Theta_n^{-1}(\boldsymbol{\mu}_n - \mathbf{r}), \tag{9.14}$$

which does not involve $\mathbf{C}_n$. So, the availability of the derivative assets does not affect the investor's optimal holdings of primary assets, which was pointed out by Lintner (1969). If the stock returns are perceived to be uncorrelated, then Θ_n is a diagonal matrix and b_{nj} is inversely proportional to the perceived variance of the return on stock j as well as directly proportional to the excess return and directly proportional to the investor's risk tolerance.

The investor's optimal portfolio of derivative assets is determined by enforcing the remaining first-order condition, namely that the risk-neutral distributions of all the investors must have the same covariances as well as the same means, i.e., $\widehat{\Theta}_n = \Theta$ for all n, where Θ denotes the market's risk-neutral covariance that is determined by prices of quadratic options as shown in (9.2). It follows that investor n's portfolio of quadratic options must satisfy:

$$\mathbf{C}_n = \tau_n(\Theta^{-1} - \Theta_n^{-1}) \tag{9.15}$$

This result could hardly be simpler. The investor's optimal matrix of quadratic option holdings is merely the difference between the market's precision (inverse covariance) matrix and her own precision matrix, scaled (like her portfolio of primary assets) in proportion to her personal risk tolerance. This equation implies that investors with more precise beliefs (those n for whom the elements of Θ_n^{-1} are larger than average) will want to hold negative amounts of quadratic options, while those with less precise beliefs will want to hold positive amounts.[4]

9.4   Market aggregation of means and covariances

In equilibrium, the market's risk-neutral covariance matrix Θ is determined from the individual true covariances by enforcing the condition that the derivative assets

---

[4]This same phenomenon is discussed by Brennan and Cao (1996) in the context of their single-risky-asset model.

must exist in zero net supply, i.e., the matrices $\{\mathbf{C}_n\}$ must sum to zero. This yields

$$\Theta^{-1} = \sum_{n=1}^{N} (\tau_n/\tau)\Theta_n^{-1} \qquad (9.16)$$

where τ is the sum of the investors' risk tolerances: the total risk tolerance of the market. Thus, *the precision matrix of the market's risk-neutral distribution is a weighted sum of the precision matrices of the investors' subjective probability distributions, with weights equal to their shares of the total risk tolerance.* In the special case where the normal/exponential investors agree on variances and covariances ($\Theta_n \equiv \Theta$), there is no trade in derivative assets ($\mathbf{C}_n \equiv \mathbf{0}$), so the absence of such trade in a complete market is an indicator of agreement on those parameters. If the investors also agree on expected returns ($\boldsymbol{\mu}_n \equiv \boldsymbol{\mu}$) then each holds a share of a single portfolio in proportion to her risk tolerance ($\mathbf{b}_n = \tau_n\Theta^{-1}(\boldsymbol{\mu} - \mathbf{r})$), and that portfolio must be the market portfolio.[5]

What's observable and what isn't: The primary assets serve to equate the mean vector of the investor's risk-neutral distribution with that of the market, while quadratic options serve to equate the covariance matrices, and investments in both kinds of assets are directly proportional to risk tolerance as shown above. $\mathbf{b}_n$, $\mathbf{C}_n$, r, and Θ are theoretically observable in financial data and portfolio statements, but there is a simultaneous lack of determination of $\boldsymbol{\mu}_n$, Θ_n and τ_n. We cannot separate the effects of the investor's beliefs (subjective means and covariances of returns) from her risk tolerance, so *we can't tell if an investor who holds large amounts of risky assets does so because her means and/or covariances are very different from those of other investors or because she has an unusually high tolerance for risk.* Another potential problem with disentangling the beliefs and risk preferences of an individual investor is that she may have *unobserved prior investments* in events that are correlated with asset returns—e.g., her job security or year-end-bonus may depend on the stock price of the company for which she works—which means that her effective true values of $\mathbf{b}_n$ and $\mathbf{C}_n$ may not be exactly known.

There is no way in general to observe a representative aggregation of the investors' subjective valuations of *mean* returns. It is natural to *define* an aggregate vector of mean returns by:

$$\boldsymbol{\mu} := \sum_{n=1}^{N} (\tau_n/\tau)\Theta\Theta_n^{-1}\boldsymbol{\mu}_n \qquad (9.17)$$

so that the following summation over investors holds

$$\tau\Theta^{-1}\boldsymbol{\mu} = \sum_{n=1}^{N} \tau_n\Theta_n^{-1}\boldsymbol{\mu}_n , \qquad (9.18)$$

---

[5]These aggregation results are discussed in Lintner (1969).

and also

$$\mathbf{b} \quad := \quad \sum_{n=1}^{N} \mathbf{b}_n = \sum_{n=1}^{N} \tau_n \Theta_n^{-1}(\boldsymbol{\mu}_n - \mathbf{r}) \tag{9.19}$$

$$\tau \sum_{n=1}^{N} (\tau_n/\tau)\Theta_n^{-1}(\boldsymbol{\mu}_n - \mathbf{r}) \tag{9.20}$$

$$= \quad \tau\Theta^{-1}(\boldsymbol{\mu} - \mathbf{r}) \tag{9.21}$$

in which the market's portfolio of primary assets has the same form as those of individual investors (9.14), merely with the subscripts n removed from all the terms.

If the investors agree on covariances ($\Theta_n = \Theta$ for all n), then the aggregate vector of expected returns is a simple weighted average of individual expected returns $\{\boldsymbol{\mu}_n\}$, with weights proportional to risk tolerances, which parallels the formula (9.16) for aggregating precision matrices:

$$\boldsymbol{\mu} := \sum_{n=1}^{N} (\tau_n/\tau)\boldsymbol{\mu}_n. \tag{9.22}$$

But if they disagree on covariances, the formula is a little more complicated (9.17): it is a weighted average of the "adjusted" expected return vectors $\{\Theta\Theta_n^{-1}\boldsymbol{\mu}_n\}$. This gives proportionally more weight to investors with lower covariances (higher precision), who will tend to hold larger quantities of risky assets, ceteris paribus, as specified in (9.14).

9.5   The subjective capital asset pricing model (CAPM)

The preceding definition of the aggregate mean return vector (9.17) gives rise to a subjective version of the CAPM (Nau and McCardle (1991)). Let b denote the vector of total dollar amounts invested in the primary assets by all investors (which is strictly positive because they exist in positive net supply), and let $\mathbf{v} := \mathbf{b}/(1 \cdot \mathbf{b})$ denote the vector of the market's portfolio weights obtained by normalization of $\mathbf{b}$, let $\mu := \boldsymbol{\mu} \cdot \mathbf{v}$ denote the expected return on the market portfolio according to the aggregate mean return vector $\boldsymbol{\mu}$, and let $R := \mathbf{R} \cdot \mathbf{v}$ denote the actual return on the market portfolio. Then let $\boldsymbol{\beta}$ denote the vector of "betas" calculated from the risk-neutral covariance matrix Θ using the market portfolio:

$$\boldsymbol{\beta} := \text{cov}(\mathbf{R}, R)/\text{var}(R) = \Theta\mathbf{v}/(\mathbf{v} \cdot \Theta\mathbf{v}) \tag{9.23}$$

Its jth element β_j is the covariance of stock j's return (R_j) with the market portfolio return (R) divided by the variance of the portfolio return. If Θ were an empirical covariance matrix computed from time series data, then β_j would be the slope coefficient in the regression of stock j's return on the portfolio return. Here, however, the parameters of the CAPM are entirely subjective: β is the vector of betas calculated from the market's *risk-neutral* covariance matrix, whose elements are directly revealed by prices of quadratic options as shown in (9.2).

In these terms the familiar CAPM relationship holds by definition (as usual):

Theorem 9.2: Subjective capital asset pricing model

The expected excess return on an asset is equal to its beta times the excess return on the market:

$$\mu_j - r = \beta_j(\mu - r) . \tag{9.24}$$

In vector form:

$$\boldsymbol{\mu} - \mathbf{r} = \Theta \mathbf{b}/\tau = \boldsymbol{\beta}(\mu - r) . \tag{9.25}$$

Proof: By definition $\mathbf{b} = \mathbf{v}b$ where b is the total amount invested in risky assets, and also $\mu = \mathbf{v} \cdot \boldsymbol{\mu}$ and $r = \mathbf{v} \cdot \mathbf{r}$. The identity $\boldsymbol{\mu} - \mathbf{r} = \Theta \mathbf{b}/\tau$ follows from 9.21, and multiplication on both sides by $\mathbf{v}$ yields

$$\mu - r = (\mathbf{v} \cdot \Theta \mathbf{v})(b/\tau) , \tag{9.26}$$

whence

$$b/\tau = (\mu - r)/(\mathbf{v} \cdot \Theta \mathbf{v}) . \tag{9.27}$$

Substituting back into the original equation yields:

$$\boldsymbol{\mu} - \mathbf{r} = \Theta \mathbf{b}/\tau = \Theta \mathbf{v}(b/\tau) = ((\Theta \mathbf{v})/(\mathbf{v} \cdot \Theta \mathbf{v}))(\mu - r) = \boldsymbol{\beta}(\mu - r) . \tag{9.28}$$

Thus, based on the revealed value of Θ (9.2) and the defined value of $\boldsymbol{\mu}$ (9.17), the CAPM holds for a representative investor whose subjective beliefs and risk preferences are aggregations of those of the real investors.

While the market's instantaneous subjective betas can be determined from the risk-neutral covariance matrix together with the vector of total wealth invested in different primary assets, *the excess return on the market portfolio cannot be determined from asset prices*. As in the individual case, there is a simultaneous lack of determination of μ and τ in the equation $\Theta b / \tau = \beta(\mu - r)$: we cannot infer the representative investor's true expected returns by observing asset prices without independently knowing the total risk tolerance of the market.

Despite the inherent subjectivity of investor beliefs and preferences, and aggregations thereof, as illustrated by the CAPM model discussed above, the finance field has mostly adhered to the notion that asset prices are random variables with objective probability measures. The risk-neutral expected value of the future price of an asset, discounted to the present, is expressed as the objective expected value of the future asset price multiplied by a so-called "pricing kernel" or "stochastic discount factor":

> "The asset pricing kernel summarizes investor preferences for payoffs over different states of the world. In the absence of arbitrage, all asset prices can be expressed as the expected value of the product of the pricing kernel and the asset payoff. Thus, the pricing kernel, when it is used with a probability model for the states, gives a complete description of asset prices, expected returns, and risk premia." (Rosenberg and Engel (2002))

Here's how this works: there is an observable risk-neutral distribution determined by contingent-claim prices, and an observable fixed discount factor determined by the risk-free rate of interest, with respect to which the current price of every asset is the discounted risk-neutral expected value of its future price. (This is the no-arbitrage condition.) Now assume a "probability model" with an objective probability distribution for future asset prices. Define the pricing kernel as the discounted risk-neutral distribution function divided by the assumed objective distribution function. By construction, then, asset prices are determined by the objective expected value of the product of the pricing kernel and the future asset price. In this equation, the pricing kernel replaces the fixed discount factor at the same time as the objective probability distribution replaces the risk-neutral one, which is why it is called the stochastic discount factor. However, it is not directly observable. There is joint indeterminacy in the pricing kernel and the assumed objective distribution. The latter can be estimated from historical frequency data or based on theory or assumptions, but it does not necessarily represent investors' true beliefs about what is happening right now in the contingent-claims market. This is another example of the foundational problem of separating probability from utility that arises in subjective expected utility theory, state-preference theory, ambiguity modeling, and noncooperative game theory. There too, as has been shown, risk-neutral probabilities have a role to play as observable parameters of individual and joint beliefs.

Chapter 10

Summary of the fundamental theorems and models

10.1 Perspectives on the foundations of rational choice theory

The fundamental representation theorems of rational choice in decisions under risk and uncertainty, noncooperative games, and asset pricing can all be framed as linear programming duality results (in fact, minor variations on a single result). The primal optimization problem is to find values of psychological variables (probabilities, utilities, state-dependent utilities, risk-neutral probabilities, equilibrium beliefs, etc.) according to which the actions of agents are individually and reciprocally optimal. The dual problem is to find a material arbitrage opportunity for an observer (a riskless gain ex-ante or ex-post from acceptable bets or trades with one or more agents). This duality relationship, which follows from separating hyperplane arguments, is what connects rational thought with rational action in quantitative terms. It allows parsimonious low-dimensional psychological models to be applied to the description, prediction, and prescription of wide ranges of behavior, as if agents are motivated by quantifiable internally consistent beliefs and desires. In this book, "rationality" refers to this property of consistency rather than correctness of beliefs or goodness of what is desired.

DOI: 10.1201/9781003527145-10

The representation of mental states by numbers that are precise and commonly knowable is best justified in environments where money is available as a tool of thought, a unit of account, a medium of exchange, a language for contracts, and a bottom line for evaluating and comparing outcomes. People almost never think in a language of "utility," and probability is not quantified in daily personal decisions either. An exception can be found in the rapidly growing industry of online sports betting and prediction markets, where (significantly) the numbers flashed on TV and computer screens are often expressed in units of 3-digit money lines[1] rather than odds or probabilities.

Money isn't everything, but where it is present, it provides a self-justifying standard of economic rationality: don't throw it away or allow your pocket to be picked. And this is not only a standard of individual rationality but also a standard of group rationality and common knowledge thereof, one that allows noncooperative game theory (in its most basic form) to be framed as a simple extension of subjective probability theory without the a priori imposition of solution concepts. Correlated equilibrium, not Nash equilibrium, emerges as the dual characterization of rational play in strategic-form games. The late development of correlated equilibrium is a curiosity in the history of game theory, as noted by Myerson in his quotation on page 7 (from Solan and Vohra (2002)).

Even if money is involved, the effects of personal probabilities, utilities, and status quo stakes in events are confounded with each other in determining behavior, and hence they may not be uniquely and separately observable. This problem is not intractable, however, and it can be finessed in some situations by converting the units of analysis to *risk-neutral probabilities* (interpretable as products of subjective probabilities and relative marginal utilities) and their derivatives (which are the carriers of information about local aversion to uncertainty and ambiguity). Risk-neutral probabilities have been well-known in the finance field since the 1970s, when the topic of asset pricing by arbitrage became central, but their relevance for modeling personal decisions and games, in principle if not in practice, has not been sufficiently recognized, and proving the case is one of the main purposes of this book.

Preferences are typically incomplete due to limits of cognition and information, time for reflection and communication, and perturbation by the process of elicitation. Measurements of mental parameters such as probabilities and utilities typically provide intervals, not points, state spaces may be coarse, and optimal decisions may not be uniquely determined.

Rational choice models are populated with individuals who arrive on the scene fully endowed with their own idiosyncratic and internally consistent beliefs and

---

[1]Money lines are defined in Chapter 3, footnote 11.

tastes, who process information independently, who interact through impersonal mechanisms, and who are equally good decision-makers in a technical sense. Real decisions have intersubjective elements, and decision-making skills are not equal, a lesson from behavioral decision theory and principles of management and leadership. Directly or indirectly, through formal or informal communication and reliance on social knowledge, decisions are informed and guided by groups. They draw upon norms, laws, information sets, and choice sets that have been determined by others. They may also reflect beliefs and tastes that are manipulated by others. The modeling approach that has been explored in this book, which emphasizes the perspective of an observer, does not draw a bright line between the rationality of an individual and the rationality of a group.

By the time at which a decision problem has been reduced to numerical form, much of the hardest and most important work has already been done, namely identifying and then quantifying the relevant events and outcomes and attributes of preferences and imagining the opportunities for action, often with help from others. "Consequences" (states of a person or group) do not exist in isolation and cannot be assigned arbitrarily to events, contrary to standard practice (Savage (1954), Anscombe and Aumann (1963)). They arise from conjunctions of real events and feasible actions that the decision-maker chooses to consider, which in general is a creative process.

We violate the assumptions of rational choice theory in ways that are obvious by introspection and have been documented by decades of behavioral experiments. Sometimes these violations can be finessed by as-if arguments that agents will approximately satisfy rational choice assumptions within a social context where they can communicate with others and take advantage of norms and customs and decision tools and market institutions—and where their pockets really will be picked if they are not careful. And sometimes these violations can be turned around and used as a basis for arguing that agents ought to deliberately use rational choice models to straighten out their thinking (as is the case in applied decision analysis) or that mechanisms should be designed to channel their thinking along such lines when it is profitable or socially useful (another important success story).

But not only do we fail to live up to the assumptions of rational choice theory: the theory itself is only a small-world projection of a complex and evolving large world in which there are important considerations other than consistency. Agents in rational choice models do not subjectively experience the passage of time and they do not imagine the future or learn or create anything new or believe in things or disagree with others or outwit them or change their minds or lead them in the usual sense of those terms. They differ from each other only in regard to the information and preferences that have been dealt to them from a common deck, they strategize only in predictable ways, and they lead only in the sense of being awarded the first move. Even in dynamic models they are assumed to have complete and correct views of the possible sequences of choices and events through which the present will unfold into the future. Improbable things may happen, but no discoveries are made and no one is ever genuinely surprised. When agents receive information as anticipated, they instantiate a branch in an already-drawn and already-solved decision tree or game tree or price lattice. Whereas, in the large world, unexpected events happen and unexpected opportunities arise, and more so with the passage of more time. The experience of time is measured by an accumulation of genuinely new experiences and new options and new thoughts, not just traveling farther along paths already mapped. Bayes' rule is not a model of "learning" over time, and there may be option value in having flexibility to deviate from contingent strategies drawn up in advance. Prices of futures and options provide market-based estimates of the cumulative effects of unforeseen events that occur in financial markets in real time, but they don't capture the truly unexpected ("black swans").

Notwithstanding those caveats, the remaining sections of this chapter will provide a not-too-technical review of some of the highlights from Chapters 3 through 9.

10.2   Axioms for preferences and acceptable bets

The SP, EU, and state-dependent SEU models are all linear models in which subjective parameters of the decision-maker (probabilities, utilities, or state-dependent utilities) multiply the objective parameters of acts that are being evaluated. They have the same axioms, merely applied to different vector-valued objects of choice: assignments of money to states of the world in the SP model, assignments of objective probabilities to consequences in the EU model, and assignments of objective lotteries over consequences to states of the world in the SEU model. The axioms can be given either in terms of the properties of preferences among acts or the properties of acceptable bets, where a bet has the interpretation of what the decision-maker experiences when one act is exchanged for another. Thus, for example, if f and g are acts and $x := f - g$ is a bet, then x is preferable if f is weakly preferred to g, and conversely $f + x$ is weakly preferred to f if x is preferable, A preferable bet x is (furthermore) *acceptable* if the decision-maker is publicly willing to accept αx for any small positive α chosen by an observer regardless of her status quo position. The acceptance of a bet establishes a material connection between the decision-maker and the observer which makes its parameters common knowledge and which makes arbitrage opportunities visible. It will be tacitly assumed throughout that all preferable bets are acceptable, but it should be kept in mind that these concepts are not quite the same, and the difference is important. In plain language the axioms are as follows

Axioms for preferences:

Axiom 1a (Completeness): For any acts f and g, either f is weakly preferred to g, or g is weakly preferred to f, or both.

Axiom 2a (Continuity): The set of acts that are weakly preferred to a given act is closed.

Axiom 3a (Transitivity): If f is weakly preferred to g, and g is weakly preferred to h, then f is weakly preferred to h.

Axiom 4a (Independence): If one act is weakly preferred to another, the direction of preference is preserved when both are linearly combined in the same way with the same alternative act.

Axiom 5a (Strict monotonicity): If one act strictly dominates another, it is strictly preferred.

Axioms for acceptable bets:

Axiom 1b (Completeness): For any bet x, either x is acceptable or $-x$ is acceptable or both.

Axiom 2b (Continuity): The set of acceptable bets is closed

Axiom 3b (Convexity): A linear interpolation between two acceptable bets is acceptable.

Axiom 4b (Linear extrapolation): A positive multiple of an acceptable bet is acceptable

Axiom 5b (No-arbitrage): A strictly negative bet is not acceptable.

Axioms 1, 2, and 5 in both sets are direct translations of each other, and the combination of the independence and transitivity axioms for preferences corresponds to the combination of the convexity and linear extrapolation axioms for acceptable bets, as discussed in regard to Lemmas 3.1 and 3.2.

The equivalence of these sets of axioms in the SP model was very clearly pointed out by de Finetti (1937), as quoted at the beginning of Chapter 3. (He is quite explicit about the role of the independence axiom, although he is usually not given credit for it.) Despite this formal equivalence, there are important epistemic and practical differences, namely that the acceptability of a bet is common knowledge between an agent and an observer who takes the other side of it because it materially affects both of them in a way that they both understand.

In subjective expected utility theory, an additional pair of axioms is invoked to decompose the representation into unique probabilities and state-independent utilities:

> **Axiom 6a (Separability of beliefs and tastes among preferences):** A constant SEU-act[2] is weakly preferred to another constant SEU-act if and only if the same preference holds when both are conditioned on the same non-null event.

> **Axiom 6b (Separability of beliefs and tastes among acceptable bets):** A constant SEU-bet is weakly acceptable if and only if it is weakly acceptable when conditioned on a non-null event.

10.3 Subjective probability theory

Subjective probability theory is the theory of rational beliefs, and here (following de Finetti) it has been axiomatized in terms of preferences among acts and acceptance of bets with monetary payoffs. An SP-act is a mapping from a finite set of states to real-valued amounts of money, and the corresponding notion of a bet, called a $-bet, is a difference between two such acts (a vector of gains and losses in units of money assigned to states). A $-bet is acceptable if the agent is willing to have it added to her status quo act at the discretion of an observer. Importantly, the status quo act need not be observable.

> **Fundamental theorem of subjective probability (Theorem 3.2).** A collection of preferences among SP-acts satisfies the axioms of completeness, continuity, transitivity, independence, and coherence if and

---

[2]SEU-acts and SEU-bets (referred to as v-acts and v-bets in Chapter 5) are said to be constant if they yield the same objective lotteries over consequences in every state of the world, and to condition them on an event means to equalize or zero-out their payoffs in states in which the event does not occur. A non-null event is one whose effects are non-trivial in at least some such constructions, which is to say, its revealed probability is not zero.

only if the corresponding collection of acceptable $-bets satisfies the axioms of completeness, continuity, convexity, linear extrapolation, and no-arbitrage, and these conditions hold if and only if there exists a unique supporting probability distribution with respect to which every preference is determined by maximizing expected value and every acceptable bet has non-negative expected value.

The proof is simple: the no-arbitrage condition is satisfied if and only if the closed convex set of acceptable bets and the open convex set of strictly negative bets are separated by a hyperplane, and the coordinates of the normal vector of that hyperplane, scaled to sum to 1, are the supporting probability distribution. In the case of a finite state space and a finite number of bets that are directly asserted to be acceptable, the supporting probability distribution can be found by linear programming, as shown in the next chapter. This theorem can be generalized to the case of continuous probability distributions via the application of an infinite-dimensional separating hyperplane theorem, as shown in Section 3.7.

If the *completeness axiom* is dropped, the unique probability distribution is replaced by a convex set thereof (a convex polytope in the case of a finite state space and finite number of bets), and a bet is acceptable if and only if it has non-negative expectation with respect to every probability distribution in the set. In this representation, beliefs are described *by lower and upper probabilities* (Theorem 3.3).

A collection of acceptable bets is *ex-post-coherent* in state m if it does not contain an *ex-post-arbitrage opportunity* in state m, which is an acceptable bet that yields a negative payoff in state m and non-positive payoffs in all other states. The requirement that acceptable bets should be ex-post-coherent is a stronger rationality standard that is appropriate to use in situations where the agent may have some amount of control or otherwise certain knowledge of events, and it leads to the:

Ex-post fundamental theorem of subjective probability (Theorem 3.6). A collection of $-bets that satisfies the assumptions of the subjective probability model is ex-post-coherent in state m if and only if there is a probability distribution on states that assigns non-negative expected value to every acceptable bet *and* assigns strictly positive probability to state m.

This means the state that occurs must not be one that was implicitly or explicitly assigned a probability of zero. The agent should have known better or should have acted otherwise. This ex-post standard of rational betting, which is a simultaneous constraint on beliefs and events, can be used to directly portray noncooperative game theory (in the case of strategic-form games) as a multi-agent extension of subjective probability theory, as shown in Chapter 8.

10.4 Expected utility theory

Expected utility theory is the theory of choice under risk, in which outcomes of an arbitrary nature are received with given objective probabilities. It is merely the dual image of subjective probability theory. They have the same axioms and same separating hyperplane representation theorem except for an interchange of the primal and dual variables. In SP theory the primal variables are subjective evaluations of probabilities and the dual variables are multipliers of bets that are in units of objective amounts of money. In EU theory, in the simplest case, the primal variables are subjective evaluations of amounts of money and the dual variables are multipliers of bets that are in units of objective amounts of probability.

An EU-act is a mapping from a finite set of N "consequences" to real-valued amounts of objective probability. A consequence could be an amount of money, but in general it is a "state of the person" that could have arbitrary attributes. We are asked to imagine that states of the person can be assigned counterfactually to outputs of a random device. The corresponding notion of a bet, called a p-bet, is a difference between two such acts (a vector of gains and losses in units of probabilities assigned to consequences). A p-bet is acceptable if the agent is willing to have it added to her status quo act at the discretion of an observer, thus adding or subtracting amounts of probability from different consequences, in any multiple that is feasible (i.e., does not yield negative probabilities). One of the consequences (c_N) is assumed to be weakly worst a priori, and one of the other consequences (c_1) is assumed to be weakly best and strictly better than c_N for purposes of scaling the utility function so that c_N and c_1 can be assigned utilities 0 and 1 without loss of generality. An arbitrage opportunity is an acceptable p-bet that transfers positive amounts of probability to c_N from each of the others, all of which are weakly preferred and at least one of which is strictly preferred.

> **Fundamental theorem of expected utility (Theorem 4.2).** A collection of preferences among EU-acts satisfies the axioms of completeness, continuity, transitivity, independence, and strict monotonicity if and only if the corresponding collection of acceptable p-bets satisfies the axioms of completeness, continuity, convexity, linear extrapolation, and no-arbitrage, and these conditions hold if and only if there exists a unique[3] utility function with respect to which every preference is determined by maximizing expected utility and every acceptable p-bet has

---

[3]Up to positive affine transformations.

non-negative expected utility. If the completeness axiom is dropped, the unique utility function is replaced by a convex set thereof.

This is the same as the fundamental theorem of subjective probability, except for the interpretations of the variables. The only formal difference is the need to distinguish a weakly worst and a strictly better consequence a priori in the EU model, which corresponds to assuming that more money is preferred to less in the SP model. This point-by-point correspondence in assumptions of the SP and EU models in the case of finite sets of states and consequences is generally not mentioned in standard ax-iomatizations of these models that go back to original sources, where they are not presented side by side and where the role of separating hyperplanes is not stressed or sometimes not even mentioned (e.g., in Kreps' *Notes on the Theory of Choice* (1988)). The role of separating hyperplanes in both theorems *is* noted by Gilboa (2009).

The complete-preference version of the duality theorem can be generalized to acts that are *continuous* probability distributions over an *interval* of payoffs via ap-plication of an infinite-dimensional separating hyperplane theorem. (Theorem 4.4) In this setting, risk aversion can be modeled in the usual way as aversion to mean-preserving spreads (Figures 4.4 and 4.5), and risk premia are quantified in the usual way by the Arrow-Pratt measure in which the variance of the payoff is multiplied by a ratio of second derivatives to first derivatives of the utility function (equation 4.14).

The expected utility model is also applied in social choice theory as a basis for cardinal measurements of utility. In Harsanyi's (1955) "social aggregation theo-rem" (also called the "utilitarian cardinal welfare theorem") there are N individuals for whom "society" chooses among a set of M alternatives and objective lotteries over them. Such lotteries (which are merely thought-experiments in this setting) are represented by M-vectors of probabilities, generically denoted by $\mathbf{f}$ or $\mathbf{g}$ here. All individuals have vNM utility functions for the lotteries, represented by M-vectors $\mathbf{u}_n$, $n = 1, ..., N$, and society also has a vNM utility function $\mathbf{u}_0$. The expected utility for lottery $\mathbf{f}$ can then be expressed as $\mathbf{f} \cdot \mathbf{u}_n$ for $n = 0, ...N$. Let society's and the individuals' indifference and weak/strong preference relations be denoted by $\sim_n$, $\succsim_n$, and $\succ_n$ respectively. Thus, $\mathbf{f} \sim_n \mathbf{g}$ [resp. $\mathbf{f} \succsim_j \mathbf{g}$, $\mathbf{f} \succ_n \mathbf{g}$] means $\mathbf{f} \cdot \mathbf{u}_n = \mathbf{g} \cdot \mathbf{u}_n$ [resp. $\mathbf{f} \cdot \mathbf{u}_n \geq \mathbf{g} \cdot \mathbf{u}_n$, $\mathbf{f} \cdot \mathbf{u}_n > \mathbf{g} \cdot \mathbf{u}_n$]. Let $\mathbf{v}$ be an M-vector that is pro-portional to the difference between a pair of lotteries, i.e., $\mathbf{v} = \alpha(\mathbf{f} - \mathbf{g})$ for some acts $\mathbf{f}$ and $\mathbf{g}$ and constant $\alpha > 0$. This is a "social p-bet," a change in risk profile that results from a proportional swap of one lottery for another. Then $\mathbf{v} \cdot \mathbf{u}_n$ is individ-ual n's gain in expected utility that follows from the acceptance of the social p-bet $\mathbf{v}$, and $\mathbf{v} \cdot \mathbf{u}_0$ is society's own gain in expected utility. A social p-bet is [*strictly, indifferently*] *acceptable* to an individual or to society if its gain in expected utility for them is non-negative [positive, zero]. Consider the following standard Pareto optimality principles:

Pareto indifference: if $\mathbf{f} \sim_n \mathbf{g}$ for all $n > 0$, then $\mathbf{f} \sim_0 \mathbf{g}$.

Semistrong Pareto: if $\mathbf{f} \succsim_n \mathbf{g}$ for all $n > 0$, then $\mathbf{f} \succsim_0 \mathbf{g}$.

Strong Pareto: if $\mathbf{f} \succsim_n \mathbf{g}$ for all $n > 0$ and $\mathbf{f} \succ_j \mathbf{g}$ for some j, then $\mathbf{f} \succ_0 \mathbf{g}$.

In terms of acceptable social p-bets, Pareto indifference means that if $\mathbf{v}$ is indifferently acceptable to every individual, then it is indifferently acceptable to society; semistrong Pareto means that if $\mathbf{v}$ is acceptable to every individual, then it is acceptable to society; and strong Pareto means that if $\mathbf{v}$ is acceptable to every individual and strictly acceptable to at least one, then it is strictly acceptable to society. The utilitarian implications of these principles are given by:

> **Fundamental theorem of utilitarianism/social aggregation (Theorem 4.5)**
>
> (i) Pareto indifference is satisfied if and only if society's utility function is a weighted sum of the individual utility functions, with weights unrestricted in sign.
>
> (ii) Semistrong Pareto is satisfied if and only if society's utility function is a non-negatively weighted sum of the individual utility functions.
>
> (iii) Strong Pareto and semistrong Pareto are satisfied if and only if society's utility function is a positively weighted sum of the individual utility functions.

This result is proved in each case by constructing a primal/dual pair of linear programs in which the primal program seeks an agreeing set of weights and the dual program tests for the existence of an "arbitrage" opportunity in the form of a social p-bet that violates a Pareto optimality condition.

10.5 Subjective expected utility theory

Subjective expected utility is the theory of choice under uncertainty, in which outcomes with subjectively evaluated utilities are received in states with subjectively evaluated probabilities. The version of the theory introduced by Anscombe and Aumann (1963), which has been studied here, has almost exactly the same structure and representation theorem as SP and EU theory. An SEU-act is a synthesis of an SP-act and an EU-act: a state-dependent mapping from a finite set of states of nature to objective probability distributions over a finite set of consequences. This construction raises the counterfactualism of the expected utility model to the Mth power, so to speak. In standard English, a "consequence" is what happens when an agent chooses a course of action for herself and others (nature and/or other agents) take their own course. By definition, consequences cannot be mapped to choices and states in an arbitrary fashion, but we proceed as if this can be imagined. The corresponding notion of a bet in this setting, called a v-bet, is a difference between two SEU-acts (a state-dependent p-bet). A v-bet is acceptable if the agent is willing to have it added to her status quo act at the discretion of an observer, in any multiple that is feasible. As in the EU model, one particular consequence is assumed a priori to be weakly worse than the others and strictly worse than one in particular, and here those assumptions apply in every state of nature with respect to the same two consequences.

Fundamental theorem of subjective expected utility (Theorem 5.2).
A collection of preferences among SEU-acts satisfies the axioms of completeness, continuity, transitivity, independence, and strict monotonicity if and only if the corresponding collection of acceptable v-bets satisfies the axioms of completeness, continuity, convexity, linear extrapolation, and no-arbitrage, and these conditions hold if and only if there exists a unique *state-dependent* utility function with respect to which every preference is determined by maximizing expected utility and every acceptable v-bet has a non-negative expected utility. If, in addition, the axioms of *separability of beliefs and tastes* are satisfied, the state-dependent utility function can be decomposed as the product of a unique subjective probability distribution and a *state-independent* utility function.

The state-dependent version of the SEU model is formally the same as the SP and EU models, apart from the fact that its objects of choice involve both states and consequences, and (again) it has the same set of axioms. In the version that has a unique state-independent utility function in the representation (via the separability axioms), the utility function cannot actually be *verified* to be state-independent. In principle it could still have state-dependent scale factors that would interact with the probabilities.

If the completeness axiom is dropped from the state-dependent version of the model, the unique state-dependent utility function is merely replaced by a convex set thereof, as in the SP and EU models with lower and upper probabilities and utilities. If it is dropped from the state-*independent* version of the model and replaced by the same weaker axioms of monotonicity, the result is that there must exist a subjective probability distribution and a state-independent utility function (a "probability/utility pair") which are consistent with asserted preferences and acceptable bets. (Theorem 5.3) This does *not* say, however, that *lower and upper bounds* on the agent's subjective expected utilities are achieved by probability/utility pairs. A strengthening of the state-dependence. axioms (SEU6^{a*}, SEU6^{b*}) is needed for that.[4] (Theorem 5.4) Insofar as a particular probability distribution may be paired up with (only) a particular utility function in this representation, it does not fully separate beliefs about events from utilities for consequences from the viewpoint of an observer. Other authors (most notably Galaabaatar and Karni (2012, 2013) have pursued a further refinement of the model that achieves such a separation, i.e., which leads to separate sets of probabilities and utilities whose elements can be paired up arbitrarily in establishing bounds on subjective expected utility.

10.6   State-preference theory and risk-neutral probabilities

The state-preference framework (Arrow (1953), Debreu (1959)) provides a pathway for modeling choice under uncertainty that is more operational and less counterfactual than subjective expected utility theory. It is expressly designed to deal with concrete problems such as insurance and financial investments, and its units of analysis are (or can be) risk-neutral probabilities. Rather than generalizing the expected-utility model to deal with subjective probability, it generalizes the subjective probability model to deal with nonlinear and state-dependent utility. Consequences are measured in units of money, or else money is at least one attribute of a consequence. An act f is a vector of amounts of money assigned to M states: $f = (f_1, ..., f_M)$. Agents may or may not display expected-utility preferences for such acts. The amounts of money that are the coordinates of f do not necessarily represent an agent's total wealth in the given states, whatever that might mean, nor is money the only thing that the agent cares about. The value that the agent may attach to other attributes of outcomes is revealed indirectly through its effects on marginal utility for money.

---

[4]Seidenfeld et al. (1995) present an alternative axiomatization of the probably/utility pairs model.

Fundamental theorem(s) of state-preference theory (Theorem 6.1 and corollaries): A preference relation among monetary acts satisfies axioms of completeness, continuity, transitivity, strong monotonicity and strict payoff-convexity if and only if it is represented by a continuous and strictly monotonic and strictly quasiconcave utility function $U(\mathbf{f})$. If, in addition, preferences satisfy a "coordinate independence" condition (Wakker (1989)), the utility function is additive across states: $U(\mathbf{f}) = \sum_{m=1}^{M} v_m(f_m)$ where v is a state-dependent utility function. This can be viewed as a model of subjective expected utility without separation of probabilities and utilities. Uniquely determined probabilities and state-independent utilities can be achieved by the further addition of an axiom of "tradeoff consistency." This yields the familiar representation $U(\mathbf{f}) = \sum_{m=1}^{M} p_m u(f_m)$ where p_m is the subjective probability of state m, and u is a state-independent utility function for money. However, the existence of the latter representation still does not imply that the probabilities measure the decision-maker's beliefs nor that utility for money is really state-independent because of the possibility that utility for money could have unknown state-dependent scale factors.

An agent's local *risk-neutral probability distribution* is the probability distribution with respect to which she is willing to accept any sufficiently small bet whose expected value is positive. It is her apparent subjective probability distribution as revealed by low-stakes betting (de Finetti's method). In general the risk-neutral distribution depends on the agent's status quo portfolio of investments and other risks, which need not be observable. (The agent herself might find it difficult to conduct a detailed self-audit!) In the subjective expected utility environment, risk-neutral probabilities are the product of the agent's subjective probabilities and her local state-dependent relative marginal utilities for money. The risk-neutral probability of state m satisfies $\pi_m(\mathbf{f}) \propto p_m u'_m(f_m)$. In the general state-preference environment, the local risk-neutral distribution at act $\mathbf{f}$ is the normalized gradient of the utility function there: $\boldsymbol{\pi}(\mathbf{f}) = DU(\mathbf{f})/(\mathbf{1} \cdot DU(\mathbf{f}))$. It is uniquely determined by preferences (given the axioms above) despite the fact that U is unique only up to monotonic transformations. Some of the most fundamental modeling issues in rational choice under uncertainty (e.g., state-dependent utility, ambiguity aversion, game-theoretic interactions, asset pricing) can be framed in terms of risk-neutral probabilities (which are observable and which have direct implications for interactions with other agents) rather than pure probabilities and utilities (which in general are not observable because they are determined in part by preferences among acts that are counterfactual and/or they are contaminated by state-dependent utility and unmeasurable background uncertainty).

Suppose that a decision-maker in possession of an act $\mathbf{f}$ contemplates taking on some additional exposure to uncertainty in the form of a small bet whose payoff vector is $\mathbf{x}$. The *marginal price* of the bet $\mathbf{x}$ in the vicinity of act $\mathbf{f}$ is its *risk-neutral expected value* $\boldsymbol{\pi}(\mathbf{f}) \cdot \mathbf{x}$. The bet is *neutral* in the vicinity of act $\mathbf{f}$ if its risk-neutral expected value is zero. The *buying price* for $\mathbf{x}$ in its entirety in the vicinity of act $\mathbf{f}$, denoted by $B(\mathbf{x}; \mathbf{f})$, is the maximum amount the decision-maker would be willing to pay for $\mathbf{x}$, i.e., it satisfies $\mathbf{f} + \mathbf{x} - B(\mathbf{x}; \mathbf{f}) \sim \mathbf{f}$. The *uncertainty premium* for $\mathbf{x}$ at $\mathbf{f}$, denoted by $b(x; \mathbf{f})$, is the difference between the risk-neutral expected value of x and its buying price: $b(\mathbf{x}; \mathbf{f}) := \boldsymbol{\pi}(\mathbf{f}) \cdot \mathbf{x} - B(\mathbf{x}; \mathbf{f})$. If the decision-maker is strictly uncertainty-averse, this quantity is strictly positive for all $\mathbf{x} \neq \mathbf{0}$. The uncertainty premium is a generalization of Pratt's definition of a risk premium that extends it from the expected-utility framework to the non-expected utility framework of state-preference theory.

The *matrix of derivatives of the risk-neutral probabilities* (a novel construct) is the carrier of information about local aversion to uncertainty and ambiguity. It's the matrix $D\boldsymbol{\pi}(\mathbf{f})$ whose jk^{th} element is $[D\boldsymbol{\pi}(\mathbf{f})]_{jk} := \partial\pi_j(\mathbf{f})/\partial f_k$. It measures the relative curvature of the utility function in all directions at $\mathbf{f}$. When the vector $\mathbf{f}$ undergoes a small change $d\mathbf{x}$, the corresponding vector change in the local risk-neutral distribution is $d\boldsymbol{\pi} = D\boldsymbol{\pi}(\mathbf{f})d\mathbf{x}$. $D\boldsymbol{\pi}$ is the basis of a simple and elegant generalization of the Arrow-Pratt measure of risk aversion to a setting of non-expected utility preferences and unobservable background uncertainty:

> **Theorem on the quadratic approximation of the uncertainty premium (Theorem 6.2):** The uncertainty premium associated with a small neutral asset $\mathbf{x}$ at observable wealth distribution $\mathbf{f}$ has the local second-order approximation:
>
> $$b(\mathbf{x}; \mathbf{f}) \approx -\frac{1}{2}\mathbf{x} \cdot D\boldsymbol{\pi}(\mathbf{f})\mathbf{x} \qquad (10.1)$$

If the coordinate independence axiom also holds, then the utility function has an additive representation, $U(\mathbf{f}) = \sum_{m=1}^{M} u_m(f_m)$ as noted earlier. In this case, the uncertainty premium formula can be further simplified. Let $\mathbf{r}(\mathbf{f})$ be the vector whose m^{th} element is $r_m(f_m) := -u_m''(f_m)/u_m'(f_m)$. It is a state-dependent version of the usual Arrow-Pratt risk aversion measure. The quadratic approximation of the uncertainty premium now has the form:

$$b(\mathbf{x}; \mathbf{f}) \approx \frac{1}{2} \sum_{m=1}^{M} \pi_m(\mathbf{f}) r_m(\mathbf{f}) x_m^2 \qquad (10.2)$$

This is the usual premium formula except for the fact that a risk-neutral distribution replaces the unobservable true subjective probability distribution, and the local risk aversion measure is state-dependent. These results underscore the fact that *risk-neutral probabilities, rather than pure subjective probabilities, are the fundamental units of analysis in quantifying aversion to uncertainty in measurable terms, as well as in communication and transactions among agents.* The two kinds of probability happen to coincide only in the special case of state-independent utility and zero background uncertainty.

The matrix of derivatives of the risk-neutral probabilities is closely related to the *Slutsky matrix* of consumer theory as applied to the situation where commodities are Arrow securities. Let $S(\mathbf{f})$ denote the Slutsky matrix evaluated at observable state-dependent wealth $\mathbf{f}$. Then $D\pi(\mathbf{f})$ and $S(\mathbf{f})$ are linked by the following equivalence:

$$S(\mathbf{f})d\pi = d\mathbf{f} \quad \Longleftrightarrow \quad D\pi(\mathbf{f})d\mathbf{f} = d\pi. \tag{10.3}$$

The quantity $d\mathbf{f}$ in the equality on the left is the change in $\mathbf{f}$ that the decision-maker will seek (via purchases and sales) if risk-neutral probabilities (state prices) change by $d\pi$. (This is the defining property of $S(\mathbf{f})$.) The quantity $d\pi$ in the equality on the right is the change in π that the decision-maker will experience if her wealth changes by $d\mathbf{f}$. (This is the defining property of $D\pi(\mathbf{f})$.) Each matrix can be derived from the other as follows:

Theorem on the relation of the Slutsky matrix and the matrix of derivatives of the risk-neutral probabilities (Theorem 6.3): $S(\mathbf{f})$ and $D\pi(\mathbf{f})$ have an almost-inverse relation to each other, but they are not literally inverses because both are singular with rank $M-1$. Rather, *bordered versions of them are inverses of each other*:

$$\begin{bmatrix} D\pi(\mathbf{f}) & \pi(\mathbf{f}) \\ \pi(\mathbf{f})^T & 0 \end{bmatrix}^{-1} = \begin{bmatrix} S(\mathbf{f}) & \mathbf{w}(\mathbf{f}) \\ 1^T & 0 \end{bmatrix} \tag{10.4}$$

where $\mathbf{w}(\mathbf{f})$ is the *wealth expansion path* at $\mathbf{f}$, i.e., the path along which prices (risk-neutral probabilities) do not change. (See equation 6.42.)

This is a variation on the classic bordered-Hessian-matrix formula for obtaining the Slutsky matrix (equations 6.43 and 6.44).

10.7   Ambiguity and source-dependent uncertainty aversion

The phenomenon of aversion to ambiguity illustrated by Ellsberg's (1961) paradoxes may be explained in variety of ways, some of which involve a relaxation of the completeness axiom (thus allowing probability distributions to be set-valued, as in Section 3.5 of this book) and others of which involve relaxation of the independence axiom (thus allowing representations with source-dependent uncertainty aversion or status–quo-dependent subjective probabilities in the SEU and state-preference modeling environments). The independence axiom has the non-obvious property that it requires an uncertainty-averse decision-maker to be equally averse to all sources of uncertainty, thus placing constraints on the shape of her indifference curves: they must, in a sense, curve similarly in all directions. This property is behaviorally unrealistic in consumer theory, and insofar as bets may be considered as special cases of consumer goods, it ought not to be surprising to find it violated in settings of choice under uncertainty.[5] The modeling approach studied here is that of state-preference theory, as developed in Chapter 6, which descends directly from consumer theory. Beliefs are modeled via risk-neutral probabilities (marginal state prices) and concepts of subjective probability and utility are not necessarily separated. The decision-maker's indifference curves in state-payoff space are allowed to bend arbitrarily along directions corresponding to different sources of uncertainty.

A symmetrized version of Ellsberg's 2-urn problem is shown in Table 10.1. There are two urns containing a total of 100 balls each, which may be red or black. Urn 1 contains unknown proportions of red and black while urn 2 contains 50 of each. A ball will be independently drawn from each urn, and bets may be placed on the results. The letter A [B] denotes the event in which the ball drawn from urn 1[2] is red. A bet whose payoff depends only on A [B] will be called an A-bet [B-bet]. $x_A(\Delta)$ denotes an A-bet which yields a payoff of Δ (which could be positive or negative) if A occurs and $-\Delta$ otherwise. The B-bet called $x_B(\Delta)$ yields a payoff of Δ if B occurs and $-\Delta$ otherwise.

---

[5]The traditional argument in favor of the independence axiom in the context of choice under uncertainty is that from an ex post perspective, consequences experienced in different states of the world cannot serve as complements or substitutes. Only one state will occur in the end, and the others are just might-have-beens. Solution of a decision tree by backward induction relies on this argument: when you are evaluating one subtree, you ought not to look at others that are disjoint from it. But from an ex ante perspective, when the state of the world is not yet known, that is an arbitrary imposition on the decision-maker's preferences and her way of framing the problem for herself.

TABLE 10.1: A-bets and B-bets with symmetric payoffs

	Bet $\mathbf{x}_A(\Delta)$			Bet $\mathbf{x}_B(\Delta)$	
	B	$\overline{B}$		B	$\overline{B}$
A	$+\Delta$	$+\Delta$	A	$+\Delta$	$-\Delta$
$\overline{A}$	$-\Delta$	$-\Delta$	$\overline{A}$	$+\Delta$	$-\Delta$

If the independence axiom holds, then changing the sign of Δ should change the direction of preference between these two bets. To see this, start by assuming $\mathbf{x}_A(\Delta) \prec \mathbf{x}_B(\Delta)$. The independence axiom implies that the direction of preference is unchanged if the common payoff of $+\Delta$ in state AB is replaced by $-\Delta$ and the common payoff of $-\Delta$ in state $\overline{AB}$ is replaced by $+\Delta$. This combination of substitutions changes $\mathbf{x}_A(\Delta)$ to $\mathbf{x}_B(-\Delta)$ and it changes $\mathbf{x}_B(\Delta)$ to $\mathbf{x}_A(-\Delta)$. Therefore, according to the independence axiom, $\mathbf{x}_A(\Delta) \prec \mathbf{x}_B(\Delta)$ ought to imply $\mathbf{x}_A(-\Delta) \succ \mathbf{x}_B(-\Delta)$, i.e., changing the signs of the payoffs ought to flip the direction of preference. Yet the typical pattern of behavior is $\mathbf{x}_A(\Delta) \prec \mathbf{x}_B(\Delta)$ for both positive and negative Δ. It is suggestive to say that the agent is averse to the ambiguity concerning the composition of urn 1, and by extension, averse to ambiguity in general. This is an artificial laboratory example, however. By assumption the agent has no prior monetary stakes in these events nor any other interest in them. In a more general economic setting, where the events are naturally occurring and are of personal importance, that might not be true. The agent could have unknown prior stakes in the events (background uncertainty) and/or state-dependent marginal utility for money, whether or not she has expected-utility preferences. This possibility can be addressed by adjusting the payoffs in the 4 states by dividing them by the agent's corresponding risk-neutral probabilities for them, denoted π_{AB}, $\pi_{A\overline{B}}$, $\pi_{\overline{A}B}$, and $\pi_{\overline{A}\,\overline{B}}$.[6] Bets of this nature will be called $A{:}B$-bets and $B{:}A$-bets with payoff functions $\mathbf{x}_{A:B}(\Delta)$ and $\mathbf{x}_{B:A}(\Delta)$ as shown here:

TABLE 10.2: $A{:}B$-bets and $B{:}A$-bets with payoffs adjusted by risk-neutral probabilities

	Bet $\mathbf{x}_{A:B}(\Delta)$			Bet $\mathbf{x}_{B:A}(\Delta)$	
	B	$\overline{B}$		B	$\overline{B}$
A	$+\Delta/\pi_{AB}$	$+\Delta/\pi_{A\overline{B}}$	A	$+\Delta/\pi_{AB}$	$-\Delta/\pi_{A\overline{B}}$
$\overline{A}$	$-\Delta/\pi_{\overline{A}B}$	$-\Delta/\pi_{\overline{A}\,\overline{B}}$	$\overline{A}$	$+\Delta/\pi_{\overline{A}B}$	$-\Delta/\pi_{\overline{A}\,\overline{B}}$

---

[6]The risk neutral probabilities can be elicited in terms of preferences among bets with payoffs much smaller in magnitude than Δ. They represent first-order effects, whereas preferences among bets with payoffs on the order of Δ reveal second-order effects.

By construction these are neutral bets. Each cell contributes $\pm\Delta$ to the risk-neutral expected value, with two positive signs and two negative signs. They also have the following property. If the agent's preferences satisfy the independence axiom, her utility function is additive across states, and Theorem 6.2 then yields the result that the quadratic approximation of the agent's uncertainty premium must be the same for both bets regardless of the sign of Δ. If the agent has a strict preference for $\mathbf{x}_{B:A}(\Delta)$ over $\mathbf{x}_{A:B}(\Delta)$ regardless of the sign of Δ, she violates the independence axiom in a way that is suggestive of ambiguity aversion, regardless of background uncertainty and regardless of state-dependence of utility for money.

As shown in the chapter, it is easy to construct utility functions that exhibit ambiguity aversion via source-dependence of aversion to uncertainty and which allow the effect of ambiguity to be quantified without separating probabilities from utilities. In the simplest form of such a model there are two distinct sources of uncertainty represented by two logically independent partitions of events, $\mathcal{A} := \{A_1, ..., A_J\}$ and $\mathcal{B} := \{B_1, ..., B_K\}$, so that the state space is $\mathcal{A} \times \mathcal{B}$. Events that are measurable with respect to $\mathcal{A}$ and $\mathcal{B}$ are called $\mathcal{A}$-events and $\mathcal{B}$-events, respectively. Suppose that the decision-maker's utility function has the compound additive form:

$$U(\mathbf{f}) = \sum_{j=1}^{J} u_j \left(\sum_{k=1}^{K} v_{jk}(f_{jk}) \right), \tag{10.5}$$

where j and k refer to A_j and B_k, and the evaluation functions $\{u_j\}$ and $\{v_{jk}\}$ are strictly concave. The inner summation is a state-dependent additive utility function for monetary payoffs ("first order utility") and the outer summation is an $\mathcal{A}$-event-dependent additive utility function ("second-order utility") for first-order utilities. Concavity of the second-order functions $\{u_j\}$ amplifies the decision-maker's aversion to uncertainties that depend on the $\mathcal{A}$-events. The local risk-neutral probabilities at $\mathbf{f}$ are determined by $\pi_{jk}(\mathbf{f}) \propto u_j'(v_j)v_{jk}'$, where $v_j := \sum_{k=1}^{K} v_{jk}$. Let $\mathbf{s}$ and $\mathbf{t}$ denote JK-vectors uniquely defined by:

$$s_{jk} \quad : \quad = -\frac{u_j''}{u_j'}v_{jk}' \tag{10.6}$$

$$t_{jk} \quad : \quad = -\frac{v_{jk}''}{v_{jk}'},$$

which are positive numbers that measure local *second-order and first-order uncertainty aversion* in state jk. Note that they are familiar-looking ratios of second derivatives to first derivatives as in the measurement of pure risk aversion, and they are measured in real units of 1/money. In particular t_{jk} is a first-order state-dependent Arrow-Pratt risk aversion measure for money received in state jk. The multiplier v'_{jk} in the formula for s_{jk} converts it from units of 1/utility to 1/money.

Let $\mathbf{x}$ be the payoff vector of a small neutral asset, and let $\overline{\mathbf{x}}$ be the J-vector whose j^{th} element is the conditional risk-neutral expectation of x given event A_j. Let s_j denote the marginal sum of s_{jk}:

$$s_j := \sum_{k=1}^{K} s_{jk}, \tag{10.7}$$

which is a second-order state-dependent Arrow-Pratt risk aversion measure for the first-order expected utility received in event j. In these terms we have the following:

Theorem on the decomposition of the uncertainty premium in the two-source model (Theorem 7.1):

uncertainty premium for $\mathbf{x}$ $\approx$ $\mathcal{A}$-*premium* + $\mathcal{A}\mathcal{B}$-*premium*

where

$$\mathcal{A}\text{-}premium = \frac{1}{2} \sum_{j=1}^{J} \pi_j s_j \overline{x}_j^2 \tag{10.8}$$

$$\mathcal{A}\mathcal{B}\text{-}premium = \frac{1}{2} \sum_{j=1}^{J} \sum_{k=1}^{K} \pi_{jk} t_{jk} x_{jk}^2, \tag{10.9}$$

For every $\mathcal{B}$-event, the $\mathcal{A}$-premium (which may be interpreted as an ambiguity premium) is *zero* to a second-order approximation. If $\{u_j\}$ are all strictly concave, then the $\mathcal{A}$-premium of an $\mathcal{A}$-bet is *positive* to a second-order approximation, while if $\{u_j\}$ are all linear, then the $\mathcal{A}$-premia are zero for all bets. Meanwhile, the $\mathcal{A}\mathcal{B}$-premium is determined from the decision-maker's joint risk-neutral probability distribution exactly as in the case of state-dependent expected utility preferences. Hence, the compound cardinal utility function (10.5) yields measurable ambiguity-averse behavior even in a generalized 2-urn problem where probabilities and utilities need not be separated and unobservable background uncertainty may be present.

Another approach to modeling Ellsberg's paradox that has been widely studied is one in which "ambiguity" is considered as uncertainty about the agent's subjective probability distribution. In the simplest case of such a model the agent has 2 or more "credal states," indexed by i, each of which is characterized by its own subjective probability distribution p_i, and they have a common von Neuman-Morgenstern utility function v representing a shared aversion to uncertainty. Uncertainty about the credal state is modeled with a 2nd-order probability probability distribution μ and aversion to the uncertainty about the credal state is modeled by a 2nd-order utility function u. The agent's overall utility function $U(\mathbf{f})$ is the second-order-expected-utility-of-the-first-order-expected-utility:

$$U(\mathbf{f}) = \sum_{i=1}^{I} \mu_i u \left(\sum_{j=1}^{J} \sum_{k=1}^{K} p_{jk|i} v(x_{jk}) \right), \qquad (10.10)$$

where μ_i is the probability of credal state i and $p_{jk|i}$ is the probability of first-order event jk in credal state i. This is the discrete form of the "KMM model" (Klibanoff et al. (2005, 2011ab, 2014, 2022)), which they axiomatized in a Savage-type framework with the set of consequences being an interval of real numbers (like amounts of money). The KMM model has become perhaps the most widely used model of ambiguity aversion in microeconomics. It is discussed in Section 7.5 (and in Nau (2001, 2006a)), and it yields a decomposition of an uncertainty premium into the sum of a risk premium and an ambiguity premium, similar in principle to that of the source-dependent model.

An appealing feature of the KMM model is that its parameterization separates risk from ambiguity and also separates attitude-toward-risk from attitude-toward-ambiguity, which tells a nice story and provides flexibility in theoretical modeling. However, its axiomatization involves yet another ratcheting-up of the counterfactuality of acts that the agent is implicitly able to compare. It is necessary to imagine the contemplation of acts whose counterfactual payoffs depend on credal states that by definition are impossible to observe. Also there is a problem in determining how to parameterize the second-order probability distribution even in the simplest cases. How should "unknown" proportions of red and black balls be modeled in the 2-urn problem? Should a uniform distribution be used? A binary distribution? A tri-angular or beta distribution? How would an agent approach that problem from a decision analysis perspective? There is an interaction between the variance of the assumed distribution and the risk tolerance of the assumed 2nd-order utility function, which makes it hard to separate their implications for concrete behavior even if they are conceptually distinct. By comparison, the compound utility function discussed above (10.5) requires only the introduction of a second-order utility function, which might have only a single parameter to assess, in addition to an assignment of ordinary subjective probabilities to observable events. In its application to the 2-urn problem, it allows the agent to sum up her preferences in this way: "By symmetry I believe that red and black are equally likely to be drawn from either urn. However, I would be willing to pay a small premium z to draw from urn 2 regardless of the winning color." All that's needed is to take that statement as a primitive description of the agent's thought process without asking "why."

10.8   Noncooperative game theory

A noncooperative game is a joint decision problem in which there are two or more players who have some degree of control over the outcomes of events via their own strategies. They might have mutual interests or opposing interests in what happens. There could be incentives for communication and cooperation or incentives for privacy and deception. It is assumed that the players cannot make binding promises to each other (multilateral contracts), and in the moment of play they are free to act independently so as to maximize their expected payoffs given their beliefs about what others will do. Those beliefs are to be endogenously determined (to the extent possible) from analysis of whatever is commonly known about their payoff functions. In the conventional telling of this story, the viewpoint is that of individual players who try to reason about each others' likely moves in the construction of their own beliefs, given what everyone already knows about the payoff functions, which normally are just "given." A rational solution of the game is by definition an equilibrium (usually a Nash equilibrium) in those beliefs. In the version of the story that has been considered here, the viewpoint is that of an observer who knows exactly as much as the players commonly know, as revealed by bets they are willing to take in public (unilateral contracts), which are the medium through which some amount of information about the rules of the game becomes common knowledge. From the observer's perspective, the players are acting as one betting opponent and are subject to same standard of rationality that would apply to such an opponent, namely, don't throw money away without having had some possibility of gain.

The simplest case of a noncooperative game (the best case for quantitative analysis and illustration of general principles) is a strategic-form game in which the sets of strategies are discrete, the payoffs are monetary, and the players are risk-neutral. Rational play in such a game can be modeled via a simple extension of the subjective probability model. The payoff function of each player can be partially revealed through unilateral offers to accept conditional bets that are exactly aligned with her personal incentives. These bets are of the following form: "in the event that you observe me to play strategy s in preference to some other available strategy s' of my own, you may infer that I am willing to accept a small bet whose payoffs are proportional to those of s minus those of s', at your discretion." This does not mean that the player reveals the separate payoff vectors of the two strategies. Rather, she only reveals a bet whose conditional payoffs are proportional to *differences* between those payoff vectors. If the player's choice of a strategy is based on an assignment of personal probabilities to strategies of her opponents, the accepted bet has non-negative expected value precisely when the chosen strategy has greater or equal expected value than the alternative strategy.

Given the acceptable bets through which information about payoffs is publicly revealed, an outcome of the game is defined to be *jointly coherent* if it does not lead to ex-post-arbitrage, i.e., the players jointly offering to accept a bet that yields a negative total payoff to them in that outcome and non-positive total payoffs in all other outcomes. This is the same as the standard of ex-post-coherence for an individual's subjective probabilities that was introduced in Section 3.7. A violation of joint coherence does not necessarily mean that any one player has accepted a loss without having had a possibility of gain. Rather, their combined actions have that property. But insofar as the acceptable bets are public, the players ought to realize that they look irrational as a group when this happens. Furthermore in this case, any one of the players can exploit the same arbitrage opportunity that is being offered to an observer. The construction of the arbitrage portfolio may include a bet with herself, but that is one that she would have accepted anyway.

The revelation of acceptable bets of this nature determines the set of *correlated equilibria* (as will be shown in the theorems below) and also the set of *Nash equilibria*, but it does not provide information about external effects, i.e., how one player would benefit from changes in another's strategy, ceteris paribus. The latter information is invisible to both solution concepts—it gets subtracted out when constructing the bets—which is why noncooperative equilibria are not necessarily Pareto efficient. The set of correlated equilibria is a convex polytope, directly comparable to convex polytopes of probability distributions that arise in the theory of subjective probability minus the completeness axiom when the number of states is finite. From an observer's perspective, a set of game players who reveal information about their payoff functions via acceptable bets is equivalent to an individual whose possibly-incomplete beliefs regarding the outcome of the game are described by the set of correlated equilibria.

Fundamental theorem of noncooperative games:

Risk-neutral case (Theorem 8.2): When the rules of a noncooperative game among risk-neutral[7] players are (partially) revealed through the acceptance of monetary side bets, the outcome of the game is jointly coherent (avoids ex-post-arbitrage) if and only if there is a correlated equilibrium in which it has positive probability. This is a special case of Theorem 3.7 for ex-post-coherence of subjective probabilities. Thus,

---

[7]Risk-neutral means constant and state-independent marginal utility for money, whether or not money is the only attribute of a game's outcome.

the rationality principle of no-arbitrage leads to correlated equilibrium, not Nash equilibrium.

Risk-averse case (Theorem 8.4): When the rules of a noncooperative game among risk-averse[8] players are (partially) revealed through the acceptance of monetary side bets, the outcome of the game is jointly coherent (avoids ex-post-arbitrage) if and only if there is a risk-neutral equilibrium (a correlated equilibrium whose parameters are risk-neutral probabilities) in which it has positive probability.

Other properties of solutions of games:

Superficiality of Nash equilibria (Theorem 8.1): The Nash equilibria of a game lie on the surface of the correlated equilibrium polytope, i.e., on its boundary or relative boundary, though not necessarily at its vertices or along its edges. This fact is not obvious (hence not noticed for decades) but is trivial to prove.

Existence of correlated equilibria (Theorem 8.3) : The existence proof invokes the existence of a stationary distribution of a Markov chain, a weaker condition than existence of a fixed point. The proof is constructive and easy to describe. Suppose that it is possible for an observer to choose a combination of bets yielding a sure loss to the players as a group. For every player n and every pair of strategies s_{nj} and s_{nk} of that player, the observer's assignment of a positive weight to a bet whose payoffs to the player are those of s_{nj} minus those of s_{nk}, conditional on her choosing s_{nj}, is a signal to the player that, in an independent mixed strategy, she ought to shift some probability from s_{nj} to s_{nk} (because her playing s_{nj} is evidently better for the observer). Iteration of this adjustment process defines a Markov chain whose stationary distribution is an independent mixed strategy in which all bets accepted by the player have an expected payoff of zero, and similarly for all the other players, which contradicts the assumption that a fixed set of bets chosen by the observer could yield a sure loss for them as a group.

The vertices of the polytope (extremal correlated equilibria) have rational coordinates, while Nash equilibria could have irrational coordinates. For example, a game

---

[8]Risk averse means that marginal utility for money is strictly less in one outcome of the game than another if that outcome is strictly preferred.

could have a unique Nash equilibrium with irrational coordinates lying in the middle of an edge of the polytope, as in the 3-player example of Table 8.22. The set of Nash equilibria could also include a curve lying in a face of the polytope. It is possible for the number of vertices of the polytope to be enormous even in a modest sized game: a 2-player game with 4 strategies per player could have a correlated equilibrium polytope with over 100,000 vertices. However, in a simulation involving 250 games of this size with randomly chosen, positively correlated payoffs, it was found that around half of them had correlated equilibrium polytopes with 5 or fewer vertices. (Nau et al. (2004))

Rationalizability and correlated rationalizability, two other generalizations of Nash equilibrium that are widely used as standards of common knowledge of rationality, sometimes allow game outcomes that are jointly incoherent, i.e., they do not necessarily protect the players from ex-post-arbitrage. (Table 8.28) Also, in games with 3 or more players, rationalizability sometimes excludes dominant solutions that take advantage of opportunities for formal or informal coordination among subsets of players. (Table 8.21).

Games of incomplete information:

This entire setup generalizes straightforwardly to games of incomplete information, as formalized by Harsanyi (1967), in which player n has a finite set of "types," T_n, which are characterized by different payoff functions and/or states of information, as well as a finite set S_n of strategies. The set of joint strategies is $S := S_1 \times ... \times S_N$ and the set of joint types is $T := T_1 \times ... \times T_N$ and the set of outcomes of the game is $S \times T$. In the usual story, there is a *common prior distribution* over the set of types, which induces mutually consistent reciprocal beliefs about payoff functions and states of information. A *Bayesian equilibrium* of the game is a Nash equilibrium of the game played among the types, i.e., a pure or objectively randomized strategy for each type that is expected-payoff maximizing for that type on the assumption that all other types of all players adhere to their own recommendations. The common prior is simply "given," avoiding the question of how players might come to have such beliefs, even in principle. As we have seen, there are deep philosophical and practical problems with the idea that different players can know each other's personal probabilities, let alone agree on them, except when those probabilities are really objective and impersonal or when the players have state-independent marginal utility for money (something that is not verifiable) and they agree on betting odds. In the generalization of this theory presented in Section 8.11, the parameters of the game—payoff functions of types *and* reciprocal beliefs of types—are revealed through the acceptance of monetary bets as in subjective probability theory and in the theory of noncooperative games of complete information developed earlier in the chapter.

From the viewpoint of an observer, the state space for an incomplete-information game is the set $S \times T$, and the parameters of the game are revealed through two kinds of gambles that the players may offer to accept: "preference gambles" and "belief gambles." Preference gambles have the same structure as those in the model of games of complete information, except with the additional feature of conditioning on types. Each such gamble is of the form: "if you observe me to play strategy s_{nj} of my own rather than strategy s_{nk} when my type is t_n, then I am willing to accept a conditional gamble whose payoffs are proportional to those of s_{nj} minus those of s_{nk} given the occurrence of t_n. (The absolute payoff vectors of the strategies are not actually revealed, only the vector of conditional differences in payoffs.) Belief gambles are conditional gambles on types of other players given a player's own type. An outcome of the game is defined to be jointly coherent if there is no set of gambles available to the observer that yields a strictly negative aggregate payoff to the players when that outcome occurs and a non-positive aggregate payoff in all other outcomes, i.e., if there is no ex-post-arbitrage opportunity. A probability distribution π on $S \times T$ is a *correlated Bayesian equilibrium* if it has the property that if it were used by a mediator to select types and private recommended strategies for all the players, then (a) it would be optimal for each player to comply with her own strategy recommendation for each of her types if all other players do likewise, and (b) it assigns non-negative expected value to the acceptable belief gambles of all players (i.e., serves as a common prior on types). In these terms we have:

Fundamental theorem of games of incomplete information (Theorem 8.6): In a game of incomplete information among risk-neutral players, when payoff functions and beliefs in regard to types are (partially) revealed through the acceptance of monetary side bets, the outcome of the game is jointly coherent (avoids ex-post-arbitrage) if and only if there is a correlated Bayesian equilibrium in which it has positive probability.

This model can be generalized to the case of nonlinear and/or state-dependent utility for money along the same lines as that of games of complete information. The parameters of the randomized strategies and the common prior distribution on types merely become risk-neutral probabilities, which provides an answer to the question of why and how a common prior might come to exist.

10.9　Asset pricing theory

The theory of asset pricing by arbitrage, which revolutionized finance in the 1970s, combines the elements of models discussed in Chapters 3 and 6: subjective probability and state-preference theory. Objects of choice are mappings from states to amounts of money and all agents are risk-averse. Asset prices are subjectively determined by a group of market participants rather than a single individual, multiple time periods are involved, and payoffs in future time periods are discounted at an interest rate determined by the price of a riskless bond. In addition to primary assets (stocks and bonds) there are derivative assets (various kinds of options) whose payoff functions may be arbitrary, and one of the objectives is to price each of those assets correctly in regard to others.

The term "risk-neutral probability" originated in the finance literature in this context. It is sometimes mischaracterized as the probability distribution of a "risk-neutral representative investor." The real representative investor is risk-averse, just like the individuals. (In the normal/exponential special case discussed below, the parameters of the representative investor are literally totals or averages of those of the individuals.) The risk-neutral distribution of the market is often expressed as the product of an imaginary "true" probability measure on asset price movements and a "pricing kernel" or "stochastic discount factor" that adjusts for the representative investor's state-dependent marginal utility for money, as in equations 6.12 and 6.13, but from a subjectivist perspective there is actually no such a thing as a true probability for an economic event, nor even a subjective probability that everyone would agree upon.

> **Fundamental theorem of asset pricing for a finite set of states and two time periods (Theorem 9.1):** There is no arbitrage opportunity in a complete financial market if and only if there is a risk-neutral probability distribution with respect to which the present value of every asset's expected future payoff is equal to its current price, and correspondingly the expected return on any asset is equal to the risk-free rate of return.

There is no explicit mention of utility or risk aversion here, so the proof is just a generalization of that of the fundamental theorem of subjective probability in which payoffs may be received in multiple time periods and transferred among time periods with discounting.

Two important special cases were considered: (i) a market with a discrete state space and a finite number of primary assets, and (ii) a market in which there are a finite number of primary assets with real-valued payoffs and investors are assumed to have heterogeneous normal subjective probability distributions and heterogeneous exponential utility functions. A special kind of derivative asset was introduced: a "quadratic option" whose payoff is the square of the excess return on one stock or the product of the excess returns on two stocks. (The excess return is the stock return minus the riskless rate of return.) The prices of quadratic options directly reveal the risk-neutral variances and covariances of the stock returns. The case of a finite state space and one stock is illustrated in Tables 9.3 and 9.4.

The case of a multivariate normal risk-neutral distribution of stock returns arising from the aggregate preferences of normal/exponential investors is analyzed in depth in Section 9.3. A stylized fact that emerges here is that each agent's contribution to the mean vector and covariance matrix of the market's risk-neutral distribution is proportional to her risk tolerance, the τ parameter in the exponential utility function $u(w) = -\exp(w/\tau)$. (τ is the reciprocal of the Arrow-Pratt measure of risk aversion and it is measured in units of money.) Assume that agent n's personal probability distribution for the vector of stock returns, $\mathbf{R}$, is a normal distribution with mean vector $\boldsymbol{\mu}_n$ and covariance matrix Θ_n. Its density function is:

$$p_n(\mathbf{R}) \propto \exp(-\frac{1}{2}(\mathbf{R} - \boldsymbol{\mu}_n)\Theta_n^{-1}(\mathbf{R} - \boldsymbol{\mu}_n)), \qquad (10.11)$$

and the value of her portfolio of wealth is a quadratic function of returns:

$$w_n(\mathbf{R}) = a_n r + \mathbf{b}_n \cdot \mathbf{R} + \frac{1}{2}(\mathbf{R} - \mathbf{r}) \cdot \mathbf{C}_n(\mathbf{R} - \mathbf{r}) \qquad (10.12)$$

where $\mathbf{b}_n$ is the vector of her holdings of stocks and $\mathbf{C}_n$ is the (symmetric) matrix of her holdings of quadratic options, and a_n is an irrelevant zero point.

Assume that the agent's utility function for wealth is exponential:

$$u_n(w_n(\mathbf{R})) = -\exp(-w_n(\mathbf{R})/\tau_n).$$

Her local risk-neutral density function is the (normalized) product of her probability density function and her *marginal* utility function:

$$\pi_n(\mathbf{R}) \propto p_n(\mathbf{R})u_n'(w_n(\mathbf{R})), \qquad (10.13)$$

which is a normal distribution with mean vector $\widehat{\boldsymbol{\mu}}_n$ and covariance matrix $\widehat{\Theta}_n$, where these are shifted relative those of her true subjective distribution according to:

$$\widehat{\Theta}_n = (\Theta_n^{-1} + \mathbf{C}_n/\tau_n)^{-1} \tag{10.14}$$

$$\widehat{\mu}_n = \mathbf{r} + \widehat{\Theta}_n\Theta_n^{-1}(\mu_n - \mathbf{r}) - \widehat{\Theta}_n\mathbf{b}_n/\tau_n. \tag{10.15}$$

In equilibrium (no-arbitrage), these must be the same for all investors and for the market as a whole, which yields the following conditions:

$$\mathbf{b}_n = \tau_n\Theta_n^{-1}(\mu_n - \mathbf{r}), \tag{10.16}$$

$$\mathbf{C}_n = \tau_n(\Theta^{-1} - \Theta_n^{-1}) \tag{10.17}$$

$$\Theta^{-1} = \sum_{n=1}^{N}(\tau_n/\tau)\Theta_n^{-1} \tag{10.18}$$

Note that the investor's vector of holdings of stocks $\mathbf{b}_n$ is proportional to the difference between her own expected return vector and the risk-free rate vector, and $\mathbf{C}_n$ is proportional to the difference between her own inverse covariance matrix and that of the market, and both are scaled in proportion to risk tolerance. More risk tolerance implies larger investments, ceteris paribus. $\mathbf{b}_n$, $\mathbf{C}_n$, r, and Θ are theoretically observable in financial data and portfolio statements, but there is a simultaneous lack of determination of μ_n, Θ_n and τ_n. We cannot separate the effects of the investor's beliefs (subjective means and covariances of returns) from her risk tolerance. An investor who holds comparatively large amounts of risky assets could do because she has a high tolerance for risk and/or because she believes their expected returns to be higher than are assessed by other investors and/or because she believes that variances of returns are smaller. *This is exactly the same inseparability issue that arises in state-dependent utility theory and state-preference theory.* We also cannot directly observe the total risk tolerance of the market, τ, which is of the same order of magnitude as total money invested in the primary securities.

It is natural to define an aggregate vector of true mean returns by:

$$\mu := \sum_{n=1}^{N}(\tau_n/\tau)\Theta\Theta_n^{-1}\mu_n \tag{10.19}$$

The contribution of each agent is weighted in proportion to her risk tolerance and also in proportion to her inverse covariance matrix: more weight is (again) given to those with higher risk tolerance or lower estimates of market volatility, who will tend to have larger portfolios of risky assets as shown above. As a special case, when investors agree on covariances ($\Theta_n \equiv \Theta$, which is not unreasonable given that everyone knows the latter), this reduces to a simple weighted average with weights proportional to risk tolerances:

$$\boldsymbol{\mu} := \sum_{n=1}^{N} (\tau_n/\tau) \boldsymbol{\mu}_n. \tag{10.20}$$

However, it too is unobservable.

Let $\mathbf{b}$ denote the vector of total dollar amounts invested in the primary securities by all investors (which is strictly positive because they exist in positive net supply), and let $\mathbf{v} := \mathbf{b}/(1 \cdot \mathbf{b})$ denote the vector of the market's portfolio weights obtained by normalization of $\mathbf{b}$, let $\mu := \boldsymbol{\mu} \cdot \mathbf{v}$ denote the expected return on the market portfolio according to the aggregate mean return vector $\boldsymbol{\mu}$, and let $R := \mathbf{R} \cdot \mathbf{v}$ denote the actual return on the market portfolio. Then let $\boldsymbol{\beta}$ denote the vector of "betas" calculated from the risk-neutral covariance matrix Θ using the market portfolio:

$$\boldsymbol{\beta} := \mathrm{cov}(\mathbf{R}, R)/\mathrm{var}(R) = \Theta \mathbf{v}/(\mathbf{v} \cdot \Theta \mathbf{v}) \tag{10.21}$$

Its jth element β_j is the covariance of stock j's return (R_j) with the market portfolio return (R) divided by the variance of the portfolio return. If Θ were an empirical covariance matrix computed from time series data, then β_j would be the slope coefficient in the regression of stock j's return on the portfolio return. Here, however, $\boldsymbol{\beta}$ is the vector of betas calculated from the market's *risk-neutral* covariance matrix, whose elements are directly revealed by prices of quadratic options as shown in (9.2). In these terms we obtain the familiar CAPM formula with subjectively determined parameters:

> **Subjective capital asset pricing model (Theorem 9.2):** The expected excess return on an asset is equal to its beta times the excess return on the market:
>
> $$\mu_j - r = \beta_j(\mu - r). \tag{10.22}$$

In vector form:

$$\boldsymbol{\mu} - \mathbf{r} = \Theta \mathbf{b}/\tau = \boldsymbol{\beta}(\mu - r). \tag{10.23}$$

Because $\boldsymbol{\beta}$ and r and Θ and $\mathbf{b}$ are observable, this equation allows the vector of true expected returns on individual stocks, $\boldsymbol{\mu}$, to be computed from an assumed value of the true expected return on the market portfolio, μ, or an assumed value of the total risk tolerance of the market, τ. This does not solve the problem of unobservability of true expected returns, but it reduces it to a single parameter.

Chapter 11

Linear programming models for seeking arbitrage opportunities

11.1   LP models for arbitrage in subjective probability theory

Computational applications of all of the fundamental theorems can be carried out by linear programming. The LP models for different theorems are all minor variations on the single model introduced in Section 2.6. For the subjective probability version, let $\pi = (\pi_1, ..., \pi_M)$ denote the individual's hypothetical probability distribution on the states, and let $x_1, ..., x_K$ denote the payoff vectors of her acceptable \$-bets. Let $\mathbf{W}$ be the $K \times M$ matrix of which those vectors are the rows. Then $\mathbf{W}\pi$ is the K-vector whose elements are the expected values of the bets according to π, and $\mathbf{W} \cdot \alpha$ is the M-vector whose elements are the payoffs of a weighted sum of the acceptable bets, where α is an K-vector of non-negative weights.[1] These bets could be offered directly or else their acceptability could be inferred from subjective probability assessments ($x_k = (\mathbf{A}_k - p_k)\mathbf{B}_k$ where the conditional probability of $\mathbf{A}_k$ given $\mathbf{B}_k$ is asserted to be at least p_k) or from preferences among SP-acts ($x_k = f_k - g_k$ where f_k is asserted to be weakly preferred to g_k). The following two

---

[1]Recap of matrix multiplication notation: $\mathbf{W}\pi$ is the K-vector whose k^{th} element is $\sum_{m=1}^{M} \mathbf{W}_{km}\pi_m$, and $\mathbf{W} \cdot \alpha$ is the M-vector whose m^{th} element is $\sum_{k=1}^{K} \alpha_k \mathbf{W}_{km}$ (the so-called "dot product" of $\mathbf{W}$ and α). The latter is more commonly written as $\mathbf{W}^T\alpha$, i.e., "$\mathbf{W}$-transpose α." Dot notation is used here to avoid confusion with T as a label for other parameters used elsewhere. Also note that $1 \cdot \pi$ means the sum of the elements of the vector π.

linear programming problems, reproduced from Section 2.6, are dual to each other and hence equivalent:

- Primal problem: maximize $1 \cdot \pi$ over all $\pi \geq 0$ subject to $1 \cdot \pi \leq 1$ and $\mathbf{W}\pi \geq 0$ [look for a supporting probability distribution for the acceptable \$-bets]

- Dual problem: minimize z over all $z \geq 0$ and $\alpha \geq 0$ subject to $\mathbf{W} \cdot \alpha - z \leq -1$ [look for an arbitrage opportunity among the acceptable \$-bets]

Both problems are feasible with bounded solutions having the following properties. If the probability assessment is *coherent*, i.e., if there are no arbitrage opportunities, then π is a supporting probability distribution and $\alpha = 0$ (there is no betting). If there *is* an arbitrage opportunity, then $\pi = 0$ (there is no supporting probability distribution) and α is a non-zero vector of bet multipliers that yields strictly negative payoffs to the agent. The optimal objective values $(1 \cdot \pi, z)$ are equal to 1 in the first case and 0 in the second case. The optimal values of the decision variables in one problem are the optimal values of shadow prices in the other.

The ex-post version of the fundamental theorem can be implemented with a refined model that focuses on what happens in state m:

- Ex post primal problem: maximize π_m over all $\pi \geq 0$ such that $1 \cdot \pi \leq 1$ and $\mathbf{W}\pi \geq 0$ [look for a supporting probability distribution that assigns positive probability to state m]

- Ex post dual problem: minimize z over all $z \geq 0$, $\alpha \geq 0$, such that $\mathbf{W} \cdot \alpha - z \leq 0$ and $[\mathbf{W} \cdot \alpha]_m - z \leq -1$ [look for an ex-post-arbitrage opportunity in state m]

The probability assessment avoids ex-post-arbitrage (i.e., is ex-post-coherent) in state m if and only if the optimal objective value is positive.

These linear programming models use inequality rather than equality constraints, so they are applicable to both the case of complete preferences (unique probabilities) and the case of incomplete preferences (lower and upper probabilities). In the case of complete preferences, there will be two oppositely-signed inequality constraints for a probability that is asserted to have a unique value or for an assertion that two acts are equally preferred.

If the solution of the ex-ante version of the primal linear program yields a supporting probability distribution, then a slightly altered version of the model can be used to determine lower and upper bounds on the expected value of a bet with an arbitrary payoff vector $\mathbf{y}$. All that's needed is to change the objective function to "maximize [or minimize] $\mathbf{y} \cdot \boldsymbol{\pi}$" and to change $\mathbf{1} \cdot \boldsymbol{\pi} \leq 1$ to $\mathbf{1} \cdot \boldsymbol{\pi} = 1$. In general this might or might not lead to a unique value, depending on whether the supporting probability distribution is unique with respect to the events on which the payoffs of $\mathbf{y}$ depend.

11.2   LP model for arbitrage in expected utility theory

Insofar as the EU model is formally identical to the SP model except for labeling of the working parts, the same linear programming models apply to the primal and dual problems of finding a supporting utility function for acceptable p-bets or finding an arbitrage opportunity among them. The only small technicality is that in the EU model, the components of a p-bet sum to 0, unlike the components of a \$-bet, and a weakly best and semi-strictly worst consequence are identified a priori in the EU model, whereas there are no a priori constraints on relative probabilities of states in the SP model. These differences are unimportant, though, and can be ignored in setting up the LP models. The a priori specification of a most preferred consequence actually plays no role except to provide an arbitrary reference point for the maximum of the utility function, and its labeling could just as well wait until utilities have been elicited. In the end its determination is endogenous. As for the requirement that the elements of a p-bet must sum to 0, that can be sidestepped by merely dropping the last coordinates of the p-bet vector and utility vector, i.e., the probability and utility assigned to consequence N whose utility is 0 by assumption. Therefore, let $\widehat{\mathbf{u}} := (u_1, ..., u_{N-1})$ and $\widehat{\mathbf{x}} := (x_1, ..., x_{N-1})$ be the vectors consisting of the first $N-1$ elements of the utility function $\mathbf{u}$ and and the acceptable p-bet $\mathbf{x}$, respectively, and let $\widehat{\mathbf{W}}$ denote the matrix whose row vectors are the $\widehat{\mathbf{x}}$'s. Then the expected utility of $\mathbf{x}$ according to $\mathbf{u}$ can equally well be expressed as $\mathbf{x} \cdot \mathbf{u}$ or as $\widehat{\mathbf{x}} \cdot \widehat{\mathbf{u}}$, and the vector of expected utilities of all the acceptable p-bets can be expressed as $\mathbf{Wu}$ or as $\widehat{\mathbf{W}}\widehat{\mathbf{u}}$. Also without loss of generality, we can choose to scale the utility function so that its elements sum to 1 rather than having a maximum value of 1.

In these terms the following two linear programming problems, which have exactly the same form as the SP versions, are dual to each other and hence equivalent:

- Primal problem: maximize $\mathbf{1} \cdot \widehat{\mathbf{u}}$ over all $\widehat{\mathbf{u}} \geq \mathbf{0}$ such that $\mathbf{1} \cdot \widehat{\mathbf{u}} \leq 1$ and $\widehat{\mathbf{W}}\widehat{\mathbf{u}} \geq \mathbf{0}$ [look for a supporting utility function for the acceptable p-bets]

- Dual problem: minimize z over all $z \geq 0$, $\boldsymbol{\alpha} \geq \mathbf{0}$, such that $\widehat{\mathbf{W}} \cdot \boldsymbol{\alpha} - z \leq -1$ [look for an arbitrage opportunity among the acceptable p-bets]

In the solutions, if there are no arbitrage opportunities, then $\widehat{u}$ is a supporting utility function and $\alpha = 0$ (there is no betting). If there *is* an arbitrage opportunity, then $\widehat{u} = 0$ (there is no supporting utility function) and α is a non-zero vector of bet multipliers that subtracts positive amounts of probability from all non-worst consequences. The optimal objective value equals 1 in the first case and 0 in the second case. The optimal values of the decision variables in one problem are the optimal values of shadow prices in the other. As was the case in the SP model, the linear programming models are framed in terms of inequality constraints (one-way acceptability) and are equally applicable to complete or incomplete preferences (unique utilities or lower and upper utilities).

Thus, the SP and EU models are formally identical. They merely reverse the roles of probabilities and payoffs with respect to which ones are objective variables to manipulate and which ones are subjective parameters to measure. This is perhaps not surprising insofar as both are *linear* representations of preferences, and linear models are to some extent all the same. Still, the tight correspondence of their representations and their common proofs of the fundamental theorems via separating hyperplanes and their common implementability via linear programming are not always appreciated, especially by those whose work mainly resides in only one of the two domains and/or who go back to original sources.

11.3   LP model for arbitrage in subjective expected utility theory

The state-dependent version of the SEU model is also formally identical to the SP and EU models except that the payoffs of acts are indexed by both states and consequences. For the purposes of setting up a linear programming model to look for a supporting utility function or an arbitrage opportunity, the parametric form of a bet will be arranged as a vector rather than a matrix and denoted with a hat: $\widehat{x}$. The elements of $\widehat{x}$ are determined by successively concatenating the rows of X up to consequence $N - 1$. (As in the EU case we do not need to include a coordinate for consequence N whose utility is zero by definition in all states.) Thus, $\widehat{x} := (x_{11}, ..., x_{1,N-1}, x_{21}, ..., x_{2,N-1}, ..., x_{M1}, ..., x_{M,N-1})$, and its length is $M(N-1)$. Let $\widehat{v} := (v_{11}, ..., v_{1,N-1}, v_{21}, ..., v_{2,N-1}, ..., v_{M1}, ..., v_{M,N-1})$ denote a similarly reshaped version of a state-dependent utility matrix V. The expected utility of X that is determined by V can then be written as the inner product $\widehat{x} \cdot \widehat{v}$. As in the EU version of the LP, we can choose to scale the utility vector so that its elements sum to 1, and this makes it unnecessary to assume a priori that a particular consequence is most-preferred.

Let $\widehat{\mathbf{x}}_k$ be an acceptable reshaped v-bet vector for $k = 1, ..., K$, and let $\widehat{\mathbf{W}}$ denote the $K \times M(N-1)$ matrix whose row vectors are the $\widehat{\mathbf{x}}$'s. Then the expected utilities of all the acceptable bets according to $\widehat{\mathbf{v}}$ can be expressed as the vector $\widehat{\mathbf{W}}\widehat{\mathbf{v}}$. A weighted sum of the rows of $\widehat{\mathbf{W}}$ can be expressed as $\widehat{\mathbf{W}} \cdot \boldsymbol{\alpha}$ for some non-negative vector of weights $\boldsymbol{\alpha}$, and it is an arbitrage opportunity if its elements are strictly negative, i.e., if positive net amounts of probability are transferred to the worst consequence from each of the others in every state.

In these terms the following two linear programming problems, which have exactly the same form as the SP and EU versions, are dual to each other and hence equivalent:

- Primal problem: maximize $\mathbf{1} \cdot \widehat{\mathbf{v}}$ over all $\widehat{\mathbf{v}} \geq \mathbf{0}$ such that $\mathbf{1} \cdot \widehat{\mathbf{v}} \leq 1$ and $\widehat{\mathbf{W}}\widehat{\mathbf{v}} \geq \mathbf{0}$ [look for a supporting state-dependent utility function for the acceptable v-bets]

- Dual problem: minimize z over all $z \geq 0$, $\boldsymbol{\alpha} \geq \mathbf{0}$, such that $\widehat{\mathbf{W}} \cdot \boldsymbol{\alpha} - z \leq -1$ [look for an arbitrage opportunity among the acceptable v-bets]

11.4 LP model for ex-post-arbitrage and correlated equilibria in games

As advertised in Section 3.7, the fundamental theorem of noncooperative games (Theorem 8.2) is a special case of the ex-post version of the fundamental theorem of subjective probability (Theorem 3.6). The general acceptable-bet matrix $\mathbf{W}$ which appeared there is merely replaced by the revealed-rules-of-the-game matrix $\mathbf{G}$ whose rows are acceptable bets for different players that are determined by their payoff functions in the game. It provides the minimum information needed to compute the Nash equilibria and correlated equilibria and does so in a way that makes them common knowledge. With this substitution, the linear programs for testing for the existence of a supporting probability distribution and searching for an arbitrage opportunity are the same as those of the ex-post subjective probability model:

- Primal problem: maximize π_s over all $\boldsymbol{\pi} \geq \mathbf{0}$ such that $\mathbf{1} \cdot \boldsymbol{\pi} \leq 1$ and $\mathbf{G}\boldsymbol{\pi} \geq \mathbf{0}$ [look for a correlated equilibrium distribution that assigns positive probability to outcome s]

- Dual problem: minimize z over all $z \geq 0$, $\boldsymbol{\alpha} \geq \mathbf{0}$, such that $\mathbf{G} \cdot \boldsymbol{\alpha} - z \leq \mathbf{0}$ and $[\mathbf{G} \cdot \boldsymbol{\alpha}]_s - z \leq -1$ [look for an ex-post-arbitrage opportunity in outcome s]

Outcome s is jointly coherent (avoids ex-post-arbitrage) if and only if the optimal objective value in both problems (which is the same) is positive. In the primal problem this is the maximum of π_s on the set of correlated equilibria. In the dual problem a positive optimal value of z means that there is no non-negative $\boldsymbol{\alpha}$ that can be chosen by an opponent to give the players a strictly negative total payoff in outcome s ($[\mathbf{G} \cdot \boldsymbol{\alpha}]_s < 0$) and a non-negative total payoff in all other outcomes ($\mathbf{G} \cdot \boldsymbol{\alpha} \leq \mathbf{0}$).

Given that a correlated equilibrium always exists, there is no loss of generality in using an equality constraint on the sum of probabilities ($\mathbf{1} \cdot \boldsymbol{\pi} = 1$), and a broader search for correlated equilibria can then be performed by using $\mathbf{c} \cdot \boldsymbol{\pi}$ as the objective function in the primal problem for some coefficient vector $\mathbf{c}$. The choice of $\mathbf{c}$ will determine which vertex of the correlated equilibrium polytope will be selected if there is more than one. If $\mathbf{c}$ is a sum or positive weighted sum of the payoffs of all the players, as in some of the examples above, the solution of the primal linear program will be a Pareto efficient vertex. For example, referring to the classic 2×2 coordination games in Table 8.13, maximization of a sum of payoffs leads to vertex $\boldsymbol{\pi}^4$ in the chicken game: a correlated mixed strategy that avoids a collision and places a maximal probability of 0.5 on swerve-swerve. It leads to vertex $\boldsymbol{\pi}^1$ or $\boldsymbol{\pi}^2$ or randomization between them in the battle-of-sexes game (both go to the same venue somehow), and it leads to vertex $\boldsymbol{\pi}^1$ in the stag hunt game (both hunt the stag).

11.5 LP model for arbitrage in asset pricing theory

A financial market can be viewed as a single agent with state-dependent subjective expected utility preferences (which are generally incomplete) who takes bets on a set of events (namely future asset payoffs) and whose (risk-adjusted) beliefs are parameterized by risk-neutral probabilities. In the case of finite sets of states and assets, the same linear programming models for finding supporting probability distributions or arbitrage opportunities that were used in the subjective probability setting can be applied here too without modification. All that is needed is to specify the nature of the bets that are in play. Suppose that there are M states of the world at time 1 and that there are K risky assets that can be purchased at market prices at time 0. There is also a riskless asset that can be bought and sold at time 0 for p_0 per unit and which yields a time-1 payoff of 1 per unit for purposes of borrowing and lending, so that the multiplicative risk-free return factor (1 plus the risk-free rate) is $r := 1/p_0$. Let x_k denote the time-1 payoff vector of asset k and let p_k denote its time-0 buying price, which can be deferred to time 1 by borrowing at the risk-free rate. Thus, the market's offer of the opportunity for an agent (a potential arbitrageur) to purchase 1 unit of asset k at price p_k at time 0 is equivalent to offering to accept a bet whose vector of net payoffs at time 1 is $x_k - p_k r$. Purchases and sales of a given real asset are modeled separately, i.e., considered as distinct assets with oppositely signed payoffs, so that the decision variables are all non-negative and so that there could be bid-ask spreads in prices. Let W be the $K \times M$ matrix of net payoffs of unit purchases of the assets, whose k^{th} row is $x_k - p_k r$. We can test the rationality of the market prices by solving exactly the same primal/dual pair of linear programs as in the subjective probability version:

- Primal problem: maximize $1 \cdot \pi$ over all $\pi \geq 0$ such that $1 \cdot \pi \leq 1$ and $W\pi \geq 0$ [look for a supporting risk-neutral probability distribution for the asset prices]

- Dual problem: minimize z over all $z \geq 0$, $\alpha \geq 0$, such that $W \cdot \alpha - z \leq -1$ [look for an arbitrage opportunity among the asset prices]

The optimal objective value in both problems will be 1 if there is a risk-neutral distribution π which prices the assets, i.e, which satisfies $\pi \cdot x_k \geq p_k r$ for all k. This means that from the rational market's perspective p_k is a fair lower-bound buying price for 1 unit of asset k for every k. The optimal objective value will be 0 otherwise, in which case the optimal values of elements of α in the dual problem will be multiples of the assets to be purchased so as to achieve an arbitrage profit. This does not mean, however, that π is an estimate of the representative investor's subjective probability distribution for the future asset prices. It contains an adjustment for state-dependent marginal utility of money which shifts its mean downward to match that of the riskless asset.

Chapter 12

Selected proofs

Theorem 5.4: Theorem 5.3, which is based on SEU1*–SEU6 only, establishes that $\succsim$ is represented by a convex cone $\mathcal{X}$ of acceptable bets and dually by a convex set $\mathcal{V}$ of s.d.e.u. functions containing at least one probability/utility pair. Let $\mathcal{X}^*$ and $\mathcal{V}^*$ denote the corresponding sets that are obtained if the stronger axioms SEU6^{a*} and SEU6^{b*} are applied. It must be shown that $\mathcal{V}^*$ is the convex hull of a set of probability/utility pairs, which means that for any acts $\mathbf{F}$ and $\mathbf{G}$, the minimum of $E_\mathbf{V}(\mathbf{F}) - E_\mathbf{V}(\mathbf{G})$ over all $\mathbf{V} \in \mathcal{V}^*$ is achieved at some $\mathbf{V}$ which is a probability/utility pair. Let $d = \min_{\mathbf{V} \in \mathcal{V}}(E_\mathbf{V}(\mathbf{F}) - E_\mathbf{V}(\mathbf{G}))$. Since $E_\mathbf{V}(\mathbf{L}_N) := 0$ (because consequence N is worst by definition) and $E_\mathbf{V}(\mathbf{L}[u]) = u$, it follows that d is the maximum value of u for which

$$\tfrac{1}{2}\mathbf{F} + \tfrac{1}{2}\mathbf{L}_N \succsim \tfrac{1}{2}\mathbf{G} + \tfrac{1}{2}\mathbf{L}[u] \qquad (12.1)$$

for all $\mathbf{V} \in \mathcal{V}^*$, or equivalently it is the maximum value of u for which $E_\mathbf{V}(\tfrac{1}{2}\mathbf{F} + \tfrac{1}{2}\mathbf{L}_N) \geq E_\mathbf{V}(\tfrac{1}{2}\mathbf{G} + \tfrac{1}{2}\mathbf{L}[u])$, or equivalently it is the maximum value of u for which $\mathbf{F} + \mathbf{L}_N - \mathbf{G} - \mathbf{L}[u] \in \mathcal{X}^*$. To prove that there is a probability/utility pair for which this condition holds with equality, it will suffice to show that the reverse conditions, namely that $E_\mathbf{V}(\tfrac{1}{2}\mathbf{F} + \tfrac{1}{2}\mathbf{L}_N) \leq E_\mathbf{V}(\tfrac{1}{2}\mathbf{G} + \tfrac{1}{2}\mathbf{L}[d])$ and equivalently that $-(\mathbf{F} + \mathbf{L}_N - \mathbf{G} - \mathbf{L}[d])$ is acceptable, can be applied without producing incoherence. First, define $\mathbf{X}_d = \mathbf{F} + \mathbf{L}_N - \mathbf{G} - \mathbf{L}[d]$, noting that by construction it is an extremal direction of $\mathcal{X}^*$. Then include the reverse preference by adding to the convex cone $\mathcal{X}^*$ the ray whose direction is $-\mathbf{X}_d$. Then apply SEU2^b– SEU4^b, which convexifies the set, yielding an expanded cone. Finally apply SEU6^b (the special case of SEU6^{b*} with $\alpha = 1$), which means to (i) determine which *new* conditional directions of the form $\mathbf{A}_k\mathbf{X}_k$, where $\mathbf{A}_k$ is a not-potentially-null event and $\mathbf{X}_k$ is constant, are in the expanded cone; (ii) then propagate the same conditional preferences to all possible conditioning events by adding the directions $\{\mathbf{E}_m\mathbf{X}_k\}$ for every k and every state m; and (iii) take the convex hull again. Thus, the expanded cone consists of all vectors whose directions are $\mathbf{Y} - \gamma_d\mathbf{X}_d + \sum_{m=1}^{M} \sum_{k=1}^{K} \gamma_{mk}\mathbf{E}_m\mathbf{X}_k$, where $\mathbf{Y} \in \mathcal{X}^*$, $\gamma_d \geq 0$, and $\gamma_{mk} \geq 0$.

DOI: 10.1201/9781003527145-12

Suppose that one such vector is in the open negative orthant $\mathcal{X}^-$, in violation of SEU5^b, i.e., suppose that $\mathbf{Y} - \gamma_d \mathbf{X}_d + \sum_{m=1}^{M} \sum_{k=1}^{K} \gamma_{mk} \mathbf{E}_m \mathbf{X}_k = \mathbf{Z}$, or equivalently

$$\mathbf{Y} + \sum_{m=1}^{M} \sum_{k=1}^{K} \gamma_{mk} \mathbf{E}_m \mathbf{X}_k = \gamma_d \mathbf{X}_d + \mathbf{Z}, \qquad (12.2)$$

where $\mathbf{Y} \in \mathcal{X}^*$ and $\mathbf{Z} \in \mathcal{X}^-$. It will be shown that this leads to a contradiction. Because the conditional direction $\mathbf{A}_k \mathbf{X}_k$ is in the convex hull of $\mathcal{X}^*$ *plus* the additional direction $-\mathbf{X}_d$, but not in $\mathcal{X}^*$ alone, it satisfies $\mathbf{X}_d + \mathbf{A}_k \mathbf{X}_k \in \mathcal{X}^*$. Because $\mathbf{X}_d \in \mathcal{X}^*$ and each $\mathbf{A}_k$ is not-potentially-null, and $\mathcal{X}^*$ satisfies SEU6^{b*}, there exist positive constants $\{\delta_{mk}\}$ such that $\mathbf{X}_d + \delta_{mk} \mathbf{E}_m \mathbf{X}_k \in \mathcal{X}^*$ for all m and k. [This is the step at which the stronger property of SEU6^{b*} is needed: $\mathbf{X}_d$ is the background noise in the presence of which the state-independence condition is applied.] Multiplying through by γ_{mk}/δ_{mk}, summing, and invoking the convexity of $\mathcal{X}^*$, yields $\left(\sum_{m=1}^{M} \sum_{k=1}^{K} \gamma_{mk}/\delta_{mk} \right) \mathbf{X}_d + \sum_{m=1}^{M} \sum_{k=1}^{K} \gamma_{mk} \mathbf{E}_m \mathbf{X}_k \in \mathcal{X}^*$, whence also $\mathbf{Y} + \left(\sum_{m=1}^{M} \sum_{k=1}^{K} \gamma_{mk}/\delta_{mk} \right) \mathbf{X}_d + \sum_{m=1}^{M} \sum_{k=1}^{K} \gamma_{mk} \mathbf{E}_m \mathbf{X}_k \in \mathcal{X}^*$. Adding $\left(\sum_{m=1}^{M} \sum_{k=1}^{K} \gamma_{mk}/\delta_{mk} \right) \mathbf{X}_d$ to both sides of (11.2) and comparing, it follows that $\left(\gamma_d + \sum_{m=1}^{M} \sum_{k=1}^{K} \gamma_{mk}/\delta_{mk} \right) \mathbf{X}_d + \mathbf{Z} \in \mathcal{X}^*$. Multiplication by $\alpha = \left(\gamma_d + \sum_{m=1}^{M} \sum_{k=1}^{K} \gamma_{mk}/\delta_{mk} \right)^{-1}$ yields $\mathbf{X}_d + \alpha \mathbf{Z} \in \mathcal{X}^*$, where $\alpha \mathbf{Z} \in \mathcal{X}^-$, contradicting the assumption that $\mathbf{X}_d$ was an extreme direction of $\mathcal{X}^*$ and constructively proving the existence of $u > d$ for which $\frac{1}{2}\mathbf{F} + \frac{1}{2}\mathbf{L}_N \succsim \frac{1}{2}\mathbf{G} + \frac{1}{2}\mathbf{L}[u]$. ∎

Theorem 5.3(c): "If" follows from the fact that the probability/utility pair $\{\mathbf{p}, \mathbf{q}\}$ determines a state-independent utility function that satisfies the same conditions as in part (a). For "only if" it suffices to show that a preference relation satisfying the given assumptions can be extended to assign a state-independent utility to any consequence between 1 and consequence N (those endpoints having utilities of 1 and 0 by definition). This extension can be performed for $n = 2, 3, ..., N - 1$ (in any order) to obtain a state-independent utility function. (The result is not necessarily unique because the order of consequences could matter.) To begin, consider some consequence c_n and let u_n denote the maximum value of u such that $\mathbf{L}_n \succsim \mathbf{L}[u]$, that is, the *greatest lower utility* for c_n. (The maximum exists by virtue of continuity.) Since SEU0^b, SEU1^{b*}, and SEU2^b–SEU5^b apply, there exists a cone of acceptable gambles representing $\succsim$. Let this cone be denoted by $\mathcal{X}^*$. The condition $\mathbf{L}_n \succsim \mathbf{L}[u_n]$ means that $\mathcal{X}^*$ contains $\mathbf{L}_n - \mathbf{L}[u_n]$, and when SEU6^b also applies, $\mathcal{X}^*$ also contains the vectors $\mathbf{E}_m (\mathbf{L}_n - \mathbf{L}[u_n])$ for all m. It remains to show that $\succsim$ can be coherently extended so as to assign consequence c_n an *upper* utility also equal to u_n. Let $\mathcal{X}^{**}$ denote the enlarged set of acceptable gambles that is obtained by adding the upper utility assessment. It is the convex hull of $\mathcal{X}^*$ and the rays whose directions are $\mathbf{E}_m(\mathbf{L}[u_n] - \mathbf{L}_n)$ for all m and n. Note that the latter vectors are opposite in sign to those previously known (via SEU6^b) to belong to $\mathcal{X}^*$. Since $\mathcal{X}^{**}$ is also a convex cone, it potentially represents a relation satisfying SEU0^b SEU1^{b*}, and SEU2^b–SEU5^b; and since it is formed by adding to $\mathcal{X}^*$ a difference between two constant acts and all the conditionalizations thereof, the relation also satisfies SEU6^b. If $\mathcal{X}^{**}$ is disjoint from the negative orthant $\mathcal{X}^-$, then the relation it represents is coherent and is the desired extension $\succsim_*$. It remains to be shown, then, that $\mathcal{X}^{**}$ is disjoint from $\mathcal{X}^-$. Suppose that this is not true. Then there must exist $\mathbf{X} \in \mathcal{X}^*$ and $\{\beta_m\} \in [0,1]$ such that[1]:

$$\mathbf{X} + \left(\sum_{m=1}^{M} \beta_m \mathbf{E}_m(\mathbf{L}[u_n] - \mathbf{L}_n) \right) \ll 0 \qquad (12.3)$$

$$\Longleftrightarrow \mathbf{X} + \left(\sum_{m=1}^{M} (1 - \beta_m)\mathbf{E}_m(\mathbf{L}_n - \mathbf{L}[u_n]) \right) - (\mathbf{L}[u_n] - \mathbf{L}_n) \ll 0$$

$$\Longleftrightarrow \mathbf{X}' + (\mathbf{L}[u_n] - \mathbf{L}_n) \ll 0$$

where

m

$$\mathbf{X}' = \mathbf{X} + \left(\sum_{m=1}^{M} (1 - \beta_m)\mathbf{E}_m(\mathbf{L}_n - \mathbf{L}[u_n]) \right) \qquad (12.4)$$

Note that $\mathbf{X}' \in \mathcal{X}^*$, because $\mathcal{X}^*$ includes all vectors of the form $\mathbf{E}_m(\mathbf{L}_n - \mathbf{L}[u_n])$ and their nonnegative linear combinations. Rearrangement of the last inequality yields $(\mathbf{L}_n - \mathbf{L}[u_n]) \gg \mathbf{X}'$, which is equivalent to $(\mathbf{L}_n - \mathbf{L}[u_n + \epsilon]) \geq \mathbf{X}'$ for some $\epsilon > 0$. Because $(\mathbf{L}_n - \mathbf{L}[u_n + \epsilon])$ weakly dominates $\mathbf{X}'$, an element of $\mathcal{X}^*$, it follows that $(\mathbf{L}_n - \mathbf{L}[u_n + \epsilon]) \in \mathcal{X}^*$, whence also $\mathbf{L}_n \succsim \mathbf{L}[u_n + \epsilon]$. Thus, $u_n + \epsilon$ is a lower utility for consequence n, contradicting the assumption that u_n was its greatest lower utility under the relation $\succsim$, and therefore $\mathcal{X}_n^*$ must be disjoint from $\mathcal{X}^-$. This exercise can then be repeated for all the other consequences strictly between 1 and N to assign them all state-independent utilities. ∎

Theorem 6.2: The first equality (6.33) is Proposition 2(a) in Nau (2003). A proof of a more general result is given there, but the short version is as follows. By assumption, $\mathbf{f} + \mathbf{x} + b(\mathbf{x}, \mathbf{f})\mathbf{1}$ lies on the same indifference curve as $\mathbf{f}$, and in principle the decision-maker could move from f to $\mathbf{f} + \mathbf{x} + b(\mathbf{x}, \mathbf{f})\mathbf{1}$ along the curve by receiving small increments of $\mathbf{x}$ and being compensated for them at their

---

[1]"$\mathbf{Y} \ll 0$" means the same thing as "$\mathbf{Y} \in \mathcal{X}^-$."

marginal prices. (Here $b(\mathbf{x}, \mathbf{f})\mathbf{1}$ is the constant vector $\mathbf{1}$ scaled by $b(\mathbf{x}, \mathbf{f})$. The marginal prices are negative and increasingly so, and the equivalent compensation has the opposite sign, yielding a positive number.) To a first-order approximation, the price vector changes linearly with the amount purchased, and in particular it changes from $\boldsymbol{\pi}(\mathbf{f})$ to approximately $\boldsymbol{\pi}(\mathbf{f}) + D\boldsymbol{\pi}(\mathbf{f})(\mathbf{x} + b(\mathbf{x}, \mathbf{f})\mathbf{1})$ from start to finish. The average value of the price vector along this linear trajectory is the midpoint, $\boldsymbol{\pi}(\mathbf{f}) + \frac{1}{2}D\boldsymbol{\pi}(\mathbf{f})(\mathbf{x} + b(\mathbf{x}, \mathbf{f})\mathbf{1})$. So the total compensation she must receive to accept $\mathbf{x}$ is approximately $-\mathbf{x} \cdot (\boldsymbol{\pi}(\mathbf{f}) + \frac{1}{2}D\boldsymbol{\pi}(\mathbf{f})(\mathbf{x} + b(\mathbf{x}, \mathbf{f})\mathbf{1}))$, which reduces to $-\frac{1}{2}\mathbf{x} \cdot D\boldsymbol{\pi}(\mathbf{f})(\mathbf{x} + b(\mathbf{x}, \mathbf{f})\mathbf{1}))$ because $\mathbf{x} \cdot \boldsymbol{\pi}(\mathbf{f}) = 0$ for a neutral asset. The exact amount of compensation she must receive is $b(\mathbf{x}, \mathbf{f})$ by definition, which yields the approximation:

$$b(\mathbf{x}, \mathbf{f}) \approx -\frac{1}{2}\mathbf{x} \cdot D\boldsymbol{\pi}(\mathbf{f})(\mathbf{x} + b(\mathbf{x}, \mathbf{f})\mathbf{1})). \tag{12.5}$$

The second term in parentheses on the RHS can be ignored in the limit as x becomes small because $b(\mathbf{x}, \mathbf{f}) = O(\|\mathbf{x}\|^2)$, which yields $b(\mathbf{x}, \mathbf{f}) \approx -\frac{1}{2}\mathbf{x} \cdot D\boldsymbol{\pi}(\mathbf{f})\mathbf{x}$ as claimed. The second equality (6.34) follows by applying (6.16) to rewrite $D\boldsymbol{\pi}(\mathbf{f})$ in terms of $DU(\mathbf{f})$ and $D^2U(\mathbf{f})$ and noting that the second term in (6.16) must yield $\mathbf{0}$ when it is premultiplied by $\mathbf{x}$ because each of its columns is proportional to $\boldsymbol{\pi}(\mathbf{f})$, and $\mathbf{x} \cdot \boldsymbol{\pi}(\mathbf{f}) = 0$ if $\mathbf{x}$ is a neutral asset. The third equality (6.35) follows from the fact that

$$\left[\frac{D^2U(\mathbf{f})}{\mathbf{1} \cdot DU(\mathbf{f})}\right]_{mj} = \frac{\partial^2 U(\mathbf{f})/\partial f_m \partial f_j}{\mathbf{1} \cdot DU(\mathbf{f})} \tag{12.6}$$

$$= \frac{\partial U(\mathbf{f})/\partial f_m}{\mathbf{1} \cdot DU(\mathbf{f})} \frac{\partial^2 U(\mathbf{f})/\partial f_j \partial f_k}{\partial U(\mathbf{f})/\partial f_m} = \pi_m(\mathbf{f})r_{mj}(\mathbf{f}). \blacksquare$$

Theorem 6.3: For a strictly uncertainty-averse decision-maker, both $D\boldsymbol{\pi}(\mathbf{f})$ and $S(\mathbf{f})$ have rank $M - 1$, while both bordered matrices have rank $M + 1$, hence they are invertible. To see that the bordered $D\boldsymbol{\pi}$ matrix has rank $M + 1$, note that $\boldsymbol{\pi}(\mathbf{f})$ is linearly independent from the columns of $D\boldsymbol{\pi}(\mathbf{f})$, because its elements do not sum to zero, so their concatenation $\begin{bmatrix} D\boldsymbol{\pi}(\mathbf{f}) & | & \boldsymbol{\pi}(\mathbf{f}) \end{bmatrix}$ has rank M. In turn, $\begin{bmatrix} \boldsymbol{\pi}(\mathbf{f})^T & | & 0 \end{bmatrix}$ is linearly independent from the rows of $\begin{bmatrix} D\boldsymbol{\pi}(\mathbf{f}) & | & \boldsymbol{\pi}(\mathbf{f}) \end{bmatrix}$, because otherwise there would exist a non-zero M-vector $\mathbf{x}$ satisfying $\mathbf{x} \cdot \begin{bmatrix} D\boldsymbol{\pi}(\mathbf{f}) & | & \boldsymbol{\pi}(\mathbf{f}) \end{bmatrix} = \begin{bmatrix} \boldsymbol{\pi}(\mathbf{f})^T & | & 0 \end{bmatrix}$, which would imply $\mathbf{x} \cdot \boldsymbol{\pi}(\mathbf{f}) = 0$ and $\mathbf{x} \cdot D\boldsymbol{\pi}(\mathbf{f}) = \boldsymbol{\pi}(\mathbf{f})^T$, which together would also imply $\mathbf{x} \cdot D\boldsymbol{\pi}(\mathbf{f})\mathbf{x} = 0$, which is impossible because $\mathbf{x} \cdot D\boldsymbol{\pi}(\mathbf{f})\mathbf{x} < 0$ for every x that satisfies $\mathbf{x} \cdot \boldsymbol{\pi}(\mathbf{f}) = 0$ if the decision-maker is strictly uncertainty-averse. Similarly for the bordered Slutsky matrix, $\mathbf{w}(\mathbf{f})$ is linearly independent from the columns of $S(\mathbf{f})$, because the latter all have a zero inner product with $\boldsymbol{\pi}(\mathbf{f})$ while the former does not, and $\begin{bmatrix} \mathbf{1}^T & | & 0 \end{bmatrix}$ is linearly independent from the rows of $\begin{bmatrix} S(\mathbf{f}) & | & \mathbf{w}(\mathbf{f}) \end{bmatrix}$

because the latter all have a zero inner product with $\left[\boldsymbol{\pi}(\mathbf{f})^T \mid 0\right]$ while the former does not.

Next, if $\boldsymbol{\pi}(\mathbf{f})$ is strictly positive, any vector $\mathbf{y}$ can be expressed as

$$\mathbf{y} = \begin{bmatrix} \alpha\boldsymbol{\pi}(\mathbf{f}) + \gamma d\boldsymbol{\pi} \\ \beta \end{bmatrix} \tag{12.7}$$

for some constants α, β, and γ, where $d\boldsymbol{\pi}$ is a vector satisfying $\mathbf{1} \cdot d\boldsymbol{\pi} = 0$ and $0 < \pi_m(f) + d\pi_m < 1$ for all m, which can be interpreted as a feasible change in relative prices. Let $df := \mathbf{S}(\mathbf{f})d\boldsymbol{\pi}$ be the compensated change in consumption induced by $d\boldsymbol{\pi}$, and note that $\mathbf{S}(\mathbf{f})\boldsymbol{\pi}(\mathbf{f}) = \mathbf{0}$ and $\boldsymbol{\pi}(\mathbf{f}) \cdot \mathbf{S}(\mathbf{f}) = \mathbf{0}$ (because a change in prices proportional to the current price vector has no effect on consumption, and $\mathbf{S}(\mathbf{f})$ is symmetric), and $\boldsymbol{\pi}(\mathbf{f}) \cdot df = 0$ (df is self-financing at relative prices $\boldsymbol{\pi}(\mathbf{f})$). It follows that

$$\begin{bmatrix} \mathbf{S}(\mathbf{f}) & \mathbf{w}(\mathbf{f}) \\ \mathbf{1}^T & 0 \end{bmatrix} \begin{bmatrix} \alpha\boldsymbol{\pi}(\mathbf{f}) + \gamma d\boldsymbol{\pi} \\ \beta \end{bmatrix} = \begin{bmatrix} \gamma df + \beta\mathbf{w}(\mathbf{f}) \\ \alpha \end{bmatrix} \tag{12.8}$$

Multiplication by the bordered $D\boldsymbol{\pi}$ matrix yields

$$\begin{bmatrix} D\boldsymbol{\pi}(\mathbf{f}) & \boldsymbol{\pi}(\mathbf{f}) \\ \boldsymbol{\pi}(\mathbf{f})^T & 0 \end{bmatrix} \begin{bmatrix} \gamma df + \beta\mathbf{w}(\mathbf{f}) \\ \alpha \end{bmatrix} = \begin{bmatrix} \alpha\boldsymbol{\pi}(\mathbf{f}) + \gamma d\boldsymbol{\pi} \\ \beta \end{bmatrix} \tag{12.9}$$

because $D\boldsymbol{\pi}(\mathbf{f})df = d\boldsymbol{\pi}$ (first-order condition for df to yield a change of $d\boldsymbol{\pi}$ in relative marginal rates of substitution), and $D\boldsymbol{\pi}(\mathbf{f}) \cdot \mathbf{w}(\mathbf{f}) = \mathbf{0}$ (marginal rates of substitution do not change along the wealth expansion path), and $\boldsymbol{\pi}(\mathbf{f}) \cdot df = 0$ (self-financing property of df). Hence,

$$\begin{bmatrix} D\boldsymbol{\pi}(\mathbf{f}) & \boldsymbol{\pi}(\mathbf{f}) \\ \boldsymbol{\pi}(\mathbf{f})^T & 0 \end{bmatrix} \begin{bmatrix} \mathbf{S}(\mathbf{f}) & \mathbf{w}(\mathbf{f}) \\ \mathbf{1}^T & 0 \end{bmatrix} \mathbf{y} = \mathbf{y} \quad \text{for all } \mathbf{y}, \tag{12.10}$$

which implies

$$\begin{bmatrix} D\boldsymbol{\pi}(\mathbf{f}) & \boldsymbol{\pi}(\mathbf{f}) \\ \boldsymbol{\pi}(\mathbf{f})^T & 0 \end{bmatrix}^{-1} = \begin{bmatrix} \mathbf{S}(\mathbf{f}) & \mathbf{w}(\mathbf{f}) \\ \mathbf{1}^T & 0 \end{bmatrix}. \quad \blacksquare \tag{12.11}$$

Corollary 6.4: $\Delta\boldsymbol{\pi}$ and $\Delta\mathbf{f}$ must satisfy $D\boldsymbol{\pi}(\mathbf{f})\Delta\mathbf{f} \approx \Delta\boldsymbol{\pi}$ (the change in $\mathbf{f}$ must yield a change in relative marginal utilities that matches the change in relative prices, to a first-order approximation) and $(\boldsymbol{\pi}(\mathbf{f}) + \Delta\boldsymbol{\pi})\Delta\mathbf{f} = 0$ ($\Delta\mathbf{f}$ must be self-financing at the new relative prices). This can be written as the system of equations:

$$\begin{bmatrix} D\boldsymbol{\pi}(\mathbf{f}) & \boldsymbol{\pi}(\mathbf{f}) + \Delta\boldsymbol{\pi} \\ (\boldsymbol{\pi}(\mathbf{f}) + \Delta\boldsymbol{\pi})^T & 0 \end{bmatrix} \begin{bmatrix} \Delta\mathbf{f} \\ 0 \end{bmatrix} \approx \begin{bmatrix} \Delta\boldsymbol{\pi} \\ 0 \end{bmatrix} \tag{12.12}$$

The bordered $D\boldsymbol{\pi}$ matrix is non-singular, by the same argument as before, and multiplication by its inverse yields the result. $\blacksquare$

Theorem 7.1(b): This is Theorem 1(d) in Nau (2006a), but the full details of the proof are not given there, so here they are. Suppose that U is the two-source utility function

$$U(\mathbf{f}) = \sum_{j=1}^{J} u_j \left(\sum_{k=1}^{K} v_{jk}(f_{jk}) \right). \tag{12.13}$$

Its gradient, whose normalization yields the local risk-neutral distribution, is

$$[DU(\mathbf{f})]_{jk} = u_j' v_{jk}', \tag{12.14}$$

where u_j' is shorthand for $u_j' \left(\sum_{k=1}^{K} v_{jk}(f_{jk}) \right)$ and v_{jk}' is shorthand for $v_{jk}'(f_{jk})$, both of which are positive. Its Hessian matrix $D^2 U(f)$ is the sum of a diagonal matrix and a block-diagonal matrix constructed as follows. Let the rows of $D^2 U(\mathbf{f})$ be indexed by jk and let the columns be indexed by mn, where row jk corresponds to event $A_j B_k$ and column mn corresponds to event $A_m B_n$.[2] The element in row jk and column mn is

$$[D^2 U(\mathbf{f})]_{jk,mn} = u_j' v_{jk}'' 1_{jk=mn} + u_j'' v_{jk}' v_{jn}' 1_{j=m}, \tag{12.15}$$

where u_j'' and v_{jk}'' are shorthands for and $u_j'' \left(\sum_{k=1}^{K} v_{jk}(f_{jk}) \right)$ and $v''(f_{jk})$ respectively, both of which are negative. ($1_{jk=mn}$ is the indicator function for terms in which $jk = mn$, and similarly for $1_{j=m}$.) The corresponding term in the risk aversion matrix $\mathbf{R}$ is by definition the quotient of (12.13) and (12.12):

$$r_{jk,mn} = \frac{u_j' v_{jk}'' 1_{jk=mn} + u_j'' v_{jk}' v_{jn}' 1_{j=m}}{u_j' v_{jk}'} = \frac{v_{jk}''}{v_{jk}'} 1_{jk=mn} + \frac{u_j''}{u_j'} v_{jn}' 1_{j=m} \tag{12.16}$$

By Theorem 6.2, and in light of the special additive structure of $\mathbf{R}(\mathbf{f})$ illustrated in (7.9), the local quadratic approximation of the uncertainty premium is:

$$\hat{b}(\mathbf{x}; \mathbf{f}) = \frac{1}{2} \mathbf{x} \cdot \mathbf{\Pi}(\mathbf{f}) \mathbf{R}(\mathbf{f}) \, \mathbf{x} \tag{12.17}$$

$$= \frac{1}{2} \sum_{j=1}^{J} \sum_{k=1}^{K} \sum_{m=1}^{J} \sum_{n=1}^{K} \pi_{jk} r_{jk,mn} x_{jk} x_{mn} \tag{12.18}$$

$$= \frac{1}{2} \sum_{j=1}^{J} \sum_{k=1}^{K} \pi_{jk} \frac{v_{jk}''}{v_{jk}'} x_{jk}^2 + \frac{1}{2} \sum_{j=1}^{J} \sum_{k=1}^{K} \sum_{n=1}^{K} \pi_{jk} \frac{u_j''}{u_j'} v_{jn}' x_{jk} x_{jn} \tag{12.19}$$

$$= \frac{1}{2} \sum_{j=1}^{J} \sum_{k=1}^{K} \pi_{jk} t_{jk} x_{jk}^2 + \frac{1}{2} \sum_{j=1}^{J} \sum_{k=1}^{K} \pi_{jk} \frac{u_j''}{u_j'} x_{jk} \sum_{n=1}^{K} v_{jn}' x_{jn} \tag{12.20}$$

[2]Here the range of m is $\{1, ..., J\}$ and the range of n is $\{1, ..., K\}$.

The first term on the RHS is the $\mathcal{B}$-premium. To unpack the second term, let $\alpha = (\sum_{j=1}^{J} \sum_{k=1}^{K} u_j' v_{jk}')^{-1}$ and $v_j' = \sum_{n=1}^{K} v_{jn}'$, in terms of which $\pi_{jn} = \alpha u_j' v_{jn}'$ and $\pi_j = \sum_{n=1}^{K} \pi_{jn} = \alpha u_j' \sum_{n=1}^{K} v_{jn}' = \alpha u_j' v_j'$, whence $v_{jn}' = v_j'(\pi_{jn}/\pi_j)$. Then:

$$\frac{1}{2} \sum_{j=1}^{J} \sum_{k=1}^{K} \pi_{jk} \frac{u_j''}{u_j'} x_{jk} \sum_{n=1}^{K} v_{jn}' x_{jn}$$

$$= \frac{1}{2} \sum_{j=1}^{J} \pi_j \frac{u_j''}{u_j'} \sum_{k=1}^{K} \frac{\pi_{jk}}{\pi_j} x_{jk} \sum_{n=1}^{K} v_{jn}' x_{jn} \tag{12.21}$$

$$= \frac{1}{2} \sum_{j=1}^{J} \pi_j \frac{u_j''}{u_j'} \bar{x}_j \sum_{n=1}^{K} v_{jn}' x_{jn} \tag{12.22}$$

$$= \frac{1}{2} \sum_{j=1}^{J} \pi_j \frac{u_j''}{u_j'} \bar{x}_j \left(v_j' \sum_{n=1}^{K} \frac{\pi_{jn}}{\pi_j} x_{jn} \right) \tag{12.23}$$

$$= \frac{1}{2} \sum_{j=1}^{J} \pi_j \frac{u_j''}{u_j'} v_j' \bar{x}_j^2 = \frac{1}{2} \sum_{j=1}^{J} \pi_j s_j \bar{x}_j^2 \tag{12.24}$$

which is the $\mathcal{A}$-premium. For parts (c) and (d), note that by construction x_{jk}^2 is the same for both bets, so they have the same $\mathcal{AB}$-premium, and $\bar{x}_j = 0$ for all j in the $\mathcal{B}{:}\mathcal{A}$ bet—hence its $\mathcal{A}$-premium is zero—while $|\bar{x}_j| > 0$ for all j in the $\mathcal{A}{:}\mathcal{B}$ bet. If u_j is strictly concave, then $s_j > 0$, so the $\mathcal{A}$-premium is strictly positive for the $\mathcal{A}{:}\mathcal{B}$ bet if this is true for one or more j. ∎

Theorem 7.3: The second-order utility model can be viewed as a special case of the source-dependent utility model in which the first source of uncertainty is the first-order payoff and the second source of uncertainty is the credal state. With a change of notation (i taking the place formerly occupied by j, and jk taking the place formerly occupied by k alone), Theorem 7.1 yields

$$\widehat{b}(\mathbf{x}; \mathbf{f}) = \frac{1}{2} \sum_{i=1}^{I} \pi_i s_i \bar{x}_i^2 + \frac{1}{2} \sum_{i=1}^{I} \sum_{j=1}^{J} \sum_{k=1}^{K} \pi_{ijk} t_{jk} x_{jk}^2. \tag{12.25}$$

and i can then be integrated out of the second term because the first-order payoffs do not depend on it:

$$\sum_{i=1}^{I} \sum_{j=1}^{J} \sum_{k=1}^{K} \pi_{ijk} t_{jk} x_{jk}^2 = \sum_{j=1}^{J} \sum_{k=1}^{K} t_{jk} x_{jk}^2 \sum_{i=1}^{I} \pi_{ijk} = \sum_{j=1}^{J} \sum_{k=1}^{K} \pi_{jk} t_{jk} x_{jk}^2. \; \blacksquare$$

$$\tag{12.26}$$

Corollary 7.1: First consider the simpler situation in which $v'_{jk} = y_j z_k$ (i.e., z_k is the same within A and $\overline{A}$). Then the joint risk-neutral probabilities satisfy

$$\pi_{ijk}(\mathbf{f}) \quad \propto \quad (\mu_i u'_i)(p_{j|i} q_k) y_j z_k \tag{12.27}$$

$$= \quad (\mu_i u'_i)(p_{j|i} y_j)(q_k z_k) \tag{12.28}$$

which implies that $\pi_{jk|i} \propto q_k z_k$ for any fixed i and j, and also $\pi_{jk} \propto q_k z_k$ for any fixed j, and therefore also $\pi_{jk} \propto \pi_{jk|i}$ for any fixed i and j. If $x = x_{B:A}(\Delta)$, then by definition $\sum_{j \in A} \sum_{k=1}^{K} \pi_{jk} x_{jk} = 0$ and also $\sum_{j \in \overline{A}} \sum_{k=1}^{K} \pi_{jk} x_{jk} = 0$. Because $\pi_{jk} \propto \pi_{jk|i}$ for fixed i and j, it follows that $\sum_{j \in A} \sum_{k=1}^{K} \pi_{jk|i} x_{jk} = 0$ and $\sum_{j \in \overline{A}} \sum_{k=1}^{K} \pi_{jk|i} x_{jk} = 0$ for every i, which together imply $\sum_{j=1}^{J} \sum_{k=1}^{K} \pi_{jk|i} x_{jk} = 0$ for every i, which means $\overline{x}_i = 0$ for every i. Hence the $\mathcal{A}$-premium, which is a weighted sum of $\{\overline{x}_i^2\}$, must be zero. The proof is completed by observing that the same chain of implication holds if z_k is allowed to depend on whether $j \in A$ or $j \in \overline{A}$, because the cases $j \in A$ and $j \in \overline{A}$ can be treated separately in proving that $\sum_{j \in A} \sum_{k=1}^{K} \pi_{jk|i} x_{jk} = 0$ and $\sum_{j \in \overline{A}} \sum_{k=1}^{K} \pi_{jk|i} x_{jk} = 0$. Note that this argument does not apply to $\mathbf{x} = \mathbf{x}_{A:B}(\Delta)$ in general, because π_{jk} need not be proportional to $\pi_{jk|i}$ for fixed i and k. ∎

Bibliography

[1] Abdellaoui, M., A. Baillon, L. Placido, and P. Wakker (2011) The rich domain of uncertainty: source functions and their experimental implementation. *American Economic Review* 101: 695–723.

[2] Afriat, S. (1980) *Demand Theory and the Slutsky Matrix*. Princeton University Press.

[3] Albers, D., C. Reid, and G. Dantzig (1986) An interview with George B. Dantzig: the father of linear programming. *The College Mathematics Journal* 17(4): 292–314.

[4] Allais, M. (1953) Le comportement de l'homme rationnel devant le risque: critique des postulats et axiomes de l'ecole Américaine. *Econometrica* 21: 503–546.

[5] Anscombe, F., and R. Aumann (1963) A definition of subjective probability. *Annals of Mathematical Statistics* 34: 199–205.

[6] Arrow, K. (1951a) An extension of the basic theorems of classical welfare economics. *Proceedings of the Second Berkeley Symposium on Mathematical Statistics and Probability*. University of California Press.

[7] Arrow, K. (1951b) *Social Choice and Individual Values*. John Wiley and Sons, New York.

[8] Arrow, K. (1951c) Alternative approaches to the theory of choice in risk-taking situations. *Econometrica* 19: 404–437.

[9] Arrow, K. (1953) The role of securities in the optimal allocation of risk-bearing. *Quarterly Journal of Economics* 31: 91–96.

[10] Arrow, K. (1965) *Aspects of the Theory of Risk Bearing*. Yrjo Jahnssonin Saatio, Helsinki.

[11] Arrow, K. (1971) *Essays in the Theory of Risk Bearing*. Markham Publishing Co., Chicago.

293

[12] Arrow, K. (1974) Optimal insurance and generalized deductibles. *Scandinavian Actuarial Journal* 1: 1–42.

[13] Arrow, K. (1986) Rationality of self and others in an economic system. *The Journal of Business* 59(4): S385–S399.

[14] Arrow, K. (1987) Economic theory and the hypothesis of rationality. In J. Eatwell, M. Milgrave, and P. Newman, eds., *The New Palgrave: A Dictionary of Economics*. MacMillan Press Limited, New York.

[15] Arrow, K., and G. Debreu (1954) Existence of an equilibrium for a competitive economy. *Econometrica* 22(3): 265–290.

[16] Augustin, T., F. Coolen, G. de Cooman, C. Matthias, and M. Troffaes (2014) *Introduction to Imprecise Probabilities*. John Wiley and Sons.

[17] Aumann, R. (1962) Utility theory without the completeness axiom. *Econometrica* 30: 445–462.

[18] Aumann, R. (1971) Letter to L. J. Savage. In: Drèze, J. (1987) *Essays on Economic Decisions Under Uncertainty*: 76–81. Cambridge University Press.

[19] Aumann, R. (1974) Subjectivity and correlation in randomized games. *Journal of Mathematical Economics* 1: 67–96.

[20] Aumann, R. (1976) Agreeing to disagree. *Annals of Statistics* 4: 1236–1239.

[21] Aumann, R. (1987) Correlated equilibrium as an expression of Bayesian rationality. *Econometrica* 55: 1–18.

[22] Baccelli, J. (2017) Do bets reveal beliefs? *Synthese* 194: 3393–3419.

[23] Baccelli, J. (2019) The problem of state-dependent utility. *The British Journal for the Philosophy of Science* 72(2): 617–634.

[24] Baccelli, J. (2021) Moral hazard, the Savage framework, and state-dependent utility. *Erkenntnis* 86: 367–387.

[25] Bachelier, L. (1900) Théorie de la spéculation. Ph.D. Thesis, Ecole Normale Supérieure de Paris, Paris.

[26] Ben-El-Mechaiekh, H., and R. Dimand (2011) A simpler proof of the von Neumann minimax theorem. *The American Mathematical Monthly* 118(7): 636–641.

[27] Bergemann, D. and S. Morris (2016) Bayes correlated equilibrium and the comparison of information structures in games. *Theoretical Economics* 11: 487–522.

[28] Bernheim, D. (1984) Rationalizable strategic behavior. *Econometrica* 52 (4): 1007–1028.

[29] Bernoulli, D. (1738) Specimen Theoriae Novae de Mensura Sortis. Commentarii Academiae Scientiarum Imperialis Petropolitanae 5: 175–192. Translation by L. Sommer (1954) Exposition of a new theory on the measurement of risk. *Econometrica* 22: 23–36.

[30] Bewley, T. (1986) Knightian decision theory Part I. Cowles Foundation Discussion Paper No. 807. Reprinted (2002) in *Decisions in Economics and Finance* 25: 79–110.

[31] Binmore, K. (2009) *Rational Decisions*. Princeton University Press.

[32] Black, F., and M. Scholes (1973) The pricing of options and corporate liabilities. *Journal of Political Economy* 81: 637– 654.

[33] Blackorby, C., D. Donaldson, and J. Weymark (1999) Harsanyi's social aggregation theorem for state-contingent alternatives. *Journal of Mathematical Economics* 32: 365–387.

[34] Blackwell, D. and M. Girshick (1954) *Theory of Games and Statistical Decisions*. John Wiley and Sons, New York.

[35] Bleichrodt, H., C. Li, I. Moscati, and P. Wakker (2016) Nash was a first to axiomatize expected utility. *Theory and Decision* 81: 309–312.

[36] Bonnesen, T., and W. Fenchel. (1934) *Theorie der konvexen Körper.*Pp. vii, 164. RM. 18.80. 1934. Ergebnisse der Mathematik, Band III, Heft 1. Springer.

[37] Border, K. (1985) More on Harsanyi's utilitarian cardinal welfare theorem. *Social Choice and Welfare* 1(4): 279–281.

[38] Brandenburger, A., and E. Dekel (1989) *The Role of Common Knowledge Assumptions in Game Theory*. In: Hahn, F., ed., *The Economics of Missing Markets, Information, and Games*: 46–61. Oxford University Press, Oxford.

[39] Breeden, D. and R. Litzenberger (1978) Prices of state contingent claims implicit in options prices. *Journal of Business* 51: 621–652.

[40] Brennan, M., and H. Cao (1996) Information, trade, and derivative securities. *Review of Financial Studies* 9(1): 163–208.

[41] Brennan, M. (1979 The pricing of contingent claims in discrete time markets, *Journal of Finance* 24(1): 53–68.

[42] Brillouin, L. (1964) *Scientific Uncertainty and Information*. Academic Press, New York.

[43] Brouwer, L. (1912) Uber abbildung von mannigfaltigkeiten. *Mathematische Annalen* 71: 97–115.

[44] Buehler, R. (1976) Coherent preferences. *Annals of Statistics* 4(6): 1051–1064.

[45] Cao, H., and H. Ou-Yang (2008) Differences of opinion of public information and speculative trading in stocks and options. *Review of Financial Studies* 22(1): 299–335.

[46] Calvo-Armengól, A. (2003) The set of correlated equilibria of 2×2 games. Barcelona Economics WP no. 79, Universitat Autònamo de Barcelona.

[47] Cassel, G. (1924) The theory of social economy. (tr. J. McAbe) *Economic Journal* 34(134): 235–240.

[48] Chew, S.H. (1983) A generalization of the quasilinear mean with applications to the measurement of income inequality and decision theory resolving the Allais paradox. *Econometrica* 51(4): 1065–1092.

[49] Chew, S.H., and J. Sagi (2008) Small worlds: modeling attitudes toward sources of uncertainty. *Journal of Economic Theory* 139: 1–24.

[50] Christensen, D. (1991) Clever bookies. *Philosophical Review* 100(2): 229–247.

[51] Conitzer, V., and T. Sandholm (2008) New complexity results about Nash equilibria. *Games and Economic Behavior* 63: 621–641.

[52] Cornfield, J. (1969) The Bayesian outlook and its application. *Annals of Statistics* 4: 1051–1064.

[53] Cox, J., and S. Ross (1976) The valuation of options for alternative stochastic processes. *Journal of Financial Economics* 3(1/2): 145–166.

[54] Dahl, G. (2012) Linear optimization and mathematical finance. University of Oslo, Department of Mathematics, Pure Mathematics, No. 4, https://www.uio.no/studier/emner/matnat/math/MAT2700/h12/dualitetmat 2700.pdf.

[55] Danan, E., T. Gajdos, and J.-M. Tallon (2015) Harsanyi's aggregation theorem with incomplete preferences. *American Economic Journal: Microeconomics* 7(1): 61–69.

[56] Dantzig, G. (1963) *Linear Programming and Extensions*. RAND Corporation, Santa Monica, CA https://www.rand.org/pubs/reports/R366.html.

[57] de Finetti, B. (1931) Sul significato soggettivo della probabilita. *Fundamenta Mathematicae* 17: 298–329.

[58] de Finetti, B. (1937) La prévision: ses lois logiques, ses sources subjectives. *Ann. Inst. Henri Poincaré* 7: 1–68. Translation reprinted in H. Kyburg and H. Smokler, eds. (1980) *Studies in Subjective Probability,* 2nd ed., Robert Krieger, New York, 53–118.

[59] de Finetti, B. (1952) Sulla preferibilità. *Giornale degli Economisti e Annali di Economia* 11: 658–709

[60] de Finetti, B. (1974) *Theory of Probability, Vol. 1.* John Wiley and Sons, New York.

[61] de Finetti, F. (2011) Bruno de Finetti, an Italian on the border. *Proceedings of 7th International Symposium on Imprecise Probability: Theories and Applications*, Innsbruck, Austria, 2011. https://isipta11.sipta.org/bruno_de_finetti/f001.pdf.

[62] De Meyer, B., and P. Mongin (1995) A note on affine aggregation. *Economics Letters* 47: 177–183.

[63] Debreu, G. (1951) The coefficient of resource utilization. *Econometrica* 19: 273–292.

[64] Debreu, G. (1954) Valuation equilibrium and Pareto optimum. *Proceedings of the National Academy of Sciences* 40(7): 588–592.

[65] Debreu, G. (1959) *Theory of Value: An Axiomatic Analysis Of Economic Equilibrium.* Yale University Press.

[66] Debreu, G. (1960) Une économique de l'incertain [Economics under uncertainty]. *Économie Appliquée* 13 (1): 111–116.

[67] Debreu, G. (1984) Economic theory in the mathematical mode. *American Economic Review* 74(3): 267–278.

[68] Denti, T., and L. Pomatto (2022) Model and predictive uncertainty: a foundation for smooth ambiguity preferences. *Econometrica* 90(2): 551–584.

[69] Diecidue, E., and P. Wakker (2001) On the intuition of rank-dependent expected utility. *Journal of Risk and Uncertainty* 23(3): 281–298.

[70] Diecidue, E., and P. Wakker (2002) Dutch books: avoiding strategic and dynamic complications, and a comonotonic extension. *Mathematical Social Sciences* 43: 135–149.

[71] Dimand, R., and H. Ben-El-Mechaiekh (2010) Von Neumann, Ville, and the minimax theorem. *International Game Theory Review* 12: 115–137.

[72] Dirac, P. (1930) *The Principles of Quantum Mechanics*. Oxford University Press.

[73] Dooley, P. (1983) Slutsky's equation is Pareto's solution. *History of Political Economy* 15: 513–517.

[74] Drèze, J. (1970) Market allocation under uncertainty. *European Economic Review* 2: 133–165.

[75] Drèze, J. (1987) *Essays on Economic Decision under Uncertainty*. Cambridge University Press.

[76] Drèze, J., and A. Rustichini (2004). State-dependent utility and decision theory. In: Barberà, S., Hammond, P.J., Seidl, C., eds., *Handbook of Utility Theory*. Springer, Boston.

[77] Dubra, J., F. Maccheroni, and E. Ok (2004) Expected utility theory without the completeness axiom. *Journal of Economic Theory* 115: 118–133.

[78] Duncan, G. (1977) A matrix measure of multivariate local risk aversion. *Econometrica* 45(4): 893–903.

[79] Dyer, J., and R. Sarin (1979) Measurable multi-attribute value functions. *Operations Research* 27: 810–820.

[80] Edgeworth, F. (1881) *Mathematical Psychics: An Essay on the Application of Mathematics to the Moral Sciences*. Kegan Paul, London.

[81] Edwards, W. (1953) Probability-preferences in gambling. *American Journal of Psychology* 66: 349–364.

[82] Edwards, W. (1954) The theory of decision-making. *Psychological Bulletin* 51: 380–417.

[83] Edwards, W. (1955) The prediction of decisions among bets. *Journal of Experimental Psychology* 50: 201–214.

[84] Edwards, W. (1962) Subjective probabilities inferred from decisions. *Psychological Review* 69: 109–135.

[85] Einstein, A. (1905) On the movement of small particles suspended in stationary liquids required by the molecular-kinetic theory of heat. *Annals of Physics* 322: 549–560.

[86] Ellsberg, D. (1954) Classic and current notions of 'measurable utility'. *Economic Journal* 64(255): 528–556.

[87] Ellsberg, D. (1961) Risk, ambiguity and the Savage axioms. *Quarterly Journal of Economics* 75: 643–669.

[88] Epstein, L. (2010) A paradox for the 'smooth ambiguity' model of preference. *Econometrica* 78: 2085–2099.

[89] Epstein, L. and M. Schneider (2003) Recursive multiple-priors. *Journal of Economic Theory* 113(1): 1–31.

[90] Ergin H., and F. Gul (2009) A subjective theory of compound lotteries. *Journal of Economic Theory* 144: 899–929.

[91] Etner, J., M. Jeleva, and J.M. Tallon (2012) Decision theory under ambiguity. *Journal of Economic Surveys* 26(2): 234–270.

[92] Fine, T. (1973) *Theories of Probability*. Academic Press, New York, London.

[93] Farkas, J. (1902) Theorie der Einfachen Ungleichungen. *Journal für die reine und angewandte Mathematik* 124: 1–27.

[94] Fishburn, P. (1970) *Utility Theory for decision-making*. John Wiley and Sons.

[95] Fishburn, P. (1975a) A theory of subjective expected utility with vague preferences. *Theory and Decision* 6(3): 287–310.

[96] Fishburn, P. (1975b) Separation theorems and expected utilities. *Journal of Economic Theory* 11: 16–34.

[97] Fishburn, P. (1978) On Handa's "New Theory of Cardinal Utility" and the maximization of expected return. *Journal of Political Economy* 86(2): 321–324.

[98] Fishburn, P. and P. Wakker (1995) The invention of the independence condition for preferences. *Management Science* 41: 1130–1144.

[99] Fleurbaey, M. (2018) Welfare economics, risk, and uncertainty. *Canadian Journal of Economics* 51(1): 5–40.

[100] Forges, F. (1986) Correlated equilibria in games with incomplete information: a model with verifiable types. *International Journal of Game Theory* 15: 65–82.

[101] Forges, F. (1993) Five legitimate definitions of correlated equilibria in games with incomplete information. *Theory and Decision* 35: 277–310.

[102] Forges, F. (2006) Correlated equilibrium in games with incomplete information revisited. *Theory and Decision* 61: 329–344.

[103] Freedman, D. and R. Purves: (1969) Bayes' method for bookies. *Annals of Mathematical Statistics* 40: 1419–1429.

[104] Friedman, J., ed. (1996) *The Rational Choice Controversy*. Yale University Press.

[105] Frisch, R. (1926) Kvantitativ formulering av den teoretiske økonomikks lover [Quantitative formulation of the laws of economic theory]. *Statsøkonomisk Tidsskrift* 40: 299–334.

[106] Fudenberg, D., and J. Tirole (1991) *Game Theory*. MIT Press.

[107] Galaabaatar, T., and E. Karni (2012) Expected multi-utility representations. *Mathematical Social Sciences* 64: 242–246.

[108] Galaabaatar, T., and E. Karni (2013) Subjective expected utility with incomplete preferences. *Econometrica* 81(1): 255–284.

[109] Galbraith, K. (1975) *Money: Whence It Came, Where It Went*. Princeton University Press.

[110] Gale, D. (1960) *The Theory of Linear Economic Models*. McGraw-Hill, Inc., New York.

[111] Gale, D., H. Kuhn, and A.Tucker (1951) Linear programming and the theory of games. In: T. Koopmans, ed., *Activity Analysis of Production and Allocation*: 317–329. John Wiley and Sons, New York.

[112] Garman, M. (1980) A synthesis of the pure theory of arbitrage. Research Program in Finance Working Papers 98, University of California at Berkeley.

[113] Gelfand, A., and A. Smith (1990) Sampling-based approaches to calculating marginal densities. *Journal of the American Statistical Association* 85(410): 398–409.

[114] Gilboa, I. (2009) *Theory of Decision under Uncertainty*. Cambridge University Press.

[115] Gilboa, I. (2010) *Rational Choice*. MIT Press.

[116] Gilboa, I., and M. Marinacci (2016) Ambiguity and the Bayesian paradigm. In: Arló-Costa, H., V. Hendricks, and J. van Benthem, eds., *Readings in Formal Epistemology:* 385–439. Springer, Berlin.

[117] Gilboa, I., and D. Schmeidler (1989) Maxmin expected utility with a non-unique prior. *Journal of Mathematical Economics* 18: 141–153.

[118] Goedel, K. (1931) Über formal unentscheidbare Sätze der Principia Mathematica und verwandter Systeme, I. In: Solomon Feferman, ed. (1986) *Kurt Gödel Collected Works, Vol. I*: 144–195. Oxford University Press.

[119] Goetzmann, W. (2016) *Money Changes Everything: How Finance Made Civilization Possible*. Princeton University Press.

[120] Goldstein, M. (1983) The prevision of a prevision. *Journal of the American Statistical Association* 78: 817–819.

[121] Goldstein, M. (1985) Temporal Coherence. In: J. Bernardo, M DeGroot, D. Lindley, A. Smith, eds., *Bayesian Statistics 2*: 231–248. North-Holland.

[122] Good, I. (1962) Subjective probability as the measure of a non-measurable set. In: E. Nagel, P. Suppes, and A. Tarski, eds., *Logic, Methodology, and Philosophy of Science*. Stanford University Press. Reprinted in H. Kyburg and H. Smokler, eds. (1980) *Studies in Subjective Probability*, 2nd ed.: 133–146. Robert Krieger, New York.

[123] Grant, S., A Kajii, and B. Polak (1998) Intrinsic preference for information. *Journal of Economic Theory* 83: 233–259.

[124] Green, D., and I. Shapiro (1994) *Pathologies of Rational Choice Theory: A Critique of Applications in Political Science*. Yale University Press.

[125] Hacking, I. (1967) Slightly more realistic personal probability. *Philosophy of Science* 34(4): 311–325.

[126] Hacking, I. (1975) *The Emergence of Probability: A Philosophical Study of Early Ideas About Probability, Induction and Statistical Inference.* Cambridge Series on Statistical and Probabilistic Mathematics.

[127] Hájek, A. (2005) Scotching Dutch books? *Philosophical Perspectives* 19, Epistemology: 139–151.

[128] Hájek, A. (2009) Dutch book arguments. In: P. Anand, P. Pattanaik, C. Puppe., eds., *The Handbook of Rational and Social Choice*: 173–195. Oxford University Press.

[129] Handa, J. (1977) Risk, probabilities, and a new theory of cardinal utility. *Journal of Political Economy* 85(1): 97–122.

[130] Harrison, J. and D. Kreps (1979) Martingales and arbitrage in multiperiod securities markets. *Journal of Economic Theory* 2(3): 381–408.

[131] Harsanyi, J. (1955) Cardinal welfare, individualistic ethics, and interpersonal comparisons of utility. *Journal of Political Economy* 63(4): 309–321.

[132] Harsanyi, J. (1967) Games with incomplete information played by "Bayesian" players, I–III. Part I. The basic model. *Management Science* 14(3): 159–182.

[133] Harsanyi, J. (1968a) Games with incomplete information played by "Bayesian" players, I–III. Part II. Bayesian equilibrium points. *Management Science* 14(5): 320–334.

[134] Harsanyi, J. (1968b) Games with incomplete information played by "Bayesian" players, I–III. Part III. The basic probability distribution of the game. *Management Science* 14(7): 486–502.

[135] Harsanyi, J. (1982a) Subjective probability and the theory of games: comments on Kadane and Larkey's paper. *Management Science* 28(2): 120–124.

[136] Harsanyi, J. (1982b) Rejoinder to Professors Kadane and Larkey. *Management Science* 28(2): 124–125.

[137] Hart, S. and David Schmeidler (1989) Existence of correlated equilibria. *Mathematics of Operations Research* 14(1): 18–25.

[138] Heath, D. and W. Sudderth (1972) On a theorem of de Finetti, odds making, and game theory. *Annals of Mathematical Statistics* 43: 2072–2077.

[139] Hilbert, D. (1902) *Grundlagen der geometrie.* Teubner, Leipzig. Translated by E. Townsend (1950): https://math.berkeley.edu/~wodzicki/160/Hilbert.pdf.

[140] Hirshleifer, J. (1965) Investment decision under uncertainty: choice-theoretic approaches. *Quarterly Journal of Economics* 74: 509–536.

[141] Hirshleifer, J. (1966) Investment decision under uncertainty: applications of the state-preference approach. *Quarterly Journal of Economics* 80(2): 252–277.

[142] Howard, R. (1968) The foundations of decision analysis. *IEEE Transactions on Systems Science and Cybernetics* SSC-4(3): 211–219.

[143] Howard, R. (1988) Decision analysis: practice and promise. *Management Science* 34(6): 679–695.

[144] Jahoda, G. (2007) *A History of Social Psychology: From the Eighteenth-Century Enlightenment to the Second World War.* Cambridge University Press.

[145] Jose, V., R. Nau, and R. Winkler (2008) Scoring rules, generalized entropy, and utility maximization. *Operations Research* 56(5): 1148–1157.

[146] Kadane, J., and P. Larkey (1982a) Subjective probability and the theory of games. *Management Science* 28: 113–120.

[147] Kadane, J., and P. Larkey (1982b) Reply to Professor Harsanyi. *Management Science* 28: 124.

[148] Kadane, J., and P. Larkey (1983) The confusion of is and ought in game theoretic contexts. *Management Science* 29(12): 1365–1379.

[149] Kahneman, D., and A. Tversky (1979) Prospect theory: an analysis of decision under risk. *Econometrica* 47: 263–291.

[150] Kakutani, S. (1941) A generalization of Brouwer's fixed point theorem. *Duke Mathematical Journal* 8(3): 457–459.

[151] Karni, E (1979) On multivariate risk aversion. *Econometrica* 47: 1391–1401.

[152] Karni, E. (1985) *Decision-Making under Uncertainty: The Case of State-Dependent Preferences.* Harvard University Press.

[153] Karni, E. (1993) Subjective expected utility theory with state-dependent preferences. *Journal of Economic Theory* 60: 428–438.

[154] Karni, E. (1996) Probabilities and beliefs. *Journal of Risk and Uncertainty* 13: 249–262.

[155] Karni, E. (2003) On the Representation of beliefs by probabilities. *Journal of Risk and Uncertainty* 26: 17–38.

[156] Karni, E. (2008) On optimal insurance in the presence of moral hazard. *The Geneva Risk and Insurance Review* 33(1): 1–18.

[157] Karni, E. (2011a) Bayesian decision-making and the representation of beliefs. *Economic Theory* 48: 125–146.

[158] Karni, E. (2011b) Subjective probabilities on a state space. *American Economic Journal: Microeconomics* 3: 172–185.

[159] Karni, E. (2011c) Continuity, completeness and the definition of weak preferences. *Mathematical Social Sciences* 62(2): 123–125.

[160] Karni, E. and P. Mongin (2000) On the determination of subjective probabilities by choices. *Management Science* 46: 233–248.

[161] Karni, E., and D. Schmeidler (1993) On the uniqueness of subjective probabilities. *Economic Theory* 3: 267–277.

[162] Karni, E., and D. Schmeidler (2016) An expected utility theory for state-dependent preferences. *Theory and Decision* 81: 467–478.

[163] Karni, E., D. Schmeidler, and K. Vind (1983) On state dependent preferences and subjective probabilities. *Econometrica* 51: 1021–1031.

[164] Karni, E., and J. Weymark (1998) An informationally parsimonious impartial observer theorem. *Social Choice and Welfare* 15: 321–332.

[165] Keeney, R., and R. Nau (2011) A theorem for Bayesian group decisions. *Journal of Risk and Uncertainty* 43: 1–17.

[166] Kennedy, R., and C. Chihara (1979) The Dutch Book Argument: its logical flaws, its subjective sources. *Philosophical Studies* 36: 19–33.

[167] Keynes, J. (1921) *A Treatise on Probability*. Dover Publications, Mineola, New York.

[168] Kjeldsen, T. (2001) John von Neumann's conception of the minimax theorem: a journey through different mathematical contexts. *Archive for History of Exact Sciences* 56: 39–68.

[169] Klibanoff, P., M. Marinacci, and S. Mukerji (2005) A smooth model of decision-making under ambiguity. *Econometrica* 73:1849–1892.

[170] Klibanoff, P., M. Marinacci, and S. Mukerji (2011a) On the smooth ambiguity model: a reply. *Econometrica* 80(3): 1303–1321.

[171] Klibanoff, P., M. Marinacci, and S. Mukerji (2011b) Definitions of ambiguous events and the smooth ambiguity model. *Economic Theory* 48(2/3): 399-424. Symposium on the 50th Anniversary of the Ellsberg Paradox.

[172] Klibanoff, P., S. Mukerji, and K. Seo (2014) Perceived ambiguity and relevant measures. *Econometrica* 82(5): 1945–1978.

[173] Klibanoff, P., S. Mukerji, K. Seo, and L. Stanca (2022) Foundations of ambiguity models under symmetry: alpha-MEU and smooth ambiguity. *Journal of Economic Theory* 199: 105202.

[174] Knight, F. (1921) *Risk, Uncertainty and Profit.* University of Chicago Press.

[175] Köbberling, V., and P. Wakker (2003) Preference foundations for nonexpected utility: a generalized and simplified technique. *Mathematics of Operations Research* 28: 395–423.

[176] Koopman, B. (1940) The axioms and algebra of intuitive probability. *Annals of Mathematics* 41: 269–292.

[177] Krantz, D., D. Luce, P. Suppes, and A. Tversky (1971) *Foundations of Measurement, Vol. 1: Additive and Polynomial Representations.* Academic Press, New York.

[178] Kreps, D. (1988) *Notes on the Theory of Choice.* Westview Press, Boulder.

[179] Kreps, D. (1990) *Game Theory and Economic Modeling.* Clarendon Press.

[180] Kreps, D. and E. Porteus (1979) Temporal von Neumann-Morgenstern and induced preferences. *Journal of Economic Theory* 20: 89–109.

[181] Kyburg, H. (1974) *The Logical Foundations of Statistical Inference.* Reidel.

[182] Kyburg, H. (1978) Subjective probability: Criticisms, reflections, and problems. *Journal of Philosophical Logic* 7(1): 157–180.

[183] Laraki, R., J. Renault, and S. Sorin (2019) *Mathematical Foundations of Game Theory.* Springer.

[184] Lehman, R. (1955) On confirmation and rational betting. *Journal of Symbolic Logic* 20(3): 251–262.

[185] Lewis, D. (1969) *Convention: A Philosophical Study*. Harvard University Press, Cambridge, MA.

[186] Leonard, R. (2010) *Von Neumann, Morgenstern, and the Creation of Game Theory: From Chess to Social Science, 1900-1960*. Cambridge University Press.

[187] Leontief, W. (1947) A note on the interrelation of subsets of independent variables of a continuous function with continuous first derivatives. *Bulletin of the American Mathematical Society* 53: 343–350.

[188] Levi, I. (1974) On indeterminate probabilities. *Journal of Philosophy* 71(13): 391–418.

[189] Levi, I. (1980) *The Enterprise of Knowledge*. MIT Press.

[190] Lindley, D. (1980) L. J. Savage–his work in probability and statistics. *Annals of Statistics* 8(1): 1–24.

[191] Lintner, J. (1969) The aggregation of investor's diverse judgments and preferences in purely competitive security markets. *Journal of Financial and Quantitative Analysis* 4(4): 347–400.

[192] Luce, R., and H. Raiffa (1957) *Games and Decisions*. John Wiley and Sons, New York.

[193] Maccheroni, F., M. Marinacci, and A. Rustichini (2006) Ambiguity aversion, robustness, and the variational representation of preferences. *Econometrica* 74: 1447–1498.

[194] Machina, M. (1982) 'Expected utility' analysis without the independence axiom. *Econometrica* 50(2): 277–323.

[195] Machina, M. (2004) Almost-objective uncertainty. *Economic Theory* 24: 1–54.

[196] Machina, M. (2005) 'Expected utility/subjective probability' analysis without the sure-thing principle or probabilistic sophistication. *Economic Theory* 26: 1–62.

[197] Machina, M., and D. Schmeidler (1992) A more robust definition of subjective probability. *Econometrica* 60: 745–780.

[198] Malinvaud, E. (1952) Note on von Neumann-Morgenstern's strong independence Axiom. *Econometrica* 20(4): 679–679.

[199] Mandler, M. (2005) Harsanyi's utilitarianism via linear programming. *Economics Letters* 88: 85–90.

[200] Marschak, J. (1950) Rational behavior, uncertain prospects, and measurable
utility. *Econometrica* 18(2): 111–141.

[201] Mas-Colell, A., M. Whinston, and J. Green (1995) *Microeconomic Theory*.
Oxford University Press.

[202] Maschler, M., E. Solan, and S. Zamir (2020) *Game Theory, 2nd ed.*. Cambridge University Press.

[203] Merton, R. (1973) The theory of rational option pricing. *Bell Journal of Economics and Management Science* 4: 141–183.

[204] Minkowski, H. (1911) Theorie der Konvexen Korper, Insbesondere Begrun-
dung ihres Oberflachenbegriffs. In: *Gesammelte Abhandlungen von Hermann
Minkowski* (D. Hilbert, A. Speiser and H. Weyl, Eds.), Vol. 2: 131–229. Teubner, Leipzig.

[205] Mirowski, P. (1989) *More Heat than Light: Economics as Social Physics,
Physics as Nature's Economics*. Cambridge University Press.

[206] Mongin, P. (1994) Harsanyi's Aggregation Theorem: multi-profile version
and unsettled questions. *Social Choice and Welfare* 11: 331–354.

[207] Mongin, P. (1995) Consistent Bayesian aggregation. *Journal of Economic
Theory* 66: 313–351.

[208] Mongin, P., (1998) The paradox of the Bayesian experts and state-dependent
utility theory. *Journal of Mathematical Economics* 29:331–361.

[209] Mongin, P., and M. Pivato (2016) Social Evaluation under Risk and Uncertainty. In: Adler, M. and M. Fleurbaey, eds., *The Oxford Handbook of Well-
Being and Public Policy*. Oxford University Press.

[210] Moulin, H., and J.-P. Vial (1978) Strategically zero-sum games: the class of
games whose completely mixed equilibria cannot be improved upon. *International Journal of Game Theory* 7(3–4): 201–221.

[211] Myerson, R. (1985) Bayesian equilibrium and incentive compatibility In: L.
Hurwicz, D. Schmeidler, and H. Sonnenschein, eds., *Social Goals and Social
Organization*: 229–259. Cambridge University Press.

[212] Myerson, R. (1986) Multistage games with communication. *Econometrica* 54: 323–358.

[213] Myerson, R. (1999) Nash equilibrium and the history of economic theory. *Journal of Economic Literature* Vol. XXXVII: 1067–1082.

[214] Nasar, S. (1998) *A Beautiful Mind: A Biography of John Forbes Nash Jr., Winner of the Nobel Prize in Economics.* Simon & Schuster.

[215] Nash, J. (1950a) Equilibrium points in *n*-person games. *Proceedings of the National Academy of Sciences* 36(1): 48–49.

[216] Nash, J. (1950b) The bargaining problem. *Econometrica* 18(2): 155–162.

[217] Nash, J. (1951) Non-cooperative games. *Annals of Mathematics* 54(2): 286–295.

[218] Nash, J. (1953) Two-person cooperative games. *Econometrica* 21(1): 128–140.

[219] Nau, R. (1989) Decision analysis with indeterminate or incoherent probabilities. *Annals of Operations Research* 19: 375–403.

[220] Nau, R. (1992a) Joint coherence in games of incomplete information. *Management Science* 38(3): 374–387.

[221] Nau, R. (1992b) Indeterminate probabilities on finite sets. *Annals of Statistics* 20(4): 1737–1767.

[222] Nau, R. (1995a) Coherent decision analysis with inseparable probabilities and utilities. *Journal of Risk and Uncertainty* 10: 71–91.

[223] Nau, R. (1995b) The incoherence of agreeing to disagree. *Theory and Decision* 39: 219–239.

[224] Nau. R. (2001) De Finetti was right: probability does not exist. *Theory and Decision* 51: 89–124.

[225] Nau, R. (2002) The aggregation of imprecise probabilities. *Journal of Statistical Planning and Inference* 105: 265–282.

[226] Nau, R. (2003) A generalization of Pratt-Arrow measure to non-expected-utility preferences and inseparable probability and utility. *Management Science* 49(8): 1089–1104.

[227] Nau, R. (2006a) Uncertainty aversion with second-order utilities and probabilities. *Management Science* 52(1): 136–145.

[228] Nau, R. (2006b) The shape of incomplete preferences. *Annals of Statistics* 34(5): 2430–2448.

[229] Nau, R. (2011a) Imprecise probabilities in noncooperative games. *Proceedings of the Seventh International Symposium on Imprecise Probability: Theories and Applications.*

[230] Nau, R. (2011b) Risk, ambiguity, and state-preference theory. *Economic Theory* 49(1): 437–467.

[231] Nau, R. (2015) Risk-neutral equilibria of noncooperative games. *Theory and Decision* 78(2): 171–188.

[232] Nau, R., and K. McCardle (1990) Coherent behavior in noncooperative games. *Journal of Economic Theory* 50(2): 424–444.

[233] Nau, R., and K. McCardle (1991) Arbitrage, rationality, and equilibrium. *Theory and Decision* 31: 199–240.

[234] Nau, R., S. Gomez Canovas, and P. Hansen (2004) On the geometry of Nash equilibria and correlated equilibria. *International Journal of Game Theory* 32: 443–453.

[235] Nau, R., V. Jose, and R. Winkler (2007) Scoring rules, entropy, and imprecise probabilities. *Proceedings of the Fifth International Symposium on Imprecise Probabilities and Their Applications.*

[236] Nau, R., V. Jose, and R. Winkler (2009) Duality between maximization of expected utility and minimization of relative entropy when probabilities are imprecise. *Proceedings of the Sixth International Symposium on Imprecise Probability: Theory and Applications.*

[237] Ok, E., P. Ortoleva, and G. Riella (2012) Incomplete preferences under uncertainty: indecisiveness in beliefs vs. tastes. *Econometrica* 80: 1791–1808.

[238] Pareto, V. (1906) *Manuale di economia politica con una introduzione alla scienza sociale.* Translated as: *Manual of Political Economy.* Augustus M. Kelley, 1971.

[239] Peano, G. (1889) Arithmetices principia: nova methodo exposita. Translated as "The principles of arithmetic, presented by a new method" in J. van Heijenoort, 1967. *A Source Book in Mathematical Logic, 1879–1931.* Harvard University Press: 83–97.

[240] Pearce, D. (1984) Rationalizable strategic behavior and the problem of perfection. *Econometrica* 52(4): 1029–1050.

[241] Perng, C. (2017) On a class of theorems equivalent to Farkas's lemma. *Applied Mathematical Sciences* 11: 2175–2184.

[242] Peterson, M. (2017) *Introduction to Decision Theory. Cambridge University Press.*

[243] Pettigrew, R. (2020) Dutch Book Arguments. *Cambridge University Press.*

[244] Pierce, D. (1973) On some difficulties in a frequency theory of inference. *Annals of Statistics* 1: 241–250.

[245] Poundstone, W. (1992) *Prisoner's Dilemma.* Doubleday, New York.

[246] Pratt, J. (1964) Risk aversion in the small and in the large. *Econometrica* 32(1): 122–136.

[247] Preston, M. and P. Baratta (1948) An experimental study of the auction-value of an uncertain outcome. *American Journal of Psychology* 61: 183–193.

[248] Qin, Z. (2013) Speculations in option markets enhance allocation efficiency with heterogeneous beliefs and learning. *Journal of Banking & Finance* 37(12): 4675–4694.

[249] Quetelet, A. (1835) *Sur l'homme et le développement de ses facultés, ou essai de physique sociale.* Paris : Bachelier, imprimeur-libraire, quai des Augustins, no. 55.

[250] Quiggin, J. (1982) A theory of anticipated utility. *Journal of Economic Behavior and Organization* 3: 323–343.

[251] Radner, R. (1968) Competitive equilibrium under uncertainty. *Econometrica* 36(1): 31–58.

[252] Raiffa, H. (1968) *Decision Analysis: Introductory Lectures on Choices Under Uncertainty.* Addison-Wesley, Reading, Mass.

[253] Ramsey, F. (1926) Truth and Probability. In: R. Braithwaite, ed. (1931) *The Foundations of Mathematics and Other Logical Essays*: 156–198. K. Paul, Trench, Truber and Co., London.

[254] Rigotti L., and C. Shannon (2005) Uncertainty and risk in financial markets. *Econometrica* 73: 203–243.

[255] Rosenberg, J., and R. Engel (2002) Empirical pricing kernels. *Journal of Financial Economics* 64: 341–372.

[256] Ross, S. (1978) A simple approach to the valuation of risky streams. *Journal of Business* 51(3): 453–475.

[257] Rubin, H. (1987) A weak system of axioms for 'rational' behavior and the non-separability of utility from prior. *Statistics and Decisions* 5: 47–58.

[258] Rubinstein, M. (1976) The valuation of uncertain income streams and the pricing of options. *Bell Journal of Economics and Management Science* 7: 407–425.

[259] Rubinstein, M. (2006) *A History of the Theory of Investments: My Annotated Bibliography*. John Wiley and Sons.

[260] Russell, B., and A. Whitehead (1910, 1912, 1913) *Principia Mathematica*. Cambridge University Press.

[261] Samuelson, P. A. (1947) *Foundations of Economic Analysis*. Harvard University Press.

[262] Samuelson, P. (1950) Probability and the attempts to measure utility. *The Economic Review* 1(3): 169–70 [Tokyo].

[263] Samuelson, P., and R. Merton (1969) A complete model of warrant pricing that maximizes utility. *Industrial Management Review* 10: 17–46.

[264] Sarin, R., and P. Wakker (1998) Dynamic choice and nonexpected utility. *Journal of Risk and Uncertainty* 17: 87–119.

[265] Savage, L. (1954) *The Foundations of Statistics*. John Wiley and Sons, New York.

[266] Savage, L. (1971) Elicitation of personal probabilities and expectations. *Journal of the American Statistical Association* 66 (336): 783–801.

[267] Schervish, M., T. Seidenfeld, and J. Kadane (1990) State-dependent utilities. *Journal of the American Statistical Association* 85(411): 840–847.

[268] Schervish, M., T. Seidenfeld, and J. Kadane (1991) Shared preferences and state-dependent utilities. *Management Science* 37: 1575–1589.

[269] Schervish, M., T. Seidenfeld, and J. Kadane (2002) Measuring incoherence. *Sankhyā: The Indian Journal of Statistics, Series A* 64(3): 561–587.

[270] Schervish, M., T. Seidenfeld, and J. Kadane (2008) The fundamental theorems of prevision and asset pricing. *International Journal of Approximate Reasoning* 49: 148–158.

[271] Schervish, M., T. Seidenfeld, and J. Kadane (2013) The effect of exchange rates on statistical decisions. *Philosophy of Science* 80(4): 504–532.

[272] Schick, F. (1986) Dutch books and money pumps. *Journal of Philosophy* 83(2): 112–119.

[273] Schmeidler, D. (1989) Subjective probability and expected utility without additivity. *Econometrica* 57: 571–587.

[274] Scott, D. (1964) Measurement structures and linear inequalities. *Journal of Mathematical Psychology* 1: 233–247.

[275] Segal, U. (1987) The Ellsberg paradox and risk aversion: an anticipated utility approach. *International Economic Review* 28(1): 175–202.

[276] Segal, U. (1990) Two-stage lotteries without the reduction axiom. *Econometrica* 58(2): 349–377.

[277] Seidenfeld, T., J. Kadane, and M. Schervish (1989) On the shared preferences of two Bayesian decision-makers. *Journal of Philosophy* 86: 225–244.

[278] Seidenfeld, T., M. Schervish, and J. Kadane (1990a) Decisions without ordering. In: W. Sieg, ed., *Acting and Reflecting*: 143–170. Kluwer Academic Press, Boston.

[279] Seidenfeld, T., M. Schervish, and J. Kadane (1990b) When fair betting odds are not degrees of belief. *Philosophy of Science* 169(1): 517–524.

[280] Seidenfeld, T., M. Schervish, and J. Kadane (1995) A representation of partially ordered preferences. *Annals of Statistics* 23(6): 2168–2217.

[281] Seo, K. (2009) Ambiguity and second-order belief. *Econometrica* 77: 1575–1605.

[282] Shackle, G. (1959) Time and thought. *The British Journal for the Philosophy of Science* 9: 285–298.

[283] Smith, C. (1961) Consistency in statistical inference and decision. *Journal of the Royal Statistical Society, Series B*, 23: 1–37.

[284] Smith, J., and R. Nau (1995) Valuing risky projects: option pricing theory and decision analysis. *Management Science* 41(5): 795–816.

[285] Solan, E., and R. Vohra (2002) Correlated equilibrium payoffs and public signalling in absorbing games. *International Journal of Game Theory* 31: 91–121.

[286] Soltan, V. (2021) Separating hyperplanes of convex sets. *Journal of Convex Analysis*. 28: 1015–1032.

[287] Sono, M. (1943) The effect of price changes on the demand and supply of separable goods. [In Japanese] *Kokumin Keisai Zasshi* 74: 1–51.

[288] Stapleton, R., and M. Subrahmanyam (1984) The valuation of multivariate contingent claims in discrete time models. *Journal of Finance* 39(1): 207–228.

[289] Stern, N. (1986) A note on commodity taxation: the choice of variable and the Slutsky, Hessian, and Antonelli matrices (SHAM). *Review of Economic Studies* 53: 293–299.

[290] Stigler, G. (1950) The development of utility theory: I; II. *Journal of Political Economy* 58: 307–327; 373–396.

[291] Suh, S. and F. Zapatero (2008) A class of quadratic options for exchange rate stabilization. *Journal of Economic Dynamics and Control* 32(11): 3478–3501.

[292] Suppes, P. (1974) The measurement of belief. *Journal of the Royal Statistical Society B* 36: 160–175.

[293] Trautmann, S., and G. van de Kuilen (2015) Ambiguity attitudes. In: Keren, G. and G. Wu, eds., *The Wiley Blackwell Handbook of Judgment and decision-making (Ch. 3)*: 89–116. Blackwell, Oxford, UK.

[294] Turner, J., L. Beeghley, and C. Powers (2011) *The Emergence of Sociological Theory*. SAGE Publications, Inc.

[295] Tversky, A., and C. Fox (1995) Weighing risk and uncertainty. *Psychological Review* 102: 269–283.

[296] Tversky, A., and D. Kahneman (1992) Advances in prospect theory: cumulative representation of uncertainty. *Journal of Risk and Uncertainty* 5: 297–323.

[297] van Fraassen, C. (1984) Belief and the will. *Journal of Philosophy* 81(5): 235–256.

[298] Varian, H. (1987) The arbitrage principle in financial economics. *Journal of Economic Perspectives* 1: 55–72.

[299] Vicig, P., and T. Seidenfeld (2012) Bruno de Finetti and imprecision: Imprecise probability does not exist! *International Journal of Approximate Reasoning* 53: 1115–1123.

[300] Ville, J. (1938) Sur la théorie génerale des jeux où intervient Ĩhabileté des joueurs. In: É. Borel et al., eds., *Traité du Calcul des Probabilités et de ses Applications*: 105–117. Gauthier-Villars, Paris.

[301] Vineberg, S. (1997) Dutch books, Dutch strategies, and what they show about rationality. *Philosophical Studies* 86(2): 185–201.

[302] Vineberg, S. (2022) Dutch book arguments. In: E. Zalta and U. Nodelman, eds., *The Stanford Encyclopedia of Philosophy*. https://plato.stanford.edu/archives/fall2022/entries/dutch-book/.

[303] von Neumann, J. (1928) Zur theorie der gesellschaftsspiele. *Mathematische Annalen* 100: 295–320.

[304] von Neumann, J. (1932) *Mathematical Foundations of Quantum Mechanics*. Springer-Verlag, Berlin.

[305] von Neumann, J. (1932) A Model of General Economic Equilibrium. *Review of Economic Studies* 13(1): 1–9 (1945–1946).

[306] von Neumann, J. (1937) Über Ein Ökonomisches Gleichungssystem Und Eine Verallgemeinerung Des Brouwerschen Fixpunktsatzes. In: K. Menger, ed., *Ergebnisse Eines Mathematischen Kolloquiums* 8: 73–83. Franz Deuticke, Leipzig.

[307] von Neumann, J., and O. Morgenstern (1944) *Theory of Games and Economic Behavior*. Princeton University Press.

[308] Wakker, P. (1987) Subjective probabilities for state dependent continuous utility. *Mathematical Social Sciences* 14(3): 289–298.

[309] Wakker, P. (1989) *Additive Representations of Preferences: A New Foundation of Decision Analysis*. Kluwer Academic Publishers, Dordrecht.

[310] Wakker, P. (2010) *Prospect Theory for Risk and Ambiguity*. Cambridge University Press, Cambridge.

[311] Wakker, P., and A. Tversky (1993) An axiomatization of cumulative prospect theory. *Journal of Risk and Uncertainty* 7: 147–176.

[312] Wakker, P., and H. Zank (1999) State dependent expected utility for Savage's state space. *Mathematics of Operations Research* 24(1): 8–34.

[313] Walley, P. (1991) *Statistical Reasoning With Imprecise Probabilities*. Chapman & Hall, London.

[314] Weatherford, R. (1982) *Philosophical Foundations of Probability Theory*. Routledge, London.

[315] Weymark, J. (1993) Harsanyi's social aggregation theorem and the weak Pareto principle. *Social Choice and Welfare* 10: 209–221.

[316] Weymark, J. (1994) Harsanyi's social aggregation theorem with alternative Pareto principles. In: W. Eichhorn, ed., *Models and Measurement of Welfare and Inequality*. Springer-Verlag, Berlin, Heidelberg.

[317] Weymark, J. (1995) Further remarks on Harsanyi's social aggregation theorem and the weak Pareto principle. *Social Choice and Welfare* 12: 87–92.

[318] Wiener, N. (1923) Differential space. *Journal of Mathematics and Physics* 58: 131–174.

[319] Wiener, N. (1948) *Cybernetics; or Control and Communication in the Animal and the Machine*. John Wiley and Sons.

[320] Williams, P. M. (1976) Indeterminate probabilities. In: M. Prezelecki et al., eds., *Formal Methods in the Methodology of Empirical Sciences*: 229–246. Reidel, Dordrecht.

[321] Yaari, M. (1969) Some remarks on measures of risk aversion and their uses. *Journal of Economic Theory* 1: 315–329.

[322] Yaari, M. (1987) The dual theory of choice under risk. *Econometrica* 55: 95–115.

[323] Zermelo, E. (1908) Untersuchungen über die Grundlagen der Mengenlehre, I, *Mathematische Annalen* 65: 261–281. [Studies on the foundations of set theory]

Author Index

Subject Index

acceptable bet
    $-bet, 63
    common knowledge property, 64
    p-bet, *88*
    v-bet, 115
additively-separable utility, 29
affine independence, 103
ambiguity aversion
    as 2nd-order uncertainty, 164
    as source-dependent utility, 154
    Ellsberg's paradox, 150
Anscombe-Aumann model, 9, 29, 104,
        106, 112, 113
Arrow-Pratt measure of risk aversion,
        18, 96, 134
    matrix-valued (in state-preference
        theory), 139, 141
    state-dependent, 140, 141, 144
asset pricing
    finite-states model, 236, 274
    fundamental theorem, 45, 236, 274
    normal/exponential/quadratic
        model, 241, 275
    pricing kernel, 248
    stochastic discount factor, 248
    subjective CAPM, 247, 277
Aumann, R.
    common knowledge definition, 19
    common prior assumption, 196
    correlated equilibrium, 6, 195, 204,
        231

EU model with incompleteness,
        91, 171
letter to Savage, 13
relaxation of common prior
        assumption, 216
subjective correlated equilibrium,
        215, 219
subjective expected utility model
        (Anscombe-Aumann), 106
axiomatic methods in rational choice
        theory, 24
axioms
    coordinate independence, 132
    expected utility, 87, 89, 91, 93
    state-preference theory, 131
    subjective expected utility, 114,
        120
    subjective probability, 61, 64, 74,
        78
    tradeoff consistency, 133

Bachelier's Brownian motion model for
        stock price movements, 25
battle-of-sexes game, 186, 214
    Binmore's solution, 191
    extremal correlated equilibria, 189
    geometry of set of correlated
        equilibria, 190
    relation to chicken and stag hunt
        games, 188
    revealed-rules matrix (G), 188

Printed in the United States
by Baker & Taylor Publisher Services

Printed in the United States
by Baker & Taylor Publisher Services